Lab Manual

Experimental Chemistry

TENTH EDITION

Steven S. Zumdahl
University of Illinois at Urbana-Champaign

Susan Arena Zumdahl
University of Illinois at Urbana-Champaign

Donald J. DeCoste
University of Illinois at Urbana-Champaign

Prepared by

John G. Little
St. Mary's High School, Stockton, CA

❄️ Cengage

Australia • Brazil • Canada • Mexico • Singapore • United Kingdom • United States

For product information and technology assistance, contact us at **Cengage Customer & Sales Support, 1-800-354-9706 or support.cengage.com.**

For permission to use material from this text or product, submit all requests online at **www.copyright.com.**

ISBN: 978-1-305-95745-9

Cengage
200 Pier 4 Boulevard
Boston, MA 02210
USA

Cengage is a leading provider of customized learning solutions with employees residing in nearly 40 different countries and sales in more than 125 countries around the world. Find your local representative at: **www.cengage.com.**

To learn more about Cengage platforms and services, register or access your online learning solution, or purchase materials for your course, visit **www.cengage.com.**

Printed in the United States of America
Print Number: 05   Print Year: 2022

Contents

Introduction

Preface v

Laboratory Glassware and Other Apparatus vii

Safety in the Chemistry Laboratory xiv

Safety Quiz xxiii

Experiments

1. The Determination of Mass 1

2. The Use of Volumetric Glassware 9

3. Density Determinations 21

4. The Determination of Boiling Point 29

5. The Determination of Melting Point 39

6. The Solubility of a Salt 47

7. Identification of a Substance 55

8. Resolution of Mixtures 1: Filtration and Distillation 63

9. Resolution of Mixtures 2: Chromatography 73

10. Stoichiometry 1: Limiting Reactant 85

11. Stoichiometry 2: Stoichiometry of an Iron(III)-Phenol Reaction 93

12. Composition 1: Percentage Composition – The Empirical Formula of Magnesium Oxide 101

13. Composition 2: Hydrates and Thermal Decomposition 109

14. Preparation and Properties of Hydrogen and Oxygen Gases 121

15. Gas Laws 1: Molar Volume and The Ideal Gas Constant 131

16. Gas Laws 2: Graham's Law 139

17. Calorimetry 147

18. Heats of Reaction and Hess's Law 159

19. Spectroscopy 1: Spectra of Atomic Hydrogen and Nitrogen 171

20. Spectroscopy 2: Emission Spectra of Metallic Elements 183

21. Molecular Properties 1: Molecular Shapes and Structures 191

22. Properties of Some Representative Elements 201

23. Classification of Chemical Reactions 217

24. Colligative Properties 1: Freezing Point Depression and the Determination of Molar Mass 235

25. Colligative Properties 2: Osmosis and Dialysis 245

26. Rates of Chemical Reactions 253

27. Chemical Equilibrium 1: Titrimetric Determination of an Equilibrium Constant 263

28. Chemical Equilibrium 2: Spectrophotometric Determination of an Equilibrium Constants 273

29. Chemical Equilibrium 3: Stresses Applied to Equilibrium Systems 281

30. The Solubility Product of Calcium Iodate 291

31. Acids, Bases, and Buffered Systems 301

32. Acid–Base Titrations 1: Analysis of an Unknown Acid Sample 317

33. Acid–Base Titrations 2: Evaluation of Commercial Antacid Tablets 327

34. The Determination of Calcium in Calcium Supplements 335

35. Determination of Iron by Redox Titration 343

36. Determination of Vitamin C in Fruit Juices 351

37. Electrochemistry 1: Chemical Cells 361

38. Electrochemistry 2: Electrolysis 371

39. Gravimetric Analysis 1: Determination of Chloride Ion 381

40. Gravimetric Analysis 2: Determination of Sulfate Ion 391

41. Preparation of a Coordination Complex of Copper(II) 401

42. Inorganic Preparations 1: Synthesis of Sodium Thiosulfate Pentahydrate 409

43. Inorganic Preparations 2: Preparation of Sodium Hydrogen Carbonate 417

44. Qualitative Analysis of Organic Compounds 425

45. Organic Chemical Compounds 435

46. Ester Derivatives of Salicylic Acid 447

47. Preparation of Fragrant Esters 457

48. Proteins 465

49. Enzymes 473

50. Polymeric Substances 1: Amorphous Sulfur 481

51. Polymeric Substances 2: Preparation of Nylon 487

52. Qualitative Analysis of Selected Cations and Anions 493

Appendices

A. Plotting Graphs in the Introductory Chemistry Laboratory 503

B. Errors and Error Analysis in General Chemistry Experiments 505

C. Vapor Pressure of Water at Various Temperatures 509

D. Density of Water at Various Temperatures 510

E. Solubility Information 511

F. Properties of Substances 512

G. Concentrated Acid Base Reagent Data 516

Preface

Introduction: Why We Study Chemistry

On the simplest level, we say that chemistry is "the study of matter and the transformations it undergoes." On a more intellectual level, we realize that chemistry seeks to explain such things as the myriad chemical reactions in the living cell, the transmission of energy by superconductors, the working of transistors, and even how the oven and drain cleaners we use in our homes achieve their tasks.

The study of chemistry is required for students in many other fields because it is a major unifying force among these other subjects. Chemistry is the study of matter itself. In other disciplines, particular *aspects* of matter or its applications are studied, but the basis for such study rests in a firm foundation in chemistry.

Laboratory study is required in chemistry courses for several reasons. In some cases, laboratory work may serve as an *introduction* to further, more difficult laboratory work (this is true of the first few experiments in this laboratory manual). Before you can perform "meaningful" or "relevant" experiments, you must learn to use the basic tools of the trade.

Sometimes laboratory experiments in chemistry are used to *demonstrate* the topics covered in course lectures. For example, every chemistry course includes a lengthy exposition of the properties of gases and discussion of the gas laws. A laboratory demonstration of these laws may help to clarify or reinforce them.

Laboratory work can teach you the various standard *technique*s used by scientists in chemistry and in most other fields of science. For example, pipets are used in many biological, health, and engineering disciplines when a precisely measured volume of liquid is needed. Chemistry lab is an excellent place to learn techniques correctly; you can learn by practicing techniques in general chemistry lab rather than on the job later.

Finally, your instructor may use the laboratory as a means of judging how much you have learned and understood from lecture. Sometimes students can read and understand their textbook and lecture notes and do reasonably well on examinations without gaining much practical knowledge of the subject. The lab serves as a place where classroom knowledge can be synthesized and applied to realistic situations.

What Will Chemistry Lab Be Like?

Undergraduate chemistry laboratories are generally 2–3 hours long, and most commonly meet once each week. The laboratory is generally conducted by a teaching assistant—a graduate student with several years of experience in chemistry laboratory. Community colleges and smaller universities may not have a graduate program in chemistry, may employ instructors whose specific duties consist of running the general chemistry labs, or the course professor may be responsible for supervision of his or her own classes. In any event, your professor undoubtedly will visit the laboratory frequently to make sure that everything is running smoothly.

The first meeting of a laboratory section is often pretty hectic. This is perfectly normal, so do not judge your laboratory by this first meeting. Part of the first lab meeting will be devoted to checking in to the laboratory. You will be assigned a workspace and locker and will be asked to go through the locker to make sure that it contains all the equipment you will need. Your instructor will then conduct a brief orientation to the laboratory, giving you the specifics of how the laboratory period will be operated. Be sure to ask any questions you may have about the operation of the laboratory.

Your instructor will discuss with you the importance of safety in the laboratory and will point out the emergency equipment. *Pay close attention to this discussion so that you will be ready for any eventuality.*

Finally, your instructor will discuss the experiment you will be performing after check-in. Although we have tried to make this laboratory manual as explicit and complete as possible, your instructor will certainly present some sort of a pre-lab discussion during which he or she will offer tips or advice based on his or her own experience and expertise. Take advantage of this information.

How Should I Prepare for Lab?

Laboratory should be one of the most enjoyable parts of your study of chemistry. Instead of listening passively to your instructor, you have a chance to witness chemistry in action. However, if you have not suitably *prepared* for lab, you will spend most of your time wondering what is going on.

This manual has been written and revised to help you prepare for laboratory. First of all, each experiment has a clearly stated *Objective*. This is a very short summary of the lab's purpose. This Objective will suggest what sections of your textbook or lecture notes might be appropriate for review before lab. Each experiment also contains an *Introduction* that reviews explicitly some of the theory and methods to be used in the experiment. You should read through this Introduction *before* the laboratory period and make cross-references to your textbook while reading. Make certain that you understand fully what an experiment is supposed to demonstrate before you attempt that experiment.

The lab manual contains a detailed *Procedure* for each experiment. The Procedures have been written to be as clear as possible, but do not wait until you are actually performing the experiment to read this material. Study the Procedure *before* the lab period, and question the instructor before the lab period on anything you do not understand. Sometimes it is helpful to write up a summary or overview of a long Procedure. This can prevent major errors when you are actually performing the experiment. Another technique students use in getting ready for lab is to prepare a *flow chart,* indicating the major procedural steps, and any potential pitfalls, in the experiment. [Author's aside: I like to cook, and am constantly trying new recipes. I make a sort of informal flow chart from the recipe so that I can quickly look and see what I need to do next.] You can keep the flow chart handy while performing the experiment to help in avoiding any gross errors as to "what comes next." A flow chart is especially useful for planning your time most effectively: for example, if a procedure calls for heating something for an hour, a flow chart can help you plan what else you can get done during that hour.

This lab manual contains a *Laboratory Report* for each of the experiments. Each Laboratory Report is preceded by a set of *Pre-Laboratory Questions*, which your instructor will probably ask you to complete before coming to lab. Sometimes these pre-laboratory questions will include numerical problems similar to those you will encounter in processing the data you will be collecting in the experiment. Obviously, if you review a calculation before lab, things will be that much easier when you are actually *in* the lab.

I hope you will find your study of chemistry to be both meaningful and enjoyable. As always, I encourage those who use this manual—both instructors and students—to assist me in improving it for the future. Any comments or criticism will be greatly appreciated.

John G. Little

Laboratory Glassware and Other Apparatus

Introduction

When you open your laboratory locker for the first time, you are likely to be confronted with a bewildering array of various sizes and shapes of glassware and other apparatus. Glass is used more than any other material for the manufacture of laboratory apparatus because it is relatively cheap, usually easy to clean (and keep clean), and can be heated to fairly high temperatures or cooled to quite low temperatures. Most importantly, glass is used because it is impervious to and non-reactive with most reagents encountered in the beginning chemistry laboratory. Most glassware used in chemistry laboratories is made of borosilicate glass, which is relatively sturdy and safe to use at most temperatures. Such glass is sold under such trade names as Pyrex® (Corning Glass Co.) and Kimax® (Kimble Glass Co.). If any of the glassware in your locker does not have either of these trade names marked on it, consult with your instructor before using it at temperatures significantly above or below room temperature.

Many pieces of laboratory glassware and apparatus have special names – often, the name of the scientist who first devised it – that have evolved over the centuries chemistry has been studied. You should learn the names of the most common pieces of laboratory apparatus to make certain that you use the correct equipment for the experiments in this manual. Study the drawings of common pieces of laboratory apparatus that are shown on the next few pages. Compare the equipment in your locker with these drawings, and identify all the pieces of apparatus that have been provided to you. It might be helpful for you to label the equipment in your locker, at least for the first few weeks of the term.

While your locker may not contain all the apparatus in the drawings, be sure to ask your instructor for help if there is equipment in your locker that is not described in the drawings. As you examine your locker equipment, be alert to chipped, cracked, or otherwise imperfect glassware or porcelainware. *Imperfect glassware is a major safety hazard, and must be discarded.* Don't think that because a beaker has just a little crack, it can still be used. *Replace all glassware that has any cracks, chips, star fractures, or any other deformity.* Most college and university laboratories will replace imperfect glassware free of charge during the first laboratory meeting of the term, but may assess charges if breakage is discovered later in the term.

You may wish to clean the set of glassware in your locker during the first meeting of your laboratory section. This is certainly admirable, but somehow lab glassware has an annoying habit of becoming dirty again without apparent human intervention. Glassware always looks clean when wet, but tends to dry with water spots (and show minor imperfections in the glass that are not visible when wet). It is recommended that you thoroughly clean all the locker glassware at the start, and then rinse the glassware out before its first use. In the future, clean glassware before leaving lab for the day, and rinse before using. Instructions for the proper cleaning of laboratory glassware are provided after the diagrams of apparatus.

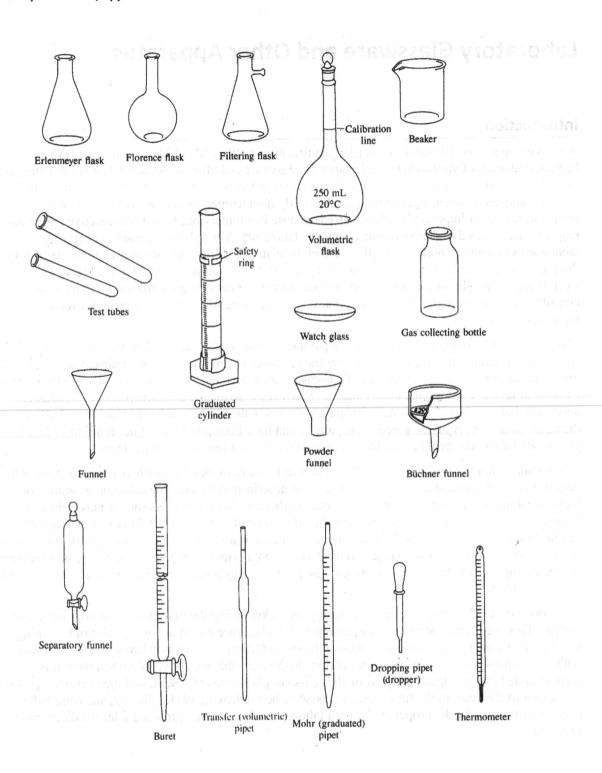

Erlenmeyer flask

Florence flask

Filtering flask

Calibration line

Beaker

250 mL
20°C

Volumetric flask

Test tubes

Safety ring

Watch glass

Gas collecting bottle

Funnel

Graduated cylinder

Powder funnel

Büchner funnel

Separatory funnel

Buret

Transfer (volumetric) pipet

Mohr (graduated) pipet

Dropping pipet (dropper)

Thermometer

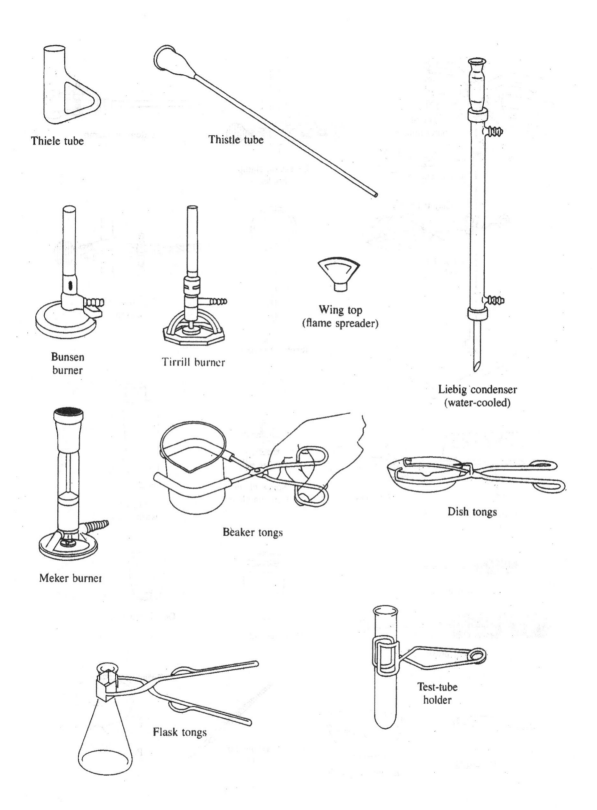

Thiele tube

Thistle tube

Wing top
(flame spreader)

Bunsen
burner

Tirrill burner

Liebig condenser
(water-cooled)

Meker burner

Beaker tongs

Dish tongs

Flask tongs

Test-tube
holder

Laboratory Glassware/Apparatus

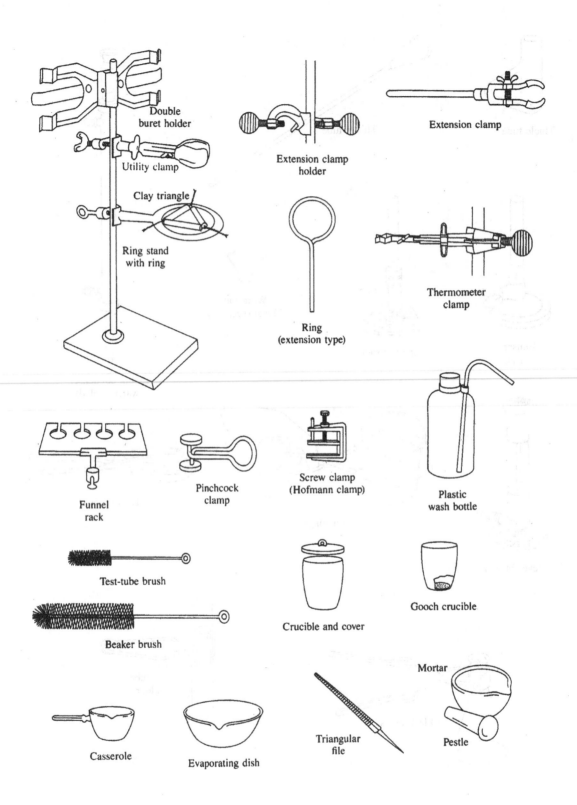

Double buret holder

Utility clamp

Clay triangle

Ring stand with ring

Extension clamp holder

Ring (extension type)

Extension clamp

Thermometer clamp

Funnel rack

Pinchcock clamp

Screw clamp (Hofmann clamp)

Plastic wash bottle

Test-tube brush

Beaker brush

Crucible and cover

Gooch crucible

Casserole

Evaporating dish

Triangular file

Mortar

Pestle

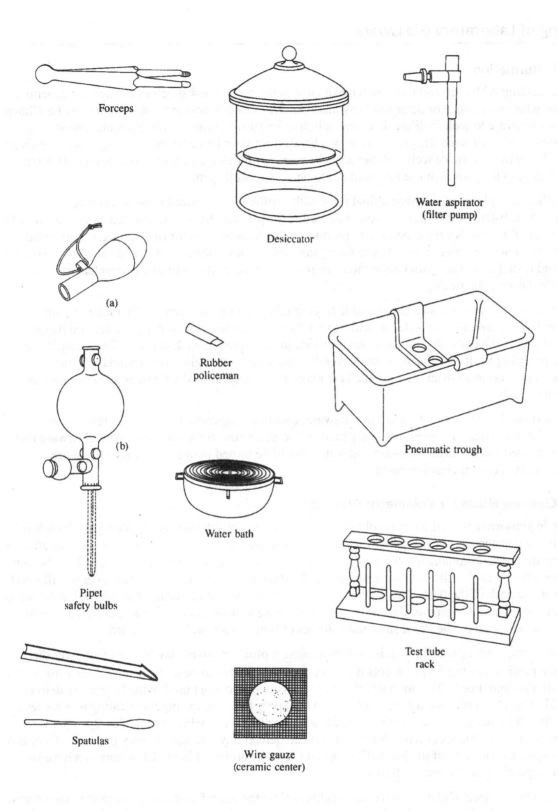

Forceps

Desiccator

Water aspirator
(filter pump)

(a)

Rubber
policeman

Pneumatic trough

(b)

Water bath

Pipet
safety bulbs

Test tube
rack

Spatulas

Wire gauze
(ceramic center)

Cleaning of Laboratory Glassware

General Information

A simple washing with soap and water will make most laboratory glassware clean enough for general use. Determine what sort of soap or detergent is available in the lab and whether the detergent must be diluted. Allow the glassware to soak in dilute detergent solution for 10–15 minutes (which should remove any grease or oil), and then scrub the glassware with a brush from your locker if there are any caked solids on the glass. Rinse the glassware well with tap water to remove all detergent. Investment in a small container of good dish soap for your own use may turn out to be money well spent!

Usually laboratory glassware is given a final rinse with distilled or deionized water to remove any contaminating substances that may be present in the local tap water. However, distilled/deionized (usually labeled, "d.H$_2$O") water is very expensive to produce and cannot be used for rinsing in great quantities. Therefore, fill a plastic squeeze wash bottle from your locker with distilled water, and rinse the previously cleaned and rinsed glassware with two or three separate 5–10 mL portions of distilled water from the wash bottle. Discard the rinsings.

If wooden or plastic drying hooks are available in your lab, you may use them to dry much of your glassware. If hooks are not available, or items do not fit on the hooks, spread out glassware on paper towels to dry. Occasionally, drying ovens are provided in undergraduate laboratories. Only simple flasks and beakers should be dried in such ovens. Never dry finely calibrated glassware (burets, pipets, graduated cylinders) in an oven, because the heat may cause the calibrated volume of the container to change appreciably.

If simple soap and water will not clean the glassware, chemical reagents can be used to remove most stains or solid materials. These reagents are generally too dangerous for student use. Any glassware that cannot be cleaned well enough with soap and water should be turned in to a person who has more experience with chemical cleaning agents.

Special Cleaning Notes for Volumetric Glassware

Volumetric glassware is used when absolutely known volumes of solutions are required to a high level of precision and accuracy. For example, when solutions are prepared to be of a particular concentration, a volumetric flask whose volume is known to ± 0.01 mL may be used to contain the solution. The absolute volume a particular flask will contain is stamped on the flask by the manufacturer, and an exact **fill mark** is etched on the neck of the flask. The symbol "TC" on a volumetric flask means that the flask is intended "to contain" the specified volume. Generally, the temperature at which the flask was calibrated (usually 20 °C) is also indicated on the flask, since the volumes of liquids vary with temperature.

If a solution sample of a particular precise size is needed, a **pipet** or **buret** may be used to deliver the sample (the precision of these instruments is also generally indicated to be to the nearest ± 0.01 mL – about $^1/_5$ of a normal drop!). The normal sort of transfer volumetric pipet used in the laboratory delivers one specific size of sample, and a mark is etched on the upper barrel of the pipet indicating to what level the pipet should be filled. Such a pipet is generally also marked "TD," which means that the pipet is calibrated "to deliver" the specified volume. A buret can deliver any size sample very precisely, from zero milliliters up to the capacity of the buret. The normal Class A buret used in the laboratory can have its volume read precisely to the nearest 0.01 mL.

Obviously, when a piece of glassware has been calibrated by the manufacturer to be correct to the nearest 0.01 milliliter, the glassware must be *absolutely clean* before use. The standard test for cleanliness of volumetric glassware involves watching a film of distilled water run down the interior sides of the glassware. Water should flow in sheets (a continuous film) down the inside of volumetric glassware, without beading up anywhere on the inside surface.

If water beads up anywhere on the interior of volumetric glassware, the glassware must be soaked in dilute detergent solution, scrubbed with a brush (except for pipets, for which no suitable brush exists), and rinsed thoroughly with both with tap water and distilled water. The process must be repeated until the glassware is absolutely clean. Narrow bristled brushes are available for reaching into the long, narrow necks of volumetric flasks, and special long-handled brushes are available for scrubbing burets. There are also brushes designed to clean the barrel of a funnel, but relatively few labs are so equipped. Since brushes cannot be fitted into the barrel of pipets, if it is not possible to clean the pipet completely on two or three attempts, the pipet should be exchanged for a new one (special pipet washers are probably available in the stockroom for cleaning pipets that have been turned in by students as uncleanable).

Volumetric glassware is generally used wet – in the case of narrow-necked vessels and slender items such as burets, drying the inside is not really practical. Rather than drying the glassware (possibly allowing the glassware to become water-spotted, possibly causing incorrect volumes readings), the user rinses the glassware with whatever it is going to be used to measure. For example, if a buret is to be filled with standard acid solution, the buret, still wet from cleaning, is rinsed with small portions of the same acid (with the rinsings being discarded). Rinsing with the solution that is going to be used with the glassware insures that no excess water will dilute the solution to be measured. *Volumetric glassware is never heated in an oven,* since the heat may destroy the integrity of the glassware's calibration. If space in your locker permits, you might wish to leave volumetric glassware, especially burets, filled with distilled water between laboratory periods. A buret that has been left filled with water will require much less time to clean on subsequent use (consult with the instructor to see if this is possible).

Safety in the Chemistry Laboratory

Introduction

Chemistry is an experimental science. You cannot learn chemistry without getting your hands dirty. Any beginning chemistry student faces the prospect of laboratory work with some apprehension. It would be untruthful to say that there is no element of risk in a chemistry lab. *Chemicals can be dangerous.* The more you study chemistry, the larger the risk will become. However, if you approach your laboratory work calmly and studiously, you will minimize the risks.

During the first laboratory meeting, you should ask your instructor for a brief tour of the laboratory room. Ask him or her to point out for you the locations of the various pieces of emergency apparatus provided by your college or university. At your bench, construct a map of the laboratory, noting the location of the exits from the laboratory and the location of all safety equipment. Close your eyes, and test whether you can locate the exits and safety equipment from memory. A brief discussion of the major safety apparatus and safety procedures follows. A Safety Quiz is provided at the end of this section to test your comprehension and appreciation of this material. For additional information on safety in your particular laboratory, consult with your laboratory instructor or course professor.

Protection for the Eyes

Government regulations, as well as common sense, demand the wearing of protective eyewear while you are in the laboratory. Such eyewear must be worn even if you personally are not working on an experiment. Figure 1 shows one common form of plastic **safety goggle.**

Figure 1. A typical student plastic safety goggle

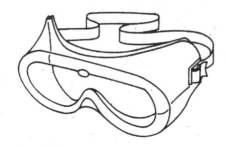

Although you may not use the particular type of goggle shown in the figure, your eyewear must include shatterproof lenses and side shields that will protect you from splashes. *Safety glasses must be worn at all times while you are in the laboratory, whether or not you are working with chemicals.* Failure to wear safety glasses may result in your being failed or withdrawn from your chemistry course, or in some other disciplinary action.

In addition to protective goggles, an **eyewash fountain** provides eye protection in the laboratory. Should a chemical splash near your eyes, you should use the eyewash fountain before the material has a chance to run in behind your safety glasses. If the eyewash is not near your bench, wash your eyes quickly but thoroughly with water from the nearest source, and then use the eyewash. A typical eyewash fountain is indicated in Figure 2:

Figure 2. Laboratory emergency eyewash fountain

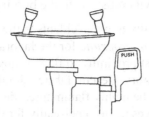

The eyewash has a panic bar that enables the eyewash to be activated easily in an emergency. If you need the eyewash, don't be modest—*use it immediately*. It is critical that you protect your eyes properly.

Protection from Fire

The danger of uncontrolled fire in the chemistry laboratory is very real, since the lab typically has a fairly large number of flammable liquids in it, and open-flame gas burners are generally used for heating (see later in this discussion for proper use of the gas burner). With careful attention, however, the danger of fire can be reduced considerably. Always check around the lab before lighting a gas burner to ensure that no one is pouring or using any flammable liquids. Be especially aware that the vapors of most flammable liquids are heavier than air and tend to concentrate in sinks (where they may have been poured) and at floor level. Since your laboratory may be used by other classes, always check with your instructor before beginning to use gas burners.

The method used to fight fires that occur in spite of precautions being taken, depends on the size of the fire and on the substance that is burning. If only a few drops of flammable liquid have been accidentally ignited, and no other reservoir of flammable liquid is nearby, the fire can usually be put out by covering it with a beaker. This deprives the fire of oxygen and will usually extinguish the fire in a few minutes. Leave the beaker in place for several minutes to ensure that the flammable material has cooled and will not flare up again (and to allow the beaker to cool!).

In the unlikely event that a larger chemical fire occurs, carbon dioxide **fire extinguishers** are available in the lab (usually mounted near one of the exits from the room). An example of a typical carbon dioxide fire extinguisher is shown in Figure 3.

Figure 3. A typical carbon dioxide fire extinguisher

(Pull the metal ring to release the extinguisher handle)

Before activating the extinguisher, pull the metal safety ring from the handle. Direct the output from the extinguisher at the base of the flames. The carbon dioxide not only smothers the flames, it also cools the flammable material quickly. If it becomes necessary to use the fire extinguisher, be sure afterward to turn the extinguisher in at the stockroom so that it can be refilled immediately. If the carbon dioxide extinguisher does not immediately put out the fire, evacuate the laboratory and call the fire department.

Carbon dioxide fire extinguishers must *not* be used on fires involving magnesium or certain other reactive metals, since carbon dioxide may react vigorously with the burning metal and make the fire worse.

One of the most frightening and potentially tragic accidents is the igniting of a person's clothing. For this reason, certain types of clothing are *not appropriate* for the laboratory and must not be worn. Since sleeves are most likely to come in closest proximity to flames, any garment that has bulky or loose sleeves should not be worn in the laboratory. Certain fabrics should also be avoided; such substances as silk and certain synthetic materials may be highly flammable. Ideally, students should be asked to wear laboratory coats with tightly fitting sleeves made specifically for the chemistry laboratory. Your particular college or university may require this clothing. Long hair also presents a clear danger if it is allowed to hang loosely in the vicinity of the flame. Long hair must be pinned back or held with a rubber band.

In the unlikely event a student's clothing or hair is ignited, his or her neighbors must take prompt action to prevent severe burns. Most laboratories have two options for extinguishing such fires: the **water shower** and the **fire blanket**. Figure 4 shows a typical laboratory emergency water shower:

Figure 4. Laboratory Emergency Deluge Shower
Use the shower to extinguish clothing fires and in the event of a large-scale chemical spill.

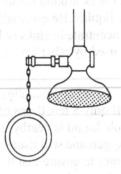

Showers such as this generally are mounted near the exits from the laboratory. In the event someone's clothing or hair is on fire, *immediately* push or drag the person to the shower and pull the metal ring of the shower. Safety showers generally dump 40–50 gallons of water, which should extinguish the flames. Be aware that the showers cannot be shut off once the metal ring has been pulled. For this reason, the shower cannot be demonstrated. (But note that the showers are checked for correct operation on a regular basis.)

Figure 5 shows the other possible apparatus for extinguishing clothing fires, the **fire blanket**. The fire blanket is intended to smother flames. Since it must be operated by the person whose clothing is on fire (he or she pulls the handle of the fire blanket and wraps it around himself or herself), it is therefore not the preferred method of dealing with such an event.

Figure 5. Fire blanket
The blanket is wrapped around the body to smother flames.

Protection from Chemical Burns

Most acids, alkalis, and oxidizing and reducing agents must be assumed to be corrosive to skin. It is impossible to avoid these substances completely, since many of them form the backbone of the study of chemistry. Usually, a material's corrosiveness is proportional to its concentration. Most of the experiments in this manual have been set up to use as dilute a mixture of reagents as possible, but this does not entirely remove the problem. Make it a personal rule to *wash your hands regularly* after using any chemical substance and to wash immediately, with plenty of water, if any chemical substance is spilled on the skin.

After working with a substance known to be particularly corrosive, you should wash your hands immediately even if you did not spill the substance. Someone else using the bottle of reagent may have spilled the substance on the side of the bottle. It is also good practice to hold a bottle of corrosive substance with a paper towel, or to wear plastic gloves, during pouring. Do not make the mistake of thinking that because an experiment calls for dilute acid, this acid cannot burn your skin. Some of the acids used in the laboratory are not volatile, and as water evaporates from a spill, the acid becomes concentrated enough to damage skin. Whenever a corrosive substance is spilled on the skin, you should inform the instructor immediately. If there is any sign whatsoever of damage to the skin, you will be sent to your college's health services for evaluation by a physician.

In the event of a major chemical spill, in which substantial portions of the body or clothing are affected, you must use the emergency water shower. Forget about modesty, and get under the shower immediately.

Protection from Toxic Fumes

Many volatile substances may have toxic vapors. A rule of thumb for the chemistry lab is, "If you can smell it, it can probably hurt you." Some toxic fumes (such as those of ammonia) can overpower you immediately, whereas other toxic fumes are more insidious. The substance may not have that bad an odor, but its fumes can do severe damage to the respiratory system. There is absolutely no need to expose yourself to toxic fumes. All chemistry laboratories are equipped with **fume exhaust hoods**. A typical hood is indicated in Figure 6.

Figure 6. A Common Type of Laboratory Fume Exhaust Hood

Use the hood whenever a reaction involves toxic fumes. Keep the glass window of the hood partially closed to provide for rapid flow of air. The hood should be marked to indicate the maximum level the window may be raised and still provide adequate exhaust airflow.

The exhaust hood has fans that draw vapors out of the hood and away from the user. The hood is also used when flammable solvents are required for a given procedure, since the hood will remove the vapors of such solvents from the laboratory and reduce the hazard of fire. The hood is equipped with a safety-glass window that can be used as a shield from reactions that could become too vigorous. Naturally, the number of exhaust hoods available in a particular laboratory is limited, but *never neglect to use the hood if it is called for*, merely to save a few minutes of waiting time. Finally, reagents are sometimes stored in a hood, especially if the reagents evolve toxic fumes. Be sure to return such reagents to the designated hood after use.

Protection from Cuts and Simple Burns

Perhaps the most common injuries to students in the beginning chemistry laboratory are simple cuts and burns. Glass tubing and glass thermometers are used in nearly every experiment and are often not prepared or used properly. Most glass cuts occur when the glass (or thermometer) is being inserted through rubber stoppers in the construction of an apparatus. Use glycerin as a lubricant when inserting glass through rubber (several bottles of glycerin will be provided in your lab). Glycerin is a natural product of human and animal metabolism; it may be applied liberally to any piece of glass. Glycerin is water soluble, and though it is somewhat messy, it washes off easily. You should always remove excess glycerin before using an apparatus since it may react with the reagents to be used.

Most simple burns in the laboratory occur when a student forgets that an apparatus may be hot and touches it. Never touch an apparatus that has been heated until it has cooled for at least five minutes, or unless specific tongs for the apparatus are available.

Report any cuts or burns, no matter how apparently minor, to the instructor immediately. If there is any visible damage to the skin, you will be sent to your college's health services for immediate evaluation by a physician. What may seem like a scratch may be adversely affected by chemical reagents or may become infected; therefore, it must be attended to by trained personnel.

Proper Use of the Laboratory Burner

The laboratory burner is one of the most commonly used pieces of apparatus in the general chemistry laboratory, and may pose a major hazard if not used correctly and efficiently.

The typical laboratory burner is correctly called a **Tirrill burner** (though the term 'Bunsen burner' is often used in a generic sense). A representation of a Tirrill burner is indicated in Figure 7. Compare the burner you will be using with the burner shown in the figure, and consult with your instructor if there seems to be any difference in construction. Burners from different manufacturers may differ slightly in appearance and operation from the one shown in the illustration.

Figure 7. A Tirrill Burner of the sort most commonly found in student laboratories

Compare this burner with the one you will use for any differences in construction or operation. The temperature of the flame is increased by allowing air to mix with the gas being burned.

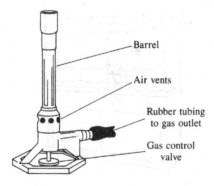

Barrel

Air vents

Rubber tubing to gas outlet

Gas control valve

Most laboratories are supplied with natural gas, which consists primarily of the hydrocarbon methane (CH_4). If your college or university is some distance from natural gas lines, your laboratory may be equipped with bottled liquefied gas, which consists mostly of the hydrocarbon propane (C_3H_8). In this case, your burner may have modifications to allow the efficient burning of propane.

A length of thin-walled rubber tubing should attach your burner to the gas main jet. If your burner has a screw-valve on the bottom (or on the side) for controlling the flow of gas, the valve should be closed (by turning in a right-hand direction) before you light the burner. The barrel of the burner should be rotated to close the air vents (or slide the air vent cover over the vent holes if your burner has this construction).

To light the burner, turn the gas main jet to the open position. If your burner has a screw-valve on the bottom, open this valve until you hear the hiss of gas. Without delay, use your striker or a match to light the burner. If the burner does not light on the first attempt, shut off the gas main jet, and make sure that all rubber tubing connections are tight. Then reattempt to light the burner. You may have to increase the flow of gas, using the screw-valve on the bottom of the burner.

After lighting the burner, the flame is likely to be yellow and very unstable (easily blown about by any drafts in the lab). The flame at this point is relatively cool (the gas is barely burning) and would be very inefficient in heating. To make the flame hotter and more stable, open the air vents on the barrel of the burner slowly to allow oxygen to mix with the gas before combustion. This should cause the flame's size to decrease and its color to change. A proper mixture of air and gas gives a pale blue flame, with a bright blue cone-shaped region in the center. The hottest part of the flame is directly above the tip of the bright blue cone. Whenever an item is to be heated strongly, it should be positioned directly above this bright blue cone. You should practice adjusting the flame to get the ideal pale blue flame with its blue inner cone, using the control valve on the bottom of the burner or the gas main jet.

Protection from Apparatus Accidents

An improperly constructed apparatus can create a major hazard in the laboratory, not only to the person using the apparatus, but to his or her neighbors as well. Any apparatus you set up should be constructed exactly as indicated in this manual. If you have any question as to whether you have set up the apparatus correctly, ask your instructor to check the apparatus before you begin the experiment.

Perhaps the most common apparatus accident in the lab is the tipping over of a flask or beaker while it is being heated or otherwise manipulated. All flasks should be *clamped securely* with an adjustable clamp to a ring support. This is indicated in Figure 8. Be aware that a vacuum flask is almost guaranteed to tip over during suction filtration if it is not clamped. This will result in loss of the crystals being filtered and will require that you begin the experiment again.

Figure 8. One Method of Supporting a Flask

Be sure to clamp all glassware securely to a ringstand before using.

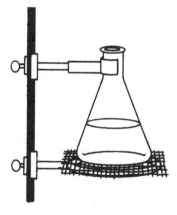

An all-too-common lab accident occurs when a liquid must be heated to boiling in a test tube. Test tubes hold only a few milliliters of liquid and require only a few seconds of heating to reach the boiling point. *A test tube MUST NOT be heated strongly in the direct full heat of the burner flame.* The contents of the test tube will super-heat and blow out of the test tube like a bullet from a gun. Ideally, when a test tube requires heating, a **boiling water bath** in a beaker should be used. If this is not possible, then hold the test tube at a 45° angle a few inches above the flame and heat only *briefly*, keeping the test tube moving constantly (from top to bottom, and from side to side) through the flame during the heating. *Aim the mouth of the test tube away from yourself and your neighbors.* See Figure 9.

Figure 9. Method for hearing a test tube containing a small quantity of liquid

Heat only for a few seconds, beginning at the surface of the liquid and moving downwards. Keep the test tube moving through the flame. Aim the mouth of the test tube away from yourself and your neighbors.

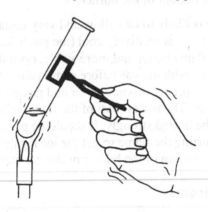

Safety Regulations

- Wear safety glasses at all times while in the laboratory.

- Do not wear short skirts, shorts, and bare-midriff shirts in the laboratory.

- Do not wear scarves and neckties in the lab, because they may be ignited accidentally in the burner flame.

- Men who have long beards must secure them away from the burner flame.

- Open-toed shoes and sandals, as well as thin canvas sneakers, are not permitted in the laboratory.

- Never leave Bunsen burners unattended when lighted.

- Never heat solutions to dryness unless this is done in an evaporating dish on a hotplate or over a boiling water bath.

- Never heat a "closed system" such as a stoppered flask.

- Never smoke, chew gum, eat, or drink in the laboratory, since you may inadvertently ingest some chemical substance.

- Always use the smallest amount of substance required for an experiment; more is never better in chemistry. *Never return unused portions of a reagent to their original bottle.*

- Never store chemicals in your laboratory locker unless you are specifically directed to do so by the instructor.

- Never remove any chemical substance from the laboratory. In many colleges, removal of chemicals from the laboratory is grounds for expulsion or other severe disciplinary action.

- Keep your work area clean, and help keep the common areas of the laboratory clean. If you spill something in a common area, remember that this substance may injure someone else.

- Never fully inhale the vapors of any substance. Use your hand to waft a tiny amount of vapor toward your nose.

- To heat liquids, always add 2–3 boiling stones to make the boiling action smoother.

- Never add water to a concentrated reagent when diluting the reagent. Always add the reagent to the water. If water is added to a concentrated reagent, local heating and density effects may cause the water to be splashed back.

- Never work in the laboratory unless the instructor is present. If no instructor is present during your assigned work time, report this to the senior faculty member in charge of your course.

- Never perform any experiment that is not specifically authorized by your instructor. Do not play games with chemicals!

- Dispose of all reaction products and unused reagents as directed by the instructor. In particular, observe the special disposal techniques necessary for flammable or toxic substances. Do not return unused reagents to the stock container from which they were obtained.

- Dispose of all glass products in the special container provided.

Information on Hazardous Substances and Procedures

Many of the pre-laboratory assignments in this manual require that you use one of the many chemical handbooks available in any scientific library to look up various data for the substances you will be working with. You should also use such handbooks as a source of information about the hazards associated with the substances and procedures you will be using. A particularly useful general reference is the booklet *Safety in Academic Laboratories*, published by the American Chemical Society (your course professor or laboratory instructor may have copies).

There are also many useful online sources available. For example, Material Safety Data Sheets (MSDS) for many chemicals are available online from the various chemical supply companies. Many chemistry journals, textbooks, and handbooks are also available online. Check the web site for your school's library: many libraries have subscriptions to online databases that you can access.

Each of the experiments in this manual includes a section entitled *Safety Precautions,* which gives important information about expected hazards. This manual also lists some of the hazards associated with common chemical substances in Appendix J. Such problems as flammability and toxicity are described in terms of a low/medium/high rating scale. If you have any questions about this material, consult with your instructor before the laboratory period. Safety is *your* responsibility.

You may now be even more hesitant about facing your chemistry laboratory experience, now that you have read all these warnings. Be cautious…be careful…be thoughtful…but don't be afraid. Every precaution will be taken by your instructor and your college or university for your protection. Realize, however, that *you* bear the ultimate responsibility for your own safety in the laboratory.

Name: _____ Section: _____

Lab Instructor: _____ Date: _____

Safety Quiz

1. Sketch below a representation of your laboratory, showing clearly the location of the following: exits, fire extinguishers, safety shower, eyewash fountain, and fume hoods.

2. Why must you wear safety glasses or goggles at all times while you are in the laboratory, even when you are not personally working on an experiment?

3. Suppose your neighbor in the lab were to spill several milliliters of a flammable substance near her burner, and the substance then ignited. How should this relatively small fire be extinguished?

4. What steps should be taken if a larger scale fire were to break out in the laboratory?

5. What measures should be taken if a student's clothing were to be ignited?

6. List five types of clothing or footwear that are *not* acceptable in the laboratory, and explain why they are not acceptable.

7. Suppose you were pouring concentrated nitric acid from a bottle into your reaction flask, and you spilled the acid down the front of your shirt. What should you do?

8. What steps should you have taken if you had spilled the nitric acid on your hand rather than on your clothing?

9. What are the most common student injuries in the chemistry lab, and how can they be prevented?

10. Where should reactions involving the evolution of toxic gases be performed? Why?

11. Why should you immediately clean up any chemical spills in the laboratory?

12. Why should you never deviate from the published procedure for an experiment?

13. Why should apparatus always be clamped securely to a metal ring stand before starting a chemical reaction?

14. Why should you never eat, drink, or smoke while you are in the laboratory?

15. Suppose the experiment you are to perform in a given lab period involves substances that are toxic or corrosive. What should you do *before* the laboratory to prepare yourself for using these substances safely?

The Determination of Mass: The Laboratory Balance

Objective

The measurement of mass is fundamentally important in any chemistry laboratory. In this experiment, you will learn how to use the balances in your particular lab.

Introduction

The accurate determination of *mass* is one of the most fundamental techniques for students of experimental chemistry. **Mass** is a direct measure of the *amount of matter* in a sample of substance; that is, the mass of a sample is a direct indication of the number of atoms or molecules the sample contains. Since chemical reactions occur in proportion to the number of atoms or molecules of reactant present, it is essential that the mass of reactant used in a process be accurately known.

There are various types of balances available in the typical general chemistry laboratory. Such balances differ in their construction, appearance, operation, and in the level of precision they permit in mass determinations. Three of the most common types of laboratory balance are indicated in Figures 1-1, 1-2, and 1-3. Determine which sort of balance your laboratory is equipped with, and ask your instructor for a demonstration of the use of the balance if you are not familiar with its operation.

Here are some general points to keep in mind when using any laboratory balance:

1. Always make sure that the balance gives a reading of 0.000 grams (or 0.00 g, or 0.0000 g, as appropriate) when nothing is present on the balance pan. Adjust the 'tare' or 'zero' knob/button if necessary. If the balance cannot be set to read zero, ask the instructor for help.

2. All balances, but especially electronic balances, are damaged by moisture. Do not pour liquids in the immediate vicinity of the balance. Clean up any spills immediately from the balance area. Unless the sample is fully contained in a spill-proof container (such as a beaker), liquids generally should not be weighed on an electronic balance.

3. No reagent chemical substance should ever be weighed directly on the pan of the balance. Ideally, reagents should be weighed directly into the beaker or flask in which they are to be used. Plastic weighing boats may also be used if several reagents are required for an experiment. Pieces of filter paper or weighing paper are *not* suitable for use in weighing reagents since the substance being weighed is easily spilled from such paper, and because filter paper is porous, not all of the substance will be transferred.

4. Procedures in this manual are generally written in such terms as "weigh 0.5 grams of substance (to the nearest milligram, or (± 0.001 g)." This does not mean that exactly 0.500 grams of substance is needed. Rather, the statement means to obtain an amount of substance between 0.450 and 0.550 grams, but to record the *actual amount of substance taken* (e.g., 0.496 grams). Unless a procedure states explicitly to weigh out an exact amount (e.g., "weigh out exactly 5.00 grams of NaCl"), you should not waste time trying to obtain an exact amount. However, *always record the amount actually taken to the full precision of the balance used.*

5. For accurate mass determinations, the object to be weighed must be at room temperature. If a hot or warm object is placed on the pan of the balance, the object causes the air around it to become heated. Warm air rises (why?), and the balance may detect this upward motion of such warm air, giving mass readings that are significantly lower than the true value.

6. For mechanical balances such as shown below, we knew to expect small errors in the absolute masses of objects determined with the balance, particularly if the balance has not been recently calibrated or had been abused. For this reason, most mass determinations were performed by *difference*; an empty container was weighed on the balance then the reagent or object whose mass is to be determined was added to the container. The resulting *difference in mass* is the mass of the reagent or object. Because of possible calibration errors, the same balance must be used throughout a procedure.

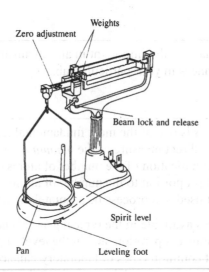

Figure 1-1. Triple Beam Balance

Never weigh a chemical directly on the pan of the balance, because it is more difficult to "zero" this type of balance once the pan becomes dirty. These are being phased out by electronic balances, such as those in Figures 1-2 and 1-3.

Today, most institutions provide electronic balances, such as those shown below. The one on the left, Figure 1-2, shows a "milligram balance," that is, it registers masses to a precision of ± 1 mg (± 0.001 g). Similar models, with other levels of precision are also used in the chemistry laboratory. Ones that are precise only to ± 0.01 gram are referred to as "centigram balances." The balance on the right (Figure 1-3) is known as an "analytical balance," and has a precision of ± 0.1 mg (± 0.0001 g). Both of these are very expensive instruments and must be handled with care and respect.

Figure 1-2: Milligram Balance

Figure 1-3: Analytical Balance

Electronic balances display the mass directly when an object is placed on the pan. For this reason, they must be calibrated and set to zero before use. Some milligram balances (such as the one shown), and all analytical balances have an enclosed chamber to stabilize the balance against air currents. In order for the chamber to be effective, the doors of the chamber must be closed before the mass is recorded. All chemicals weighed on an electronic balance must be contained in a beaker or plastic weighing boat to prevent spillage of the chemical and damage to the pan. Most electronic balances have a "tare" (zeroing) function; the container is placed on the balance, the On/Tare button is pressed, and the balance re-zeros itself. When the object or chemical to be weighed is placed in the beaker or boat, the mass may be read directly.

Safety Precautions	• **Safety eyewear approved by your institution must be worn at all times while you are in the laboratory, whether or not you are working on an experiment**

Apparatus/Reagents Required

- Unknown mass samples provided by the instructor
- Small beakers, 2 (e.g. 50-mL – large enough to hold the mass samples)

Procedure

Record all data and observations directly on the report form in ink.

Examine the balances that are provided in your laboratory. If you are not familiar with the operation of the type of balance available, ask your instructor for a demonstration of the appropriate technique.

Your instructor will provide you with two small objects whose masses you will determine. The objects are coded with an identifying number or letter. Record these identification codes on the report page.

So that you can distinguish them from one another, label two small beakers that can accommodate the objects whose masses are to be determined as **A** and **B**. Determine and record the masses of the two small beakers. The determination of the beakers' masses should be to the full level of precision permitted by the particular balance you are using. This is generally ± 0.01g, or ± 0.001g (±0.001 g = ± 1mg). See **Note**, below.

Transfer the first unknown object to **Beaker A**, and determine the combined mass of **Beaker A** and object. Record. Determine the mass of the unknown object by subtraction. Record.

Transfer the first unknown object to **Beaker B**, and determine the combined mass of **Beaker B** and object. Record. Determine the mass of the unknown object by subtraction. Record.

Although the two beakers used for the determination undoubtedly had different masses when empty, you should have discovered that the mass of the unknown object was the same, regardless of which beaker it was weighed in. Explain why this is so.

Repeat the process with the second unknown object and the two beakers.

Using a different balance from that used previously, determine the mass of each of the unknown objects on the second balance in the manner already described, using **Beaker A** only.

Compare the masses of the objects as determined on the two balances. Is there a difference in the masses determined for each object, depending on which balance was used? In future experiments always use the *same balance* for all mass determinations in a given experiment.

Show the results of your mass determinations of the unknown objects to your instructor, who will compare your mass determinations with the true masses of the unknown objects. If there is any major discrepancy, ask the instructor for help in using the balances.

Note: Most digital electronic balances have a "tare" function that permits you to place the container (beaker, etc.) on the pan then zero the balance again. This permits you to determine the mass of the contents directly.

Name: _____ Section: _____

Lab Instructor: _____ Date: _____

EXPERIMENT 1

The Determination of Mass: The Laboratory Balance

Pre-Laboratory Questions

1. Although we commonly say that we need to "weigh" a chemical in the lab, we are actually determining the **mass** of the chemical. Explain.

2. The directions in an experiment read as follows: "Weigh approximately 5 grams of NaCl to the nearest milligram." What are you expected to do?

3. When we determine the mass of an object in the lab, we commonly first determine the mass of an empty container, and then add the object of interest to the container and determine the mass of the container with the object of interest in it. Explain why this method is used.

4. Item #5 in the Introduction states that weighing objects above room temperature may introduce into a mass determination because 'warm air rises.' Why does warm air rise?

5. Why should liquids never be poured or used in the vicinity of a balance?

6. Why is it important to use the *same* balance when making several mass determinations of a given object?

EXPERIMENT 1

The Determination of Mass: The Laboratory Balance

Results/Observations

1. First Balance

	Beaker A	Beaker B
Mass of empty beaker	g	g
ID number of 1st object		
Mass of beaker + 1st object	g	g
Mass of first object alone (by difference)	g	g
Mass of empty beaker	g	g
ID number of 2nd object		
Mass of beaker + 2nd object	g	g
Mass of 2nd object alone (by difference)	g	g

2. Second balance

	Beaker A	Beaker B
Mass of empty beaker	g	g
ID number of first object		
Mass of beaker + first object	g	g
Mass of 1st object alone (by difference)	g	g
Mass of empty beaker	g	g
ID number of 2nd object		
Mass of beaker + 2nd object	g	g
Mass of 2nd object alone (by difference)	g	g

3. Difference in mass between the two balances (using **Beaker A** only).

Object ID #	Mass with 1st balance	Mass with 2nd balance	Mass Difference

Questions

1. Although the absolute masses of **Beaker A** and **Beaker B** were different, you were able to determine the correct masses of your two unknown objects using either beaker. Explain why.

2. Why is it important to keep a balance *clean*?

3. Why is it important always to use the *same balance* during the course of an experiment? Explain using examples from your own data.

EXPERIMENT 2

The Use of Volumetric Glassware

Objective

In this experiment you will investigate the precision and accuracy levels permitted by common laboratory volumetric glassware. The cleaning and care of such glassware will also be discussed.

Introduction

Most of the glassware in your laboratory locker has been marked by the manufacturer to indicate the volume contained by the glassware when filled to a certain level. The graduations etched or painted onto the glassware by the manufacturer differ greatly in the *precision* they indicate, depending on the type of glassware and its intended use. For example, beakers and Erlenmeyer flasks are marked with very *approximate* volumes, which serve merely as a rough guide to the volume of liquid in the container.[1] Other pieces of glassware, notably burets, pipets, and graduated cylinders, are marked much more carefully by the manufacturer to indicate precise volumes. It is important to distinguish when a *precise* volume determination is necessary and appropriate for an experiment and when only a *rough* determination of volume is needed.

Glassware that is intended to contain or to deliver specific precise volumes is generally marked by the manufacturer with the letters "TC" (to contain) or "TD" (to deliver). For example, a flask that has been calibrated by the manufacturer to contain exactly 500 mL of liquid at 20°C would have the legend "TC 20 500 mL" stamped on the flask. A pipet that is intended to deliver a precise 10.00 mL sample of liquid at 20°C would be stamped with "TD 20 10 mL." It is important not to confuse "TC" and "TD" glassware: such glassware may not be used interchangeably. The temperature (usually 20°C) is specified with volumetric glassware since the volume of a liquid *changes* with temperature, which causes the density of the liquid to change. While a given pipet will contain or deliver the same *volume* at any temperature, the *mass* (amount of substance present in that volume) will vary with temperature.

A. Graduated Cylinders

The most common apparatus for routine determination of liquid volumes is the *graduated cylinder*. Although a graduated cylinder does not permit as precise a determination of volume as do other volumetric devices, for many applications the precision given by the graduated cylinder is sufficient. Figures 2-1 and 2-2 show typical graduated cylinders. In Figure 2-2, notice the plastic safety ring, which helps to keep the graduated cylinder from breaking if it is tipped over. In Figure 2-1, compare the difference in graduations shown for the 10-mL and 100-mL cylinders. Examine the graduated cylinders in your lab locker, and determine the smallest graduation of volume that can be determined with each cylinder.

---

[1] They often carry the notation, "± 5%," indicating a 5% uncertainty.

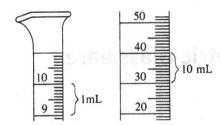

Figure 2-1. Expanded View of 10-mL and 100-mL Graduated Cylinders
Greater precision is possible with the 10-mL cylinder, since each numbered scale division represents 1 mL.

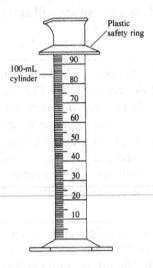

Figure 2-2. A 100-mL Graduated Cylinder
Note: The plastic ring is intended to prevent damage to the cylinder if it should tip over. This safety ring should be near the top of the cylinder during use.

When water (or an aqueous solution) is contained in a narrow glass container such as a graduated cylinder, the liquid surface is not flat, as might be expected. Rather, the liquid surface is *curved* (see Figure 2-3). This curved surface is called a **meniscus**, and is caused by an interaction between the water molecules and the molecules of the glass container wall. When reading the volume of a liquid that makes a meniscus, hold the graduated cylinder so that the meniscus is at eye level, and read the liquid level at the *bottom* of the curved surface.

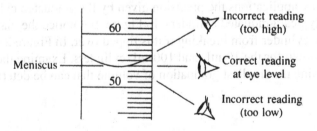

Figure 2-3. Reading A Meniscus
Read the bottom of the meniscus while holding at eye level.

B. Pipets

When a more precise determination of liquid volume is needed than can be provided by a graduated cylinder, a transfer *pipet* may be used. Pipets are especially useful if several measurements of the same volume are needed (such as in preparing uniform samples of a liquid). Two types of pipets are commonly available, as indicated in Figure 2-4. The *Mohr pipet* is calibrated at least at each milliliter and can be used to deliver *any size* sample up to the capacity of the pipet. The volumetric *transfer pipet* can deliver only *one size* sample (as stamped on the barrel of the pipet), but generally it is easier to use and is more reproducible.

Figure 2-4. A Mohr Pipet (left) and A Volumetric Transfer Pipet (right)

Note the calibration line on the upper neck of the volumetric pipet: the pipet will contain the specified volume when filled exactly to this mark.

Mohr pipets are generally labeled "TC," meaning, "*to contain*," followed by the maximum quantity of liquid the pipet can accurately handle. Volumetric pipets are generally labeled "TD," as in, "to deliver." Note that the Mohr type can be used for any volume up to its capacity, while the volumetric pipet can be used only for the volume shown on the neck.

Pipets are filled using a **rubber safety bulb** to supply the suction needed to draw liquid into the pipet. *It is absolutely forbidden to pipet by mouth in the chemistry laboratory.* Two common types of rubber safety bulb are shown in Figure 2-5.

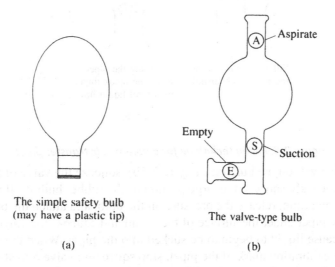

The simple safety bulb
(may have a plastic tip)

The valve-type bulb

(a)   (b)

Figure 2-5. Two common types of pipet safety bulbs: (a) the simple safety bulb (may have a plastic adapter tip); (b) the valve type bulb. Never pipet by mouth.

The *simple bulb should not actually be placed onto the barrel of the pipet.* This would most likely cause the liquid to be pulled into the bulb. Instead, squeeze the bulb first then merely press the opening of the bulb *against* the opening in the barrel of the pipet to apply the suction force, keeping the tip of the pipet under the surface of the liquid being withdrawn. Allow the suction to draw liquid into the pipet until the liquid level is 1 or 2 inches above the calibration mark on the barrel of the pipet. At this point, quickly place your index finger over the opening at the top of the pipet to prevent the liquid level from falling. By gently releasing the pressure of your index finger, the liquid level can be allowed to fall until it reaches the calibration mark of the pipet. The tip of the pipet may then be inserted into the container that is to receive the sample and the pressure of the finger removed to allow the liquid to flow from the barrel of the pipet. (See Figure 2-6.) If it is necessary to squeeze the bulb a second time to fill the pipet, remove the bulb from the pipet before squeezing it a second time.

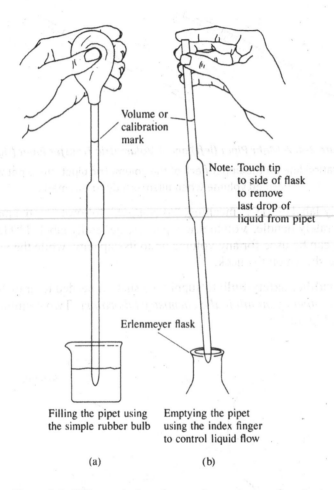

Volume or calibration mark

Note: Touch tip to side of flask to remove last drop of liquid from pipet

Erlenmeyer flask

Filling the pipet using the simple rubber bulb

Emptying the pipet using the index finger to control liquid flow

(a)   (b)

Figure 2-6. Filling technique for a volumetric transfer pipet

To use the more expensive valve-type bulb (see Figure 2-5b), squeeze the valve of the bulb marked *A (for "aspirate")*, and simultaneously squeeze the large portion of the rubber bulb itself to expel air from the bulb. Press valve *A* a second time, release the pressure on the bulb, and attach the bulb to the top of the pipet. Insert the tip of the pipet under the surface of the liquid and squeeze the valve marked *S* (*"suction"*) on the bulb, which will cause liquid to begin to be sucked into the pipet. When the liquid level has risen to an inch or two above the calibration mark of the pipet, stop squeezing valve *S* to stop the suction. Lift the tip of the pipet out of the liquid and press valve *E* (*expel*) carefully to allow the liquid to slowly drip out until the liquid meniscus is exactly on the desired mark. Transfer the pipet to the vessel that is to receive the liquid, and press valve *E* to empty the pipet. A drop or two will remain in the pipet; do not forcibly expel it. The use of this sort of bulb generally requires considerable practice to develop proficiency, but once mastered it is the fastest and easiest to use.

When using either type of pipet, observe the following rules:

1. The pipet must be scrupulously *clean* before use: wash with soap and water, and rinse with tap water and then with distilled water. If the pipet is clean enough for use, water will not bead up anywhere on the inside of the barrel.

2. To remove rinse water from the pipet (to prevent dilution of the solution to be measured), rinse the pipet with several small portions of the solution to be measured, discarding the rinsings in a waste beaker for disposal. It is not necessary to completely fill the pipet with the solution for rinsing. Draw 2–3 mL into the pipet, then place your fingers over both ends of the pipet, hold the pipet horizontally, and rotate the pipet several times to wash the walls of the pipet with the liquid.

3. The tip of the pipet must be kept *under the surface* of the liquid being measured out during the entire time suction is being applied, or air will be sucked into the pipet.

4. *When transferring* to the intended receiver vessel, the tip of the pipet must be *well above the liquid surface* of the receiver. The liquid from the pipet is allowed to fall as a stream.

5. Allow the pipet to drain for at least a minute when emptying to make certain the full capacity of the pipet has been delivered. Remove any droplets of liquid adhering to the tip of the pipet by touching the tip of the pipet to the side of the vessel that is receiving the sample.

6. If you are using the same pipet to measure out several different liquids, you should rinse the pipet with distilled water between liquids, and follow with a rinse of several small portions of the next liquid to be measured.

C. Burets

When samples of various sizes must be dispensed or measured precisely, a buret may be used. The buret consists of a tall, narrow calibrated glass tube, fitted at the bottom with a valve for controlling the flow of liquid. The valve is known as a **stopcock**. (See Figure 2-7.)

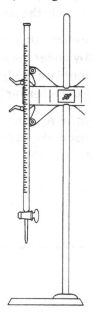

Figure 2-7. A Typical Student Buret

Most commonly, 50-mL burets are used in chemistry labs.

Like a pipet, a buret must be scrupulously *clean* before use. The precision permitted in reading a buret is on the order of 0.02 mL, but if the buret is not completely clean, this level of precision is not attainable. The buret should be cleaned at your bench, not in the sink (most sinks are too small for the length of the buret, and the buret is very easily broken). To clean the buret, first fill the buret with a small amount of soap and water using a funnel, and use a special long-handled buret brush to scrub the interior of the glass. Then rinse the buret by pouring tap water through it (allowing the water to drain from the tip), followed by several rinsings with distilled water. *Do not try to clean or rinse the pipet directly from the faucet.*

Before use, the buret should be rinsed with several small portions of the solution to be used in the buret. The buret should be tilted and rotated during the rinsings, to make sure that all rinse water is washed from it. Discard the rinsings. After use, the buret should again be rinsed with distilled water. Many of the reagent solutions used in burets may attack the glass of the buret if they are not removed, staining the glass and possibly destroying the calibration. To speed up the cleaning of a buret in future experiments, the buret may be left filled with distilled water during storage between experiments (if your locker is large enough to permit this).

A common mistake made by beginning students is to fill the buret with the reagent solution to be dispensed to exactly the 0.00 mL mark. This is *not* necessary or desirable in most experiments, and wastes time. The buret should be filled to a level that is comfortable for you to read (based on your height). The precise initial liquid level reading of the buret should be taken *before* the solution is dispensed and again *after* the liquid has been dispensed. The readings should be made to the nearest 0.02 mL. The volume of liquid dispensed is then obtained by simple subtraction of the two volume readings.

Safety Precautions	• **Safety eyewear approved by your institution must be worn at all times while you are in the laboratory, whether or not you are working on an experiment** • **When using a pipet, always use a rubber safety bulb or other pipetting device to apply the suction force. *Never pipet by mouth.*** • **Rinse the buret carefully. Do not attempt to place the top opening of the buret directly under the water tap: this usually breaks the buret. Set up the buret on the bench using a buret clamp/stand, an pour rinse water from a beaker into the buret using a funnel.** • **With caustic/corrosive reagents, use a funnel and place the buret and stand *on the floor* when filling. Never fill a buret whose openin is above your face.**

Apparatus/Reagents Required

- Graduated cylinders
- Pipets and safety bulb
- Buret and clamp
- Beakers
- Distilled water

Procedure

Record all data and observations directly on the report page in ink.

1. The Graduated Cylinder

Your instructor will set up a display of several graduated cylinders filled with different amounts of colored water. There are several sizes of cylinder available (e.g., 10-mL, 25-mL, 50-mL, and 100-mL). Examine each cylinder, paying particular attention to the marked scale divisions on the cylinder. For each graduated cylinder, to what fractional unit of volume does the smallest scale mark correspond?

Read the volume of liquid contained in each graduated cylinder and record. Make your readings to the level of precision permitted by each of the cylinders. Remember to read the liquid level as tangent to the meniscus formed by the liquid.

Check your readings of the liquid levels with the instructor before proceeding, and ask for assistance if your readings differ from those provided by the instructor.

Clean and wipe dry your 25-mL graduated cylinder and a 50-mL beaker (a rolled-up paper towel will enable you to dry the interior of the graduate). Weigh separately the graduated cylinder and beaker, and record the mass of each to the nearest milligram (0.001 gram), or to the maximum precision of the balance you are using.

Obtain about 100 mL of distilled water in a clean Erlenmeyer flask. Determine and record the temperature of the distilled water with a thermometer.

Fill the graduated cylinder with distilled water so that the meniscus of the water level lines up with the 25-mL calibration mark of the cylinder. Place distilled water in the 50-mL beaker up to the 25-mL mark. A clean plastic pipet may be used to add small portions of water to the cylinder to get the water level to exactly the calibration mark.

Weigh separately the graduated cylinder and the 50-mL beaker to the nearest milligram (0.001 gram), if your balance has that capability, and calculate the *mass* of water each contains.

Using the Table of Densities of water from the Appendix to this manual, calculate the *volume* of water present in the graduated cylinder and beaker from the exact *mass* of water present in each.

Compare the *calculated* volume of water (based on the mass of water) to the *observed* volumes of water determined from the calibration marks on the cylinder and beaker. Calculate the percentage *difference* between the calculated volume and the observed volume from the calibration marks. Why are the calibration marks on graduated cylinders and beakers taken only to be an approximate guide to volume?

2. The Pipet

Obtain a 25-mL pipet and rubber safety bulb. Clean the pipet with soap solution: draw a few mL of soap solution into the pipet, place your index fingers over the openings in either end of the pipet, hold the pipet horizontally, and rotate the pipet in your hand so as to coat all the inner surfaces with the soap solution. Rinse the pipet with tap water and then with small portions of distilled water. Practice filling and dispensing distilled water from the pipet using the rubber safety bulb until you feel comfortable with the technique. Ask your instructor for assistance if you have any difficulties in the manipulation.

Clean and wipe dry a 150-mL beaker. Weigh the beaker to the nearest milligram (0.001 gram) and record.

Obtain about 100 mL of distilled water in a clean Erlenmeyer flask. Determine and record the temperature of the water.

Pipet exactly 25 mL of the distilled water from the Erlenmeyer flask into the clean beaker you have weighed. Reweigh the beaker containing the 25 mL of water. Determine the mass of water transferred by the pipet.

Using the Table of Densities found in the Appendix to this manual, calculate the *volume* of water transferred by the pipet from the *mass* of water transferred. Compare this calculated volume to the volume of the pipet as specified by the manufacturer. Any significant difference in these two volumes is an indication that you need additional practice in pipetting. Consult with your instructor for help.

You will be asked to compare the accuracy and precision of the volume dispensed by the pipet to the volumes as determined in Part 1 using a graduated cylinder or beaker.

3. The Buret

Obtain a buret and set it up in a clamp attached to a ring stand on your lab bench. Then move the buret and ring stand to the floor. Place a large beaker under the stopcock/delivery tip of the buret.

Using a funnel, fill the buret with tap water, and check to make sure that there are no leaks from the stopcock before proceeding. If the stopcock leaks, have the instructor examine the stopcock to make sure that all the appropriate washers are present. If the stopcock cannot be made leak-proof, replace the buret.

Add a few mL of soap solution to the buret and use a long-handled buret brush to scrub the inner surface of the buret. Rinse all soap from the buret with tap water, being sure to flush water through the stopcock, as well. Rinse the buret with several small portions of distilled water.

Place the buret and stand on the floor, then use a funnel to fill the buret to above the zero mark with distilled water. Return the buret and stand to your workbench.

Open the stopcock of the buret and allow the distilled water to run from the buret into a beaker or flask. Examine the buret while the water is running from it. If the buret is clean enough for use, water will flow in sheets down the inside surface of the buret without beading up anywhere. If the buret is not clean, repeat the scrubbing with soap and water.

Once the buret is clean, refill it with distilled water to a point somewhat below the zero mark. Determine the precise liquid level in the buret to the nearest 0.02 mL.

With a paper towel, clean and wipe dry a 100- or 150-mL beaker. Determine the mass of the beaker to the nearest milligram.

Place the weighed beaker beneath the stopcock of the buret. Open the stopcock of the buret and run water into the beaker until approximately 25 mL of water have been dispensed. Determine the precise liquid level in the buret to the nearest 0.02 mL. Calculate the volume of water that has been dispensed from the buret by subtraction of the two liquid levels.

Reweigh the beaker (containing the water dispensed from the buret) to the nearest milligram, and determine the mass of water transferred to the beaker from the buret.

Use the Table of Densities in the Appendix to calculate the volume of water transferred from the mass of the water. Compare the volume of water transferred (as determined by reading the buret) with the calculated volume of water (from the mass determinations). If there is any significant difference between the two volumes, most likely you need additional practice in the operation and reading of the buret.

You will be asked to compare the volume dispensed by the buret to the volumes as determined in Part 1 using a graduated cylinder or beaker, and to the volume dispensed by the buret in Part 2?

EXPERIMENT 2

The Use of Volumetric Glassware

Pre-Laboratory Questions

1. Pipets used for the transfer of samples of solutions are always *rinsed* with a small portion of the solution to be dispensed with the pipet before the actual sample is taken. Calculate the *percentage error* likely to arise in an experiment if 1-mL, 5-mL, and 10-mL pipets are used for transfer and each pipet contains 5 drops of liquid adhering to the inside of the barrel. Assume a volume of about 0.05 mL per drop.

2. Stamped by the manufacturer on a volumetric pipet is the notation, "TD 25 mL." Explain what this means.

3. Suppose a 1.0-mL air bubble is trapped in the tip of your buret and you don't notice it before measuring out liquid samples with the buret. What percentage error would be introduced in a 35.0 mL sample of liquid if the air bubble comes out of the tip of the buret while the liquid sample is being transferred?

4. What is a *meniscus*? Why do some liquids develop a meniscus? How do we read the liquid level in a volumetric measuring device in which a solution forms a meniscus?

5. How can you tell when the inside of the barrel of a pipet or buret is *clean* enough for use?

EXPERIMENT 2

The Use of Volumetric Glassware

Results/Observations

1. **The Graduated Cylinder**

Liquid Color	Cylinder Size	Volume Contained
	mL	mL
	mL	mL
	mL	mL
	mL	mL

Mass of empty 25-mL graduated cylinder   _____

Mass of cylinder plus water sample   _____

Mass of water in cylinder   _____

Temperature of water sample   _____

Density of water at this temperature   _____

Calculated volume of water transferred   _____

Difference: observed volume and calculated volume   _____

Mass of empty 50-mL beaker   _____

Mass of beaker plus water sample   _____

Mass of water in beaker   _____

Temperature of water sample   _____

Density of water at this temperature   _____

Calculated volume of water transferred   _____

Difference: observed volume and calculated volume   _____

---

2. **The Pipet**

Mass of empty 150-mL beaker   _____

Mass of beaker plus water sample   _____

Mass of water transferred   _____

Temperature of water sample   _____

Density of water at this temperature   _____

Calculated volume of water transferred   _____

Ratio: actual/calculated volumes transferred   _____

- -

3.   The Buret

Mass of empty beaker   _____

Initial liquid level in buret   _____

Final liquid level in buret   _____

Volume of liquid transferred   _____

Mass of beaker plus water   _____

Mass of water transferred   _____

Temperature of water   _____

Density at this temperature   _____

Calculated volume of water transferred   _____

Ratio: actual/calculated volumes transferred   _____

Question

Based on your experience in this experiment, briefly discuss the *relative precision* permitted by a graduated cylinder, a pipet, and a buret. Give several circumstances, other than those used here, under which you would choose to use each instrument, in preference to the other two.

EXPERIMENT 3

Density Determinations

Objective

Density is an important property of matter that often proves useful as a method of identification. In this experiment, you will determine the densities of regularly and irregularly shaped solids, as well as the densities of pure liquids and solutions.

Introduction

The density of a sample of matter represents the mass contained within a unit volume of space within the sample. For most samples, a "unit volume" means 1.0 mL. The units of density, therefore, are quoted in units of grams per milliliter (g/mL) or grams per cubic centimeter (g/cm^3) for most samples of matter.

Since we seldom deal with exactly 1.0 mL of substance in the chemistry laboratory, we usually say that the density of a sample represents the mass of the sample divided by its volume.

$$density = \frac{mass}{volume}$$

Because the density does in fact represent a *ratio*, the mass of *any* size sample divided by the volume of that sample, in mL, gives the mass that 1.0 mL of the same sample would possess.

Densities are usually determined and reported at 20°C (around room temperature) because the volume of a sample, and hence the density, will often *vary* with temperature. This is especially true for gases, with smaller (but still often significant) changes for liquids and solids. References (such as the various chemical handbooks) always specify the *temperature* at which a density was determined.

Density can be used to determine the concentration of solutions in certain instances. When a solute is dissolved in a solvent, the density of the solution will be different from that of the pure solvent itself. Handbooks and online references list detailed information about the densities of solutions as a function of their composition. If a sample is known to contain only a single solute, the density of the solution can be measured experimentally and then the reference can be consulted to determine what concentration of solute gives rise to the observed density.

The determination of the density of certain physiological liquids is often an important screening tool in medical diagnosis. For example, if the density of urine differs from normal values, this may indicate a problem with the kidneys secreting substances that should not be lost from the body. The determination of density (specific gravity) is almost always performed during a urinalysis. Another example is the use of a hydrometer to measure the density of the coolant in an automobile radiator, to determine if it contains enough antifreeze.

There are several techniques used for the determination of density. The method used will depend on the *type of sample* and on the *precision* desired for the measurement. For example, devices that permit a quick, reliable, routine determination have been constructed for determinations of the density of urine. In general, a density determination will involve the determination of the mass of the sample with a balance, but the method used to determine the volume of the sample will differ from situation to situation. Several methods of volume determination are explored in this experiment.

For *solid* samples, there may be different methods needed for the determination of the volume, depending on whether or not the solid is regularly shaped. If a solid has a *regular* shape (e.g., cube, rectangle, cylinder), the volume of the solid may be determined by geometry:

For a cubic solid, volume = $(\text{edge})^3$

For a rectangular solid, volume = length × width × height

For a cylindrical solid, volume = $\pi \times (\text{radius})^2 \times$ height

If a solid does *not* have a regular shape, it may be possible to determine the volume of the solid making use of **Archimedes' principle**, which states that an insoluble, non-reactive solid will *displace* a volume of liquid equal to its own volume. Typically, an irregularly shaped solid is added to a liquid in a volumetric container (such as a graduated cylinder) and the *change in liquid level* determined.

For liquids, very precise values of density may be determined by pipetting an exact volume of liquid into a sealable weighing bottle (this is especially useful for highly volatile liquids) and then determining the mass of liquid that was pipeted. A more convenient method for routine density determinations for liquids is to weigh a particular volume of liquid as contained in a graduated cylinder.

Safety Precautions	• **Safety eyewear approved by your institution must be worn at all times while you are in the laboratory, whether or not you are working on an experiment.**
	• **The unknown liquids may be flammable and their vapors may be toxic. Keep the unknown liquids away from open flames and do not inhale their vapors. Dispose of the unknown liquids as directed by the instructor.**
	• **Dispose of the metal samples only in the special container designated for their collection. Do not combine different metals in the same container.**

Apparatus/Reagents Required

- Unknown liquid sample
- Unknown metal samples
- Sodium chloride
- Ruler or calipers

Procedure

Record all data and observations directly on the report page in ink.

1. Determination of the Density of Solids

Obtain a regularly shaped solid, and record its identification number. With a ruler or calipers, determine the physical dimensions of the solid to the nearest 0.2 mm. Calculate the volume of the solid. Be sure to indicate the geometry formula used for the calculation.

Determine the mass of the regularly shaped solid to at least the nearest mg (0.001 g). Calculate the density of the solid.

Obtain a sample of unknown metal pellets (metal shot) or an irregularly shaped chunk of metal and record its identification code number. Determine the mass of a sample of the metal of approximately 50 g, but record the *actual* mass of metal taken to the nearest mg (0.001 g).

Add water to your 100-mL graduated cylinder to approximately the 50-mL mark. Record the exact volume of water in the cylinder to the precision permitted by the calibration marks of the cylinder.

Carefully, to avoid splashing, pour the metal sample into the graduated cylinder, making sure that none of the pellets sticks to the walls of the cylinder above the water level. Stir/shake the cylinder to make certain that no air bubbles have been trapped among the metal pellets. (See Figure 5-1.)

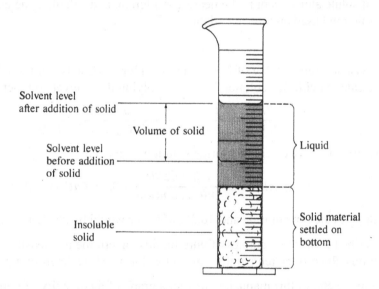

Figure 3-1. Measurement of Volume By Displacement
An insoluble object displaces a volume of liquid equal to its own volume.

Read the level of the water in the graduated cylinder, again making your determination to the precision permitted by the calibration marks of the cylinder. Assuming that the metal sample does not dissolve in or react with water, the change in water levels represents the volume of the metal pellets.

Calculate the density of the unknown metal pellets. Carefully decant as much of the water as you can, but without losing any of the metal sample. Tap the metal out onto several thicknesses of paper towels and allow the sample to dry. Turn in the sample of metal pellets to your instructor (*do not discard*).

2. Density of Pure Liquids

Clean and dry a 25-mL graduated cylinder (a rolled-up paper towel may be conveniently used to dry the interior of the cylinder). Determine the mass of the dried graduated cylinder to the nearest mg (0.001 g).

Add distilled water to the cylinder so that the water level is *above* the 20-mL mark but *below* the 25-mL mark. Determine the temperature of the water in the cylinder.

Determine the mass of the graduated cylinder and water to the nearest milligram.

Record the exact volume of water in the cylinder, to the correct level of precision permitted by the calibration marks on the barrel of the cylinder.

Calculate the density of the water. Compare the measured density of the water with the value listed online or in a handbook of chemical data for the temperature of your experiment.

Clean and dry the graduated cylinder.

Obtain an unknown liquid and record its identification number. Determine the density of the liquid, using the method just described for water. Be certain to start with a totally *clean and dry* graduate!

3.   Density of Solutions

The concentration of a solution is often conveniently described in terms of the solution's *percentage composition* on a mass basis. For example, a 5% sodium chloride solution contains 5 g of sodium chloride in every 100 g of solution (which corresponds to 5 g of sodium chloride for every 95 g of water present).

Prepare solutions of sodium chloride in distilled water consisting of the following approximate percentages by mass: 5%, 10%, 15%, 20%, and 25%. Prepare at least 25 mL of each solution. Make the mass determinations of solute and solvent to the nearest milligram, and calculate the exact percentage composition of each solution based on this mass data.

Example:

Suppose you wish to prepare 50. g of 7.5% calcium chloride ($CaCl_2$) solution. When we say we have a "7.5 % calcium chloride solution" we are implying this conversion factor

$$\frac{7.5 \ g \ CaCl_2}{100. \ g \ solution} = \frac{7.5 \ g \ CaCl_2}{7.5 \ g \ CaCl_2 + 92.5 \ g \ H_2O}$$

So, to prepare 50 g of 7.5% calcium chloride solution, you need

$$50. \ g \ solution \times \frac{7.5 \ g \ CaCl_2}{100. \ g \ solution} = 3.8 \ g \ CaCl_2 \ needed$$

Therefore, 3.8 g of calcium chloride and 46.2 g of water would be needed.

Using the method described earlier for samples of pure liquids, determine the density of each of your sodium chloride solutions. Record the temperature of each solution while determining its density.

Using graph paper from the end of this manual, construct a graph of the **density** of your solutions *versus* the **percentage of NaCl** the solution contains. What sort of relationship exists between density and the composition of the solution? Attach your graph to your report when you hand it in. Make certain the graph is properly labeled, the scales are precise and uniform, and that the plot is carefully drawn with a best-fit straight line through the points.

Use a handbook of chemical data or an online reference to determine the true density of each of the solutions you prepared. Calculate the error in each of the densities you determined.

EXPERIMENT 3

Density Determinations

Pre-Laboratory Questions

1. Define *specific gravity*. What units are used for specific gravity?

2. An insoluble, nonreactive solid metal sphere weighing 21.52 g is added to 28.7 mL of water in a graduated cylinder. The water level rises to 32.8 mL. Calculate the density of the metal.

3. An empty beaker weighs 31.9982 g. A 10-mL pipet sample of an unknown liquid is transferred to the beaker. The mass of the beaker plus the 10-mL liquid sample is 39.8521 g. Calculate the density of the unknown liquid.

4. A 25-mL pipet sample of an alcohol is found to weigh 20.71 g. What would be the mass of 35.25 mL of the same alcohol? What volume would be occupied by 125.4 g of this alcohol?

5. When determining the density of a pure liquid or a solution, the *temperature* of the sample must be measured. Why does the density of a liquid sample depend on the temperature?

6. Given your experience in mass determinations (Experiment 1) and with volumetric glassware (Experiment 2), which factor used in calculating the density – mass or volume – is likely to be measurable with greater precision? Explain.

Name: _____ Section: _____

Lab Instructor: _____ Date: _____

EXPERIMENT 3

Density Determinations

Results/Observations

1. **Density of Solids**

Regular Solid	ID # of Solid _____	Metal Pellets	ID # of Pellets _____
Dimensions of solid		Mass of pellets used	
Calculated volume of solid		Initial water level	
Mass of solid		Final water level	
Calculated density of solid		Volume of metal pellets	
----------	----------	Calculated density of metal pellets	

(Calculation Space)

2. **Density of Pure Liquids**

Water		Unknown Liquid ID # _____	
Mass of Empty Graduate		Mass of Empty Graduate	
Mass of Graduate + Water		Mass of Graduate + Unknown	
Mass of Water		Mass of Unknown	
Volume of Water		Volume of Unknown	
Density of Water		Density of Unknown	
Temperature of Water		----------	----------
Expected density at this Temperature		----------	----------

3.   Density of Solutions

% NaCl	Density as Measured	Temperature	Accepted Value	% error
5				
10				
15				
20				
25				

Questions

1. What error would be introduced into the determination of the density of the regularly shaped solid if the solid were *hollow*? Would the apparent volume of the solid be larger or smaller than the actual volume? Would the density calculated be too high or too low?

2. What error would be introduced into the determination of the density of the irregularly shaped metal pellets if you had not stirred/shaken the pellets to remove adhering air bubbles? Would the density be too high or too low?

3. Your data for the density of sodium chloride solutions should have produced a straight line when plotted. How could this plot be used to determine the density of *any* concentration of sodium chloride solution?

4. What is the difference between the *density* of a solution and its *specific gravity*?

28

EXPERIMENT 4

The Determination of Boiling Point

Objective

In this experiment, you will check your thermometer for errors by determining the temperatures of two stable reference equilibrium systems. You will then use your calibrated thermometer in determining the boiling point of an unknown substance.

Introduction

Traditionally, the most common laboratory device for the measurement of temperature has been the thermometer. The typical thermometer used in the general chemistry laboratory permits the determination of temperatures from –20° to +120°C. Most laboratory thermometers are constructed of *glass*, and so they are very fragile.

Traditionally, mercury was used as the temperature-sensing liquid in thermometers, but due to its extreme toxicity, other liquids, such as colored alcohol or hexane, have replaced mercury. Unfortunately, these liquid-filled thermometers are notoriously inaccurate; as a result, many institutions have begun using either digital thermometers (e.g., cooking thermometers) or interfaced temperature probes. If this is the case for your laboratory, you will probably be directed to skip the calibration step of the Procedure.

The typical laboratory thermometer contains a bulb (reservoir) of colored liquid at the bottom; it is this portion of the thermometer that actually senses the temperature. The glass barrel of the thermometer above the liquid bulb contains a narrow capillary opening down its center. As the liquid in the bulb gets heated, it expands, causing the liquid to rise in the capillary opening. The capillary tube in the barrel of the thermometer has been manufactured to very strict tolerances, and is very regular in cross-section along its length. This ensures that the rise in the level of liquid in the capillary tube as the thermometer is heated will be directly related to the temperature of the thermometer's surroundings.

Although the laboratory thermometer may appear similar to the sort of clinical thermometer used for determination of body temperature, the laboratory thermometer does *not* have to be shaken before use. Medical thermometers are manufactured with a *constriction* in the capillary tube that is intended to prevent the liquid level from changing once it has risen. The liquid level of a laboratory thermometer, however, changes *immediately* when removed from the substance whose temperature is being measured. For this reason, temperature readings with the laboratory thermometer must be made while the bulb of the thermometer is actually present in the material being determined.

Because the laboratory thermometer is so fragile, it is helpful to check that the thermometer provides reliable readings before any important determinations are made with it. Often, thermometers develop nearly invisible hairline cracks along the barrel, making them unsuitable for further use. This happens especially if you are not careful in opening and closing your laboratory locker.

To check whether or not your thermometer is reading temperatures correctly, you will *calibrate* the thermometer. To do this, you will determine the reading given by your thermometer in two systems whose temperature is *known with certainty*. If the readings given by your thermometer differ by more than one degree from the true temperatures of the systems measured, you should *exchange* your thermometer, and then calibrate the new thermometer. A mixture of ice and water has an equilibrium temperature of exactly 0°C, and will be used as the first calibration system. A boiling water bath, whose exact temperature can be determined from the day's barometric pressure, will be used as the second calibration system in this experiment.

Once your thermometer has been calibrated, you will use the thermometer in a simple but very important experiment: the determination of the boiling point of a pure chemical substance. The boiling points of pure substances are important because they are *characteristic* for a given substance; that is, under the same laboratory conditions, a given substance will always have the *same* boiling point. Characteristic physical properties such as the boiling point of a substance are of immense help in the identification of unknown substances and in establishing the degree of purity. Such properties are routinely reported in scientific papers when new substances are isolated or synthesized, and are compiled in tables in the various handbooks of chemical data that are available in science libraries and online. When an unknown substance is isolated from a chemical system, its boiling point may be measured (along with other characteristic properties) and then compared with tabulated data. If the experimentally determined physical properties of the unknown match those found in the literature, you may typically assume that you have identified the unknown substance.

The boiling point of a liquid is defined as the temperature at which the vapor escaping from the surface of the liquid has a pressure equal to the pressure existing above the liquid. In the most common situation of a liquid boiling in a container open to the atmosphere, the pressure above the liquid will be the day's barometric pressure. In other situations, the pressure above a liquid may be reduced by means of a vacuum pump or aspirator, which enables the liquid to be boiled at a much lower temperature than in an open container (this is especially useful in chemistry when a liquid is unstable, possibly decomposing if it were heated to its normal boiling point under atmospheric pressure). When boiling points are tabulated in the chemical literature, the *pressure* at which the boiling point determinations were made is also listed. The method to be used for the determination of boiling point is a semi-micro method that requires only a few drops of liquid.

Apparatus/Reagents Required

- Thermometer and clamp
- Beakers, 400 mL (2), 100 mL
- Melting point capillaries
- Test tube, 10 × 75 mm
- Burner and rubber tubing
- Glass-cutting file
- Medicine dropper
- Unknown sample for boiling point determination
- Ice

Safety Precautions	• Safety eyewear approved by your institution must be worn at all times while you are in the laboratory, whether or not you are working on an experiment.
	• Thermometers are often fitted with rubber stoppers as an aid in supporting the thermometer with a clamp. Inserting a thermometer through a rubber stopper must be done carefully to prevent breaking of the thermometer, which might cut you. Your instructor will demonstrate the proper technique for inserting your thermometer through the hole of a rubber stopper. Glycerin is used to lubricate the thermometer and stopper. Protect your hands with a towel during this procedure.
	• The liquid unknowns used in the experiment are *flammable*. Although only small samples of the liquids are used, the danger of fire is not completely eliminated. Keep the liquids away from all open flames!
	• Some of the liquids used in this experiment are toxic if inhaled or absorbed through the skin. The liquids should be disposed of in the manner indicated by the instructor.

Procedure

Record all data and observations directly on the report page in ink.

1. Calibration of the Thermometer – *Note: If your school uses digital thermometers or interfaced temperature probes, you may be directed to skip this step.*

Fill a 400-mL beaker with ice, and add tap water to the beaker until the ice is covered with water. Stir the mixture with a stirring rod for 30 seconds. *Do NOT stir with a glass thermometer!*

Clamp the thermometer to a ring stand so that the bottom 2–3 inches of the thermometer is dipping into the ice bath. Make certain that the thermometer is suspended freely in the ice bath and is *not* touching either the walls or the bottom of the beaker. Also, make certain that you can read the scale in the ranges of −5°C to +5°C, and 90°C to 110°C. If another part of the scale is obscured by the clamp or stopper, it will not affect your ability to complete this part of the experiment.

Allow the thermometer to stand in the ice bath for 2 minutes, and then read the temperature indicated by the thermometer to the nearest 0.2 of a degree (± 0.2°C). Remember that the thermometer must be read while still *in* the ice bath. Liquid-filled thermometers are notoriously inaccurate, but if the reading indicated by the thermometer differs from 0°C by more than two degrees, replace the thermometer and repeat the ice bath calibration.

Allow the thermometer to warm to room temperature by resting it in a safe place on the laboratory bench.

Set up an apparatus for boiling as indicated in Figure 3-1, using a 100-mL beaker containing about 75 mL of water. Add 2–3 boiling chips to the water, and heat the water to boiling.

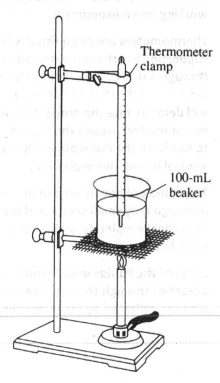

Figure 4-1. Apparatus for Calibration of the Thermometer

The thermometer must be suspended freely in the water bath and must not touch the bottom of the beaker

Using a clamp, suspend the thermometer so that it is dipping halfway into the boiling water bath. Make certain that the thermometer is *not* touching the walls or bottom of the beaker. Allow the thermometer to stand in the boiling water for 2 minutes; then record the thermometer reading to the nearest 0.2°C.

A boiling water bath has a temperature near 100°C, but the actual temperature of boiling water is dependent on the barometric pressure and changes with the weather from day to day. Your instructor will list the current barometric pressure on the chalkboard. Using a handbook of chemical data or an online source, look up the actual boiling point of water for this barometric pressure and record.

If your measured boiling point differs from the handbook value for the provided barometric pressure by more than one or two degrees, exchange your thermometer at the stockroom, and repeat the calibration of the thermometer in both the ice bath and the boiling water bath.

2. Determination of Boiling Point of an Unknown Liquid – *Note: If you are using a digital thermometer or interfaced temperature probe, the procedure is the same as is described here. The small test tube is attached to the digital thermometer or probe directly.*

Set up a beaker half-filled with water on a wire screen on an iron ring/ring stand apparatus.

Obtain a clean, dry 10×75-mm semi-micro test tube, which will contain the boiling-point sample. Attach the test tube to the lower end of your thermometer with two small rubber bands or narrow rings of rubber tubing. See Figure 3-2 on the following page.

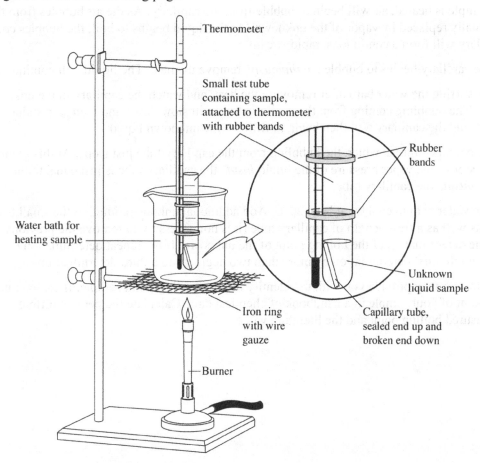

Figure 4-2. Apparatus for boiling point determination

Obtain an unknown liquid for boiling-point determination from your instructor, and record the identification code number of the unknown in your notebook and on the lab report page. Transfer part of the unknown sample to the small test tube, until the test tube is approximately half full.

Obtain a melting-point capillary. If the capillary tube is open at both ends, heat one end briefly in a flame to seal it off. Using a glass file, carefully cut the capillary about 1 inch from the sealed end (*Caution!*). Do *not* fire-polish the cut end of the capillary.

Place the small portion of capillary *sealed end up* into the boiling-point sample in the test tube.

The capillary has a rough edge at the cut end, which serves as a surface at which bubbles can form during boiling. The capillary is filled with air when inserted, sealed end up, into the liquid; the presence of this air can be used to judge when the vapor pressure of the unknown liquid reaches atmospheric pressure.

Lower the thermometer/sample apparatus into the water bath so that the bulb of the thermometer and the sample are suspended freely and not touching the bottom or sides of the beaker. Place a stirring rod in the beaker.

Begin heating the water bath in the beaker with a low flame so that the temperature rises by one or two degrees per minute – *faster heating will impair the accuracy of your results*. Continuously and gently stir the water bath with the stirring rod to make sure the heat being applied is distributed evenly. Watch the small capillary tube in the unknown sample while heating.

As the sample is heated, air will begin to bubble from the capillary. As the air bubbles from the capillary, it is gradually replaced by vapor of the unknown. As the liquid begins to boil, the bubbles coming from the capillary will form a continuous, rapid stream.

When the capillary begins to bubble *continuously*, remove the heat. The liquid will continue to boil.

Continue stirring the water bath after removing the heat, and watch the capillary in the unknown sample carefully. The bubbling coming from the capillary tube will slow down and then stop suddenly after a few moments, and the capillary will then begin to fill with the unknown liquid.

Record the temperature at which the bubbling from the capillary tube just stops. At this point (where the bubbling stops), the vapor pressure of the liquid *inside* the capillary tube is just equal to the atmospheric pressure *outside* the capillary tube.

Allow the water bath to cool by at least 20°C. Add additional unknown liquid to the small test tube if needed, as well as a fresh length of capillary tube (it is not necessary to remove the previous capillary). Repeat the determination of the boiling point of the unknown. If the repeat determination of boiling point differs from the first determination by more than two degrees, do a third determination.

Your instructor may provide you with the identity of your boiling-point sample. If so, look up the true boiling point of your sample in a handbook of chemical data. Calculate the percent difference between your measured boiling point and the literature value.

Name: _____ Section: _____

Lab Instructor: _____ Date: _____

EXPERIMENT 4

The Determination of Boiling Point

Pre-Laboratory Questions

1. Using a handbook of chemical data or an online reference, find the boiling points (at 1 atm) for the following substances:

    a. Chloroform _____

    b. 2-methyl-2-propanol _____

    c. diethyl ether _____

    d. 2-butanol _____

    e. 1-butanol _____

    f. acetone _____

    g. methanol _____

    h. (Cite reference) _____

2. What temperature should a properly working thermometer display in an ice-water slush bath? In a boiling water bath?

3. What is the purpose of the short length of broken capillary tube used in the determination of boiling point?

4. What is a *characteristic property* of a pure substance? Is the boiling point of a substance a characteristic property? Explain.

5. Describe the method to be used in this experiment for determining the boiling point of an unknown liquid sample.

EXPERIMENT 4

The Determination of Boiling Point

Results/Observations

1. **Calibration of the Thermometer**

   Reading of thermometer in ice bath, $^{\circ}C$ _____

   Reading of thermometer in boiling water, $^{\circ}C$ _____

   Barometric pressure, mm Hg   _____

   True boiling point of water at this pressure, $^{\circ}C$ _____

2. **Determination of The Boiling Point of An Unknown**

   Identification code number of unknown sample   _____

   Boiling point of unknown sample, $^{\circ}C$

          First trial   _____

          Second trial   _____

          (Third trial)   _____

Questions

1. The temperature of an ice/water mixture remains at 0 °C (until all of the ice has melted) because the combination of ice and water represents an *equilibrium system*. Explain.

2. Why were the temperatures of an *ice bath* and a *boiling water bath* chosen for the calibration of the thermometer?

3. Food products such as cake mixes often list special directions for preparation at high altitudes. Why are special directions needed in such situations? Would a food take a longer or shorter period of time to reach the desired degree of 'doneness' at high altitudes? Explain.

4. Although we nearly always fire-polish the ends of cut glass tubing (to minimize the risk of cuts), you were told not to fire-polish the capillary tube used in this experiment. What purpose did having a rough edge on the capillary serve?

EXPERIMENT 5

Recrystallization and Melting Point Determination

Objective

The separation of mixtures into their constituent components defines an entire subfield of chemistry referred to as **separation science**. In this experiment, **recrystallization**, one of the most common techniques for the resolution of mixtures of solids, will be examined.

Introduction

Mixtures occur very commonly in chemistry. When a new chemical substance is synthesized in a research lab, for example, the new substance usually must be separated from various side-products, catalysts, as well as any excess starting reagents still present. When a substance is to be isolated from a natural biological source, the substance of interest is generally found in a very complex mixture with many other substances, all of which must be removed. Chemists have developed a series of standard methods for resolution and separation of mixtures, one of which – recrystallization – will be investigated in this experiment.

Mixtures of solids often may be separated on the basis of differing *solubilities* of the components. If one of the components of the mixture is very soluble in water, for example, while the other components are insoluble, the water-soluble component may be removed from the mixture by simple *filtration* through ordinary filter paper. A more general case occurs when all the components of a mixture are soluble, but to different extents, in water or some other solvent. The solubility of substances in many cases is greatly influenced by *temperature*. By controlling carefully the temperature at which solution occurs or at which a filtration is performed, it may be possible to separate the components of the mixture. Most commonly, a sample is added to the solvent and is then heated to boiling. The hot solution is then filtered to remove completely insoluble substances. The sample is then cooled, either to room temperature or below, which causes recrystallization of those substances whose solubilities are very temperature-dependent. These crystals can then be isolated by filtration, and the filtrate remaining can be concentrated to reclaim those substances whose solubilities are not so temperature-dependent.

After a substance has been isolated as a pure solid from a mixture, it is a very common practice to determine the **melting point** of the material. When a pure solid melts during heating, the melting usually occurs quickly at one specific, *characteristic* temperature. For certain substances, especially more complicated organic substances or biological substances that tend to decompose slightly when heated, the melting may occur over a span of several degrees, called the **melting range**. Melting ranges are also commonly observed if the substance being tested is not completely pure. The presence of an impurity will broaden the melting range of the major component and will also lower the temperature at which melting starts. As with boiling points (Experiment 3), the characteristic melting points of pure, solid substances are routinely reported in the scientific literature and are tabulated in handbooks for use in the subsequent identification of unknown substances. Today, most of that information is available online.

Although this experiment assumes that melting temperatures will be determined using an oil-filled device known as a Thiele tube, many colleges and universities have adopted electrically powered melting instruments. If such is the case at your institution, your instructor will demonstrate its operation.

Safety Precautions	• Safety eyewear approved by your institution must be worn at all times while you are in the laboratory, whether or not you are working on an experiment. • The solid mixture contains benzoic acid, which may be irritating to the skin and respiratory tract. It is poisonous if ingested. • When moving hot containers, use metal tongs or a towel to avoid burns. Beware of burns from steam while solutions are being heated. • *Caution!* Oil is used as the heating fluid in the Thiele tube used for the melting point determinations that follow. Hot oil may spatter if it is heated too strongly, especially if any moisture is introduced into the oil from glassware that is not completely dry. The oil may smoke or ignite if heated above 200 °C. <u>If the oil appears cloudy, inform the instructor and do not heat it.</u> • Dispose of all solids and liquids as directed by the instructor.

Apparatus/Reagents Required

- Impure benzoic acid sample (benzoic acid that has been colored with charcoal)
- melting point capillaries
- filter paper
- rubber bands
- oil-filled Thiele tubes

Procedure

Record all observations and data directly on the report page in ink.

1. Recrystallization

Obtain a sample of impure benzoic acid for recrystallization. Benzoic acid is fairly soluble in hot water, but has a much lower solubility in cold water. The benzoic acid has been contaminated with charcoal, which is not soluble under either temperature condition. Transfer the benzoic acid sample to a clean 150-mL beaker.

Set up a short-stem gravity filter funnel in a small metal ring clamped to a ring stand. Fit the filter funnel with a piece of filter paper folded in quarters to make a cone. See Figure 5-1 (a).

Moisten the filter paper slightly so that it will remain in the funnel. Place a clean 250-mL beaker beneath the stem of the funnel.

Set up a second 250-mL beaker, about half filled with distilled water, on a wire screen over a metal ring. Heat the water to boiling.

When the water is boiling, pour about two-thirds of the water into the beaker containing the benzoic acid sample. Use a towel to protect your hands from the heat and steam.

Pour the remainder of the boiling water through the gravity funnel to heat it. If the funnel is not preheated, the benzoic acid may crystallize in the stem of the funnel rather than passing through it. Discard the water that is used to preheat the funnel.

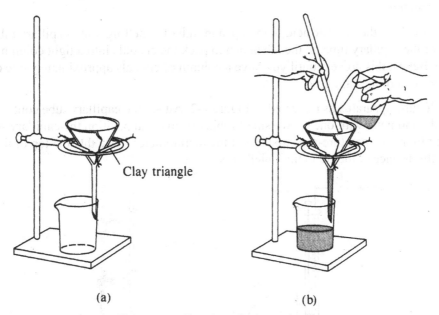

(a) (b)

Figure 5-1. Filtration of a hot solution

Use a stirring rod as a guide for running the hot solution into the funnel.
Do not fill the funnel more than half-full at a time, to prevent the solution being
lost over the rim of the filter paper cone.

Transfer the 150-mL beaker containing the benzoic acid mixture to the burner and reheat it until the mixture just begins to boil again (this should only take a few seconds). Stir the mixture to make sure that the benzoic acid dissolves to the greatest extent possible.

Using a towel to protect your hands, pour the benzoic acid mixture through the preheated funnel. Catch the filtrate in a clean beaker. See Figure 4-1 (b).

Allow the filtrate to cool to room temperature.

When the benzoic acid solution has cooled to room temperature, filter the crystals to remove water. Wash the crystals with two 10-mL portions of ice-cold water.

Transfer the filtrate from which the crystals have been removed to an ice bath to see if additional crystals will form at the lower temperature. Examine, but do not isolate this second crop of crystals.

Transfer the filter paper containing the benzoic acid crystals to a watch glass, and heat the watch glass over a 400-mL beaker of boiling water to dry the crystals. You can monitor the drying of the crystals by watching for the filter paper to dry out as it is heated on the water bath.

When the benzoic acid is completely dry, determine the melting point of the recrystallized material using the method discussed in Part 2, following.

2. Melting Point Determination – *Note: If your laboratory is equipped with digital thermometers or interfaced temperature probes, the procedure is the same. The probe or digital thermometer simply replaces the traditional glass thermometer shown in the illustrations.*

If the benzoic acid recrystallized in Part 1 is not finely powdered, grind some of the crystals on a watch glass or flat glass plate with the bottom of a clean beaker. Your instructor may also have available mortars and pestles for grinding.

Pick up a few crystals of the benzoic acid in the open mouth of a melting point capillary tube. Gently tap the sealed end of the capillary tube on the lab bench to pack the crystals into a tight column at the sealed end of the tube. Repeat this process until you have a column of crystals approximately one centimeter high at the sealed end of the capillary.

Set up the Thiele tube apparatus as indicated in Figure 4-2. Attach the capillary tube containing the crystals to the thermometer with one or two small rubber bands, and position the capillary so that the crystals are next to the temperature-sensing bulb of the thermometer. Note that the top of the capillary must be above the surface of the oil in the Thiele tube.

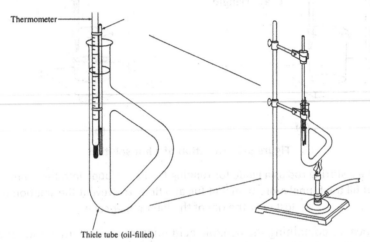

Thermometer

Thiele tube (oil-filled)

Figure 5-2.   Thiele Tube Oil Bath for Melting Point Determinations
Exercise caution when dealing with hot oil.
Replace the oil if it appears at all cloudy or discolored.

Lower the thermometer into the oil bath, and begin heating the bottom of the side arm of the Thiele tube with a very small flame. Adjust the flame as necessary so that the temperature rises only by 1 or 2 degrees per minute. Too-rapid heating can lead to erroneous results.

Watch the crystals in the capillary tube, and record the *exact temperature* at which the benzoic acid crystals *first* begin to melt, and the exact temperature at which the *last* portion finishes melting. Record these two temperatures as the melting range of the benzoic acid.

Allow the oil bath to cool by at least 20°C.

Prepare another sample of the recrystallized benzoic acid in a fresh capillary tube, and repeat the determination of the melting point. If this second determination differs significantly from the first determination, repeat the experiment a third time.

As part of the Pre-laboratory Questions, you researched the accepted value for the melting point of benzoic acid. Compare the melting point of your benzoic acid with that indicated in your reference. If the melting point you obtain is significantly lower than that reported in your reference, heat the crystals for an additional period over the hot water bath to dry them further, then repeat the melting point determination.

EXPERIMENT 5

Recrystallization and Melting Point Determination

Pre-Laboratory Questions

1. Describe the technique you will use in this experiment for determining the melting point of a crystalline substance and the manner in which you will confirm the substance's identity.

2. Suppose that, as you are preparing to use a Thiele tube for melting point determination, you discover that the oil in the tube appears cloudy. What steps should you take, and why?

3. What is meant by a *mixed melting point* determination?

4. A sample must be completely dry before its melting point is determined. What effect on the melting point would trace amounts of water have?

5. Why do we use only a very small flame to heat the Thiele tube when determining the melting point of a crystalline solid.

6. When determining the melting point of a sample, if the sample melts over a *range* of several degrees, what might this imply about the purity of the sample?

Name: _____ **Section:** _____

Lab Instructor: _____ **Date:** _____

EXPERIMENT 5

Recrystallization and Melting Point Determination

Results/Observations

1. **Recrystallization**

   Appearance of impure benzoic acid sample

   _____

   _____

   Observations of recrystallization process

   _____

   _____

   _____

   Appearance of purified benzoic acid sample

   _____

   _____

2. **Melting Point Determination**

   Observations of melting point determination

   _____

   _____

   Melting points determined

   First sample   _____

   Second sample   _____

   (Third sample)   _____

   Average melting point   _____

   Error in melting point   _____

Questions

1. It was necessary for you to *preheat* the funnel used for the gravity filtration of the benzoic acid sample? Why?

2. Why should a melting point capillary be positioned directly next to the temperature-sensing bulb of the thermometer when performing a melting point determination?

3. Why was it necessary to allow the oil bath to cool by at least 20°C before performing the second determination of the melting point of your recrystallized benzoic acid?

4. Why was it necessary to *dry* the benzoic acid crystals before determining their melting point? What would have happened to the melting point if the crystals had still been wet?

The Solubility of a Salt

Objective

In this experiment, you will determine the *solubility* of a given salt. You also will prepare the *solubility curve* for your salt.

Introduction

The term **solubility** in chemistry has both general and specific meanings. In everyday situations, we might say that a salt is "soluble," meaning that, experimentally, we are able to dissolve a sample of the salt in a particular solvent.

In a more specific sense, however, the *solubility* of a salt refers to a definite numerical quantity. Typically, the solubility of a substance is indicated as the *number of grams of the substance that will dissolve in 100 g of the solvent*. More often than not the solvent is water. In that case the solubility could also be indicated as the number of grams of solute that dissolve in 100 mL of water (the density of water is very near to 1.0 g/mL under most conditions).

Since solubility refers to a specific, experimentally determined amount of substance, it is not surprising that handbooks of chemical data and online databases contain extensive lists of solubilities of various substances. In looking at such data in a handbook or online, you will notice that the *temperature* at which the solubility was measured is always given. Solubility *changes with temperature*. For example, if you like your tea extra sweet, you have undoubtedly noticed that it is easier to dissolve two teaspoons of sugar in hot tea than in iced tea. For many solid substances, the solubility increases with increased temperature. For a number of other substances, however, the solubility decreases with increasing temperature.

For convenience, **graphs** of solubility are often used, rather than a list of solubility data. A graph of the solubility of a substance versus the temperature will clearly indicate whether or not the solubility increases or decreases as the temperature is raised. If the graph is carefully prepared, the specific numerical solubility for a particular temperature can be read from the graph.

It is important to distinguish experimentally between *whether* a substance is soluble in a given solvent, and *how fast,* and how readily the substance will dissolve. Sometimes an experimenter may wrongly conclude that a salt is not soluble in a solvent when the solute is merely dissolving at a very slow rate. The speed at which a solute dissolves has nothing to do with the final maximum quantity of solute that can enter a given amount of solvent. In practice, we use various techniques to speed up the dissolving process; grinding the solute to a fine powder or stirring/shaking the mixture, for example. Such techniques will not affect the final amount of solute that ultimately dissolves.

The solubility of a salt in water represents the amount of solute necessary to reach a state of equilibrium between saturated solution and un-dissolved additional solute. This number is a constant for a given solute/solvent combination at a constant temperature. For a salt with the general formula MX, the equilibrium may be represented as:

$$MX(s) \rightleftarrows M^+(aq) + X^-(aq)$$

Safety Precautions	• **Safety eyewear approved by your institution must be worn at all times while you are in the laboratory, whether or not you are working on an experiment.** • **Use glycerine as a lubricant when inserting the thermometer through the rubber stopper. Protect your hands with a towel.** • **Some of the salts used in this experiment may be toxic or environmental hazards. Wash your hands after use. Do not pour the salts down the drain. Dispose of the salts as directed by the instructor.**

Apparatus/Reagents Required

- 25 x 150-, or 25 x 200 mm test tube (8") fitted with 2-hole cork or slotted rubber stopper
- Copper wire for stirring
- Mortar and pestle
- Thermometer
- Buret, 50-mL
- Salt for solubility determination

Procedure

Record all data and observations directly on the report page in ink.

Obtain a salt for the solubility determination. If the salt is presented as an unknown, record the code number on the report sheet (otherwise, record the formula and name of the salt). If the salt is not finely powdered, grind it to a fine powder in a mortar.

Fit the large test tube with a 2-hole slotted rubber stopper. Protecting your hands with a towel and using glycerin as a lubricant, insert your thermometer (*Caution!*) in the slotted hole of the stopper in such a way that the thermometer can still be read from 0 to 100 °C. This may involve arranging the thermometer so that the scale can be viewed *through the slot* in the stopper.

Obtain a length of heavy-gauge copper wire for use in stirring the salt in the test tube. If the copper wire has not been prepared for you, form a loop in the copper wire in such a way that the loop can be placed around the thermometer when in the test tube. Fit the copper wire through the second hole in the stopper, making sure that the hole in the rubber stopper is big enough that the wire can be easily agitated in the test tube. (See Figure 6-1.)

Place about 300 mL of water in a 400-mL beaker, and heat the water to boiling.

While the water is heating, weigh the empty, clean, dry, test tube (*without* the stopper-thermometer-stirring assembly). Make the mass determination to the nearest milligram (0.001 g).

Add approximately 5 g of your salt for the solubility determination to the test tube, and reweigh the test tube and its contents. Again, express the mass to the nearest milligram.

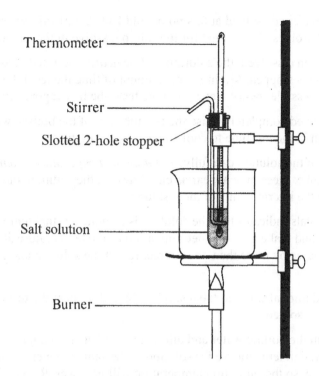

Figure 6-1. Apparatus for Stirring A Soluble Salt

Be certain the thermometer bulb dips into the solution being determined and that the scale of the thermometer can be viewed through the slot in the stopper.

Clean the 50-mL buret with soap and water, then rinse the buret with tap water, followed by several rinses with distilled water.

Fill the buret with distilled water. Make sure that water flows freely from the stopcock of the buret, but that the stopcock does not leak, especially if your buret is fitted with a glass (rather than Teflon®) stopcock.

Record the reading of the initial water level in the buret to the nearest 0.02 mL. (Recall that water forms a meniscus in glass. Read the bottom of the meniscus.).

In the following procedure, record on the report page each time a portion of water is added from the buret. It is essential to know the amount of water added at each point in the determination.

Add 3.00 ± 0.01 mL of water from the buret to the salt in the test tube. Record the precise amount of water used.

Attach the stopper with thermometer and stirrer, and clamp the test tube vertically in the boiling water bath.

Adjust the thermometer so that the bulb will be immersed in the solution in the test tube as the salt dissolves. The test tube should be set up so that the contents of the test tube are immersed fully in the boiling water, but the tube should not touch the bottom of the beaker. See Figure 6-1. Carefully stir the salt in the test tube until it dissolves completely.

If the salt does not dissolve completely after several minutes of stirring in the boiling water bath, remove the test tube and add 1.00 ± 0.01 mL additional water from the buret. Record. Return the test tube to the boiling water bath and stir.

If the salt is still not completely dissolved at this point, add 1.00 ± 0.01 mL water portions (one at a time) until the salt just barely dissolves. Allow time for the salt to dissolve in the added water. Record.

When all of the salt has been dissolved, the solution will be *nearly* saturated, and will quickly become saturated when the heating is stopped. Minimize the amount of time the test tube spends in the boiling water bath to restrict any possible loss of water from the test tube by evaporation.

After all the salt has dissolved completely, raise the test tube out of the boiling water. With constant stirring, allow the solution in the test tube to cool spontaneously in the air.

Monitor the temperature of the solution carefully, and *note the temperature when the first crystals* of salt begin to form in the test tube: these may appear as cloudiness in the solution, or crystals may begin to form on any scratches on the interior walls of the test tube.

The first formation of crystals indicates that the solution is saturated at that temperature. Reheat the test tube in the boiling water, and make a second determination of the temperature at which the first crystal forms. If your results disagree by more than one degree, reheat the solution and make a third determination.

Add 1.00 ± 0.01 mL of additional water to the test tube. Record. Reheat the test tube in boiling water until all of the solid has again dissolved.

Remove the test tube from the boiling water and allow it to cool again spontaneously. Make a determination of the saturation temperature for solution in the same manner as indicated earlier. The solution is now more dilute, so the saturation temperature will be lower. Repeat the determination of the new saturation temperature as a check on your measurement.

Repeat the addition of 1.00-mL water samples, with determination of the saturation temperatures, until you have at least six sets of data. Keep accurate records as to how much water has been added from the buret at each determination.

If the saturation temperature drops sharply on the addition of the 1.00 mL samples, reduce subsequent additions to 0.50 mL. If the saturation temperature does not change enough on the addition of 1.00-mL samples, increase the size of the water samples added to 2.00 mL. Keep accurate records of how much water is added.

From your data at each of the saturation temperatures, calculate the mass of salt that would have dissolved in 100. g of water at that temperature. Assume that the density of water is exactly 1.00 g/mL, so that your buret additions in milliliters will be equivalent to the mass of water being added.

Example

Suppose 5.101 g of a particular salt was found to begin crystallizing in 4.00 mL of water at 90°C. Assuming that the density of water is 1.00 g/mL, the solubility of this salt at 90°C could be found as follows:

$$100 \ g \ H_2O \times \frac{5.101 \ g \ salt}{4.00 \ mL \ H_2O} \times \frac{1.00 \ mL \ H_2O}{1.00 \ g \ H_2O} = 128 \ g \ salt$$

Thus, the solubility of this salt is 128 g/100. g H_2O, at 90°C.

On a piece of graph paper, plot the solubility curve for your salt, using *saturation temperature* on the horizontal axis and *solubility per 100 g of water* on the vertical axis. Attach the graph to your laboratory report.

Name: _____ Section: _____

Lab Instructor: _____ Date: _____

EXPERIMENT 6

The Solubility of a Salt

Pre-Laboratory Questions

1. Suppose it is found experimentally that 3.12 g of a salt dissolves in 4.52 mL of water at 20°C. Calculate the solubility of this salt in units of grams per 100 mL of water (g/100 mL H_2O) at 20°C. Show your calculation.

2. Stirring affects the *rate* at which a salt dissolves in water, but not the *solubility* of the salt in water. Explain.

3. What does it mean to say that the solubility of a salt represents a *dynamic equilibrium*?

4. Using your text, a handbook of chemical data, or online references, list the *formulas* of four salts, at least one of which becomes more soluble with increasing temperature, and at least one whose solubility decreases as the temperature is increased. Report their solubilities in water, at each of two temperatures at which their solubilities were measured.

Formula of Salt	Lower Temp	Solubility at Lower Temp	Higher Temp	Solubility at Higher Temp

Name: _____ Section: _____

Lab Instructor: _____ Date: _____

EXPERIMENT 6

The Solubility of a Salt

Results/Observations

Volume of Water Used (mL)	Saturation Temperature (°C)	Solubility (g/100 g H₂O)

Identity of the salt _____

Literature solubility (20°C) _____ Reference _____

Percentage error in solubility at 20°C   % _____

Attach your graph to this report page.

Questions

1.   Why was it better to determine the saturation temperature while the temperature was *dropping*, rather than while it was rising?

2.   When adding water to the salt initially, you attempted to find the minimum amount of water the salt would dissolve in at 100°C. Why was it necessary that the solution used be almost saturated?

3.   The procedure indicated that the amount of time the test tube was kept in the boiling water bath should be kept as short as possible. Why was this necessary?

EXPERIMENT 7

Identification of a Substance

Objective

Selected physical properties of an unknown substance will be measured experimentally and compared with the tabulated properties of known substances.

Introduction

Modern instrumental methods permit the routine analysis and identification of unknown substances. Because of the high cost of precision instruments, and the cost and time required for maintenance and calibration of such instruments, however, instrumental methods of analysis are primarily used for *repetitive* determinations of *similar* samples, in which case the instrumental method is relatively fast and the cost *per analysis* moderate. For a *single* sample, however, it is found that classical "wet" laboratory methods of analysis are often preferred.

Most commonly, a thorough determination of the *physical properties* of a sample will suffice for identification. The properties determined for the *unknown* sample are compared to the properties of *known* substances as tabulated in the chemical literature. If the properties match, successful identification is assumed. The physical properties most commonly used in identifications are *density* (as discussed in Experiment 3), the *boiling and melting points* (as discussed in Experiments 4 and 5), and the *solubility* of the substance in water or in some other appropriate solvent (as discussed in Experiment 6).

In general terms, one substance is likely to be soluble in another substance if the two have similar structural features (for example, a similar group of atoms) or have comparable electronic properties (hence similar intermolecular forces). Generally solutes are divided into two major classes, depending on whether they dissolve in very *polar* solvents such as water, or in very *nonpolar* solvents, such as any of the hydrocarbons (the hydrocarbon cyclohexane, $C_6H_{14}(l)$, is used in this experiment). Generally, ionic and very polar covalent substances will dissolve in water, whereas nonpolar substances will dissolve in cyclohexane. Some solutes that are intermediate in polarity may dissolve well in a solvent of intermediate polarity (such as ethyl alcohol).

The physical properties discussed above are almost always reported in the literature when new substances are prepared, and they are tabulated for previously known substances in the various handbooks of chemical data found in most laboratories and science libraries. Other gross properties that may also be helpful in identifications include unusual *color* (some transition metal ions have characteristic colors), *odor* (some types of chemical substances have very characteristic odors), *hardness* (crystals of ionic substances are very hard compared to most covalently bonded substances), and so forth.

In this experiment, you will be provided with an unknown sample chosen from those substances listed in Table 7-1. By careful determination of the physical properties of your sample, you should be able to match those properties to one of the substances listed in the table. The unknown samples have been purified and contain only a single substance. In real practice, an unknown sample may contain more than one major component and is likely also to contain minor impurities. The identification of unknown samples is a vital, important, and interesting application of modern chemistry.

55

Table 7-1. Physical Properties of Selected Substances

Liquid Substances (under normal room conditions)

Substance	Solubility			Density	Boiling Point
	W	A	C	g/mL	°C (760 mm)
Acetone	s	s	s	0.791	56
Cyclohexane	–	s	s	0.778	81
Ethyl acetate	s	s	s	0.900	78
Hexane	–	s	s	0.660	68
Methanol	s	s	s	0.793	65
1-propanol	s	s	s	0.780	97
2-propanol	s	s	s	0.786	83
2-methyl-2-propanol	s	s	s	0.789	82

Solid Substances (under normal room conditions)

Substance	Solubility			Density	Melting Point
	W	A	C	g/mL	°C
Acetamide	s	s	-	1.16	82
Acetanilide	–	s	s	1.22	113
Benzoic acid	–	s	s	1.27	123
Benzophenone	–	s	s	1.15	48
Biphenyl	–	s	s	0.867	71
Camphor	–	s	s	0.990	176
1,4-dichlorobenzene	–	s	s	1.25	53
Diphenylamine	–	s	s	1.16	53
Lauric acid	–	s	s	0.867	44
Naphthalene	–	s	s	1.03	80
Phenyl benzoate	–	s	s	1.24	71
Sodium acetate trihydrate	s	s	-	1.45	58
Stearic acid	–	s	s	0.941	72
Thymol	–	s	s	0.925	52

Solubility legend: W = in water, A = in ethyl alcohol, C = in cyclohexane

s = soluble   – = insoluble (or very low solubility)

Safety Precautions	• Protective eyewear approved by your institution must be worn at all time while you are in the laboratory.
	• Many of the unknowns used in this experiment are *flammable*; keep the unknown substances away from open flames
	• Cyclohexane, ethyl alcohol, acetone, hexane, methanol, ethyl acetate, and propanol are all *highly flammable*; keep these solvents in the fume exhaust hood.
	• Many of the unknown substances in this experiment are *toxic* if inhaled, ingested, or absorbed through the skin.
	• *Caution*: Oil is used as the heating fluid in the Thiele tube used for the determination of melting point. Hot oil may spatter if it is heated too strongly, especially if any moisture is introduced into the oil from glassware that is not completely dry. The oil may smoke or ignite if heated above 200°C. The oil should be completely transparent, with no cloudiness: if the oil appears cloudy, ask the instructor to replace the oil.

Apparatus/Reagents Required

Unknown sample, distilled water, ethyl alcohol, cyclohexane, sodium chloride, copper(II) sulfate, pentane, oleic acid, naphthalene, melting point apparatus (oil-filled Thiele tube), melting point capillaries, rubber bands, semi-micro test tubes, thermometer, assorted beakers, ring stand, ring, wire gauze, burner and tubing

Procedure

Record all data and observations directly in your notebook in ink.

Obtain an unknown substance for identification, and record its code number in your notebook and on the report page.

Record your gross observations of the color, odor, physical state, volatility, viscosity, clarity, and so forth for the unknown substance.

1. **Solubility**

   Before determining the solubility of your unknown sample, you will perform some preliminary solubility studies with *known* solutes and several solvents.

   Set up three clean, dry semi-micro test tubes in a test tube rack. Place 10 drops of distilled water in one test tube, 10 drops of ethyl alcohol in a second test tube, and 10 drops of cyclohexane (*Caution!*) in the third test tube.

   Obtain a sample of sodium chloride, and transfer a few crystals of NaCl to each of the test tubes containing the three solvents. Stopper the test tubes and shake for at least 30 seconds to attempt to dissolve the solid. Record your observations on the solubility of NaCl in the three solvents.

Repeat the solubility tests with samples of naphthalene, copper(II) sulfate, pentane, and oleic acid, determining the solubility of each solute in each of the three solvents (water, ethyl alcohol, and cyclohexane). Some of the test solutes are liquids. Add 3 drops of the liquid solute to each of the test tubes containing the solvents. If a liquid does not dissolve in a given solvent, it will form a separate layer in the test tube. Record your observations of solubility.

Repeat the solubility test, using small samples of your *unknown* material in each of the three solvents. Record your observations.

2.   Density

Techniques for the measurement of the densities of solids and liquids were discussed in detail in Experiment 3. Review the discussion given there.

If the unknown is a liquid, determine its density by measuring the mass of a specific volume of the liquid. For example, a weighed empty graduated cylinder could be filled with the liquid to a particular volume and then reweighed. Alternatively, a specific volume of liquid could be taken with a transfer pipet, and the liquid weighed in a clean, dry beaker.

If the unknown is a solid, the volume of a weighed sample of solid may be determined by displacement of a liquid in which the solid is *not soluble* (see Part 1 of this experiment). If this is performed in a graduated cylinder, the change in liquid level reflects the volume of the solid directly.

3.   Melting-Point and Boiling-Point Determinations

Semi-micro methods for the determination of boiling and melting points were discussed in detail in Experiments 4 and 5 of this manual. Review these methods.

If your unknown sample is a solid, determine its melting point as you did in Experiment 5. Fill two or three melting-point capillaries with finely powdered solid to a height of about 2-3 mm. Attach one of the capillaries to a thermometer with a rubber band, and lower the sample and thermometer into the oil bath of a Thiele tube. Heat the oil bath in such a manner (*Caution!*) that the temperature rises only one or two degrees per minute, and watch carefully for the crystals to melt. Record the melting *range* if the crystals do not melt sharply at a single temperature. Repeat the determination until you get two values that check to within one degree.

If your unknown is a liquid, determine its boiling point. Fill a semi-micro test tube to a height of 3–4 cm with the liquid, and insert a short length of melting-point capillary as described in Experiment 4. Attach the test tube to a thermometer with a rubber band, and lower into a water bath. Heat the water bath until bubbles come in a steady stream from the capillary, remove the heat, and record the exact temperature at which bubbles stop coming from the capillary. Add liquid to preserve the 3–4-cm height, add a fresh capillary tube, and repeat the determination as a check. If your results do not agree within one degree, perform a third determination.

Compare the properties determined for your unknown substance with the properties of those substances listed in Table 7-1. Make a tentative identification of your unknown substance.

EXPERIMENT 7

Identification of a Substance

Pre-Laboratory Questions

1. Define the following terms.

   a. Saturated Solution

   b. Melting point

   c. Physical properties

2. Describe the various tests you will be performing on your unknown substance, and explain what you hope/expect to learn about the substance from each of the tests.

3. An unknown solid compound is investigated, and is found to have a density of approximately 1.22–1.25 g/mL.

    a. Given the information in Table 7-1, propose some possible identities for the unknown solid.

    b. The unknown solid is found to have a melting point in the range 52–54°C. What is the most likely identity of the compound?

    c. Describe what further test or tests could be then performed to *confirm* the likely identity of the compound.

EXPERIMENT 7

Identification of a Substance

Results/Observations

Unknown identification number _____

Qualitative observations (physical state, color, etc.) of unknown

1. **Solubility Test Observations**

   Observations on solubility of known solutes

   Observations on solubility of unknown

2. **Density Determination**

   Describe the method by which you determined the density of the unknown.

   Density determined for unknown   _____

3. **Melting or Boiling Point Determination**

   Describe the method used for the melting point or boiling point determination.

   Melting or boiling (circle one) point _____

   Barometric pressure (for boiling point) _____

4. Summarize in the following table the properties determined for your unknown sample, along with the properties of the substance you believe your unknown to be. Calculate the percentage difference between the known and unknown substances' properties for each quantitative determination.

   Substance unknown is believed to be _____

	Unknown	**Known**	**% difference**
Solubility	_____	_____	_____
Density	_____	_____	_____
Melting/boiling point	_____	_____	_____

Questions

1. In this experiment, you qualitatively determined whether or not a given solute would dissolve in each of three solvents. Suggest a method by which the solubility of a solute in a solvent might be measured on a *quantitative* basis.

2. What does it mean to say that the normal boiling and melting points of a pure substance are *characteristic* properties of the substance?

EXPERIMENT 8

Resolution of Mixtures 1: Filtration and Distillation

Objective

The separation of mixtures into their constituent components defines an entire subfield of chemistry referred to as **separation science**. In this experiment, techniques for the resolution of mixtures of solids and liquids will be examined.

Introduction

Mixtures occur very commonly in chemistry. When a new chemical substance is synthesized, for example, oftentimes the new substance first must be *separated* from a mixture of various side-products, catalysts, and any excess starting reagents still present. When a substance must be isolated from a natural biological source, the substance of interest is generally found in a very complex mixture with many other substances, all of which must be removed. Chemists have developed a series of standard methods for resolution and separation of mixtures, some of which will be investigated in this experiment. Methods of separation based on the processes of chromatography are found in Experiment 9.

Mixtures of solids often may be separated on the basis of differing *solubilities* of the components. If one of the components of the mixture is very soluble in water, for example, whereas the other components are insoluble, the water-soluble component may be removed from the mixture by simple *filtration* through ordinary filter paper. A more general case occurs when *all* the components of a mixture are soluble, to different extents, in water or some other solvent. The solubility of substances in many cases is greatly influenced by *temperature*. By controlling the temperature at which the filtration is performed, it may be possible to separate the components. For water-soluble solutes, commonly a sample is added to a small amount of water and heated to boiling. The hot sample is then filtered to remove completely insoluble substances. The sample is then cooled, either to room temperature or below, which causes crystallization of those substances whose solubilities are very temperature-dependent. These crystals can then be isolated by a second filtration, and the filtrate remaining can be concentrated to reclaim those substances whose solubilities are *not* so temperature-dependent.

Mixtures of liquids are most commonly separated by *distillation*. In general, **distillation** involves heating a liquid to its boiling point, then collecting, cooling, and condensing the vapor produced into a separate container. For example, desalination of salt water may be achieved by boiling off and condensing the water. For a mixture of liquids, however, in which several of the components of the mixture are likely to be volatile, a separation is not so easy to effect. If the components of the mixture differ reasonably in their boiling points, it may be possible to separate the mixture simply by monitoring the temperature of the vapor produced as the mixture is heated. The components of a mixture will boil in turn as the temperature is gradually raised, with a sharp *rise* in the temperature of the vapor being distilled indicating when a new component of the mixture has begun to boil. Changing the receiving flask at this point accomplishes a separation. For liquids whose boiling points differ by only a few degrees, the mixture can be passed through a **fractionating column** as it is being heated. Fractionating columns generally are packed with glass beads or short lengths of glass tubing that provide a large amount of surface area to the liquid being boiled, effectively re-distilling the mixture over and over again.

Safety Precautions	• **Protective eyewear approved by your institution must be worn at all time while you are in the laboratory.**
	• **The solid mixture contains benzoic acid, which may be irritating to the skin and respiratory tract. Avoid the dust of the mixture.**
	• **When moving hot containers, use metal tongs or a towel to avoid burns. Beware of burns from steam while solutions are being heated, or while hot solutions are being poured.**
	• **The liquids used for distillation may be flammable. Keep the liquids away from open flames. Perform the distillation in the exhaust hood to help remove fumes. The liquids may be toxic if inhaled or absorbed through the skin.**
	• **Portions of the distillation apparatus may be very hot during the distillation process. Do not touch the apparatus while heating.**
	• Never heat a distillation flask to complete dryness. **The distillation flask may break if heated to dryness. Distillation to dryness also poses an explosion hazard for certain unstable organic substances, or for substances that may be contaminated with organic peroxides.**
	• **Dispose of all solids and liquids as directed by the instructor. Do not pour down the drain.**
	• **Silver nitrate, AgNO₃, stains the skin. All nitrates are strong oxidizers, are toxic, and may be carcinogenic.**

Apparatus/Reagents Required

Benzoic acid, charcoal, 1% sodium chloride solution, filter funnel and paper to fit, clay triangle, assorted beakers, ring stand, iron ring, wire gauze, towel or beaker tongs, melting point apparatus (See Experiment 5), melting point capillaries, stirring rod, 0.1 M silver nitrate, unknown mixture of two volatile liquids for fractional distillation

Procedure

Record all observations and data directly in your notebook in ink.

1. **Resolution of a Solid Mixture**

   Obtain small samples of benzoic acid and charcoal (approximately 1 g of each). Combine the benzoic acid and charcoal on a watch glass, and mix with a stirring rod until the mixture is fairly homogeneous. Charcoal is not soluble in water and will be used as a "contaminant" for the benzoic acid. Transfer the benzoic acid/charcoal sample to a clean 150-mL beaker.

   Set up a short-stem gravity funnel in a clay triangle on a small metal ring clamped to a ringstand. Fit the filter funnel with a piece of filter paper folded in quarters to make a cone. See Figure 8-1.

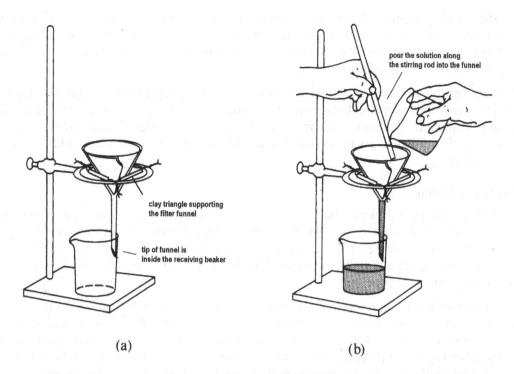

(a)   (b)

Figure 8-1. Filtration of a solution
(a)   The funnel is supported on a clay triangle.
(b)   A stirring rod serves as a guide for directing the flow of solution into the funnel

Do not fill the funnel more than halfway at a time, to avoid having solution escape over the edge of the filter paper.

Moisten the filter paper slightly so that it will remain in the funnel. Place a clean 250-mL beaker beneath the stem of the funnel.

Set up a 250-mL beaker about half filled with distilled water on a wire gauze over a metal ring. Heat the water to boiling.

When the water is boiling, pour about two-thirds of the water into the beaker containing the benzoic acid sample. Use a towel or beaker tongs to protect your hands from the heat.

Pour the remainder of the boiling water through the gravity funnel to heat it. If the funnel is not preheated, the benzoic acid may crystallize in the stem of the funnel rather than passing through it. Discard the water that is used to heat the funnel.

Transfer the beaker containing the benzoic acid mixture to the burner, and reheat it gently until the mixture just begins to boil again. Stir the mixture to make sure that the benzoic acid dissolves to the greatest extent possible.

Using a towel to protect your hands, pour the benzoic acid mixture through the preheated funnel. Catch the filtrate in a clean beaker.

Allow the benzoic acid filtrate to cool to room temperature.

When the benzoic acid solution has cooled to room temperature, filter the crystals to remove water. Wash the crystals with two 10-mL portions of ice-cold water.

Transfer the liquid filtrate from which the crystals have been removed to an ice bath to see whether additional crystals form at the lower temperature. Examine, but do not isolate, this second crop of crystals.

Transfer the filter paper containing the benzoic acid crystals to a watch glass, and dry the crystals under a heat lamp or over a 400-mL beaker of boiling water. You can monitor the drying of the crystals by watching for the filter paper to dry out as it is heated. If a heat lamp is used, do not let the paper char or the crystals melt.

When the benzoic acid has been dried, determine the melting point of the recrystallized material using the method discussed in Experiment 5. Compare the melting point of your benzoic acid with that indicated in the handbook. If the melting point you obtain is significantly lower than that reported in the handbook, dry the crystals for an additional period under the heat lamp or over the hot-water bath.

2.   Simple Distillation

Simple distillation can be used when the components of a mixture have *very different* boiling points. In this experiment, a partial distillation of a solution of sodium chloride in water will be performed (to save time, the distillation is not carried to completion). This is an extreme example, because the boiling points of water and sodium chloride differ by over 1000°C, but the technique will be clearly demonstrated by the experiment.

Your instructor has set up a simple distillation apparatus for you. (See Figure 8-2.) He or she will explain the various portions of the apparatus and will demonstrate the correct procedure for using the apparatus. The source of heat used for the distillation may be a simple burner flame, or an electrical heating device (heating mantle) may be provided. Generally, electrical heating elements are preferred for distillations, because often the substances being distilled are flammable.

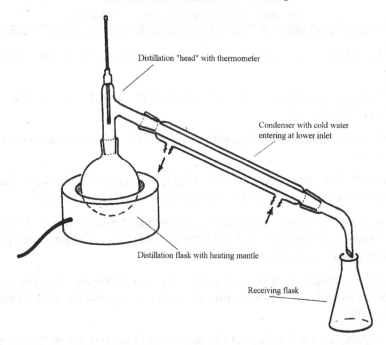

Distillation "head" with thermometer

Condenser with cold water entering at lower inlet

Distillation flask with heating mantle

Receiving flask

Figure 8-2. Simple distillation apparatus
Cold water enters the lower inlet of the condenser, cooling and condensing the distilled vapors.

Obtain about 50 mL of 1% sodium chloride solution. Place 1 mL of this solution in a small test tube, and transfer the remainder of the solution to the distilling flask.

Place a clean, dry beaker under the mouth of the condenser of the distillation apparatus to collect the water as it distills from the salt solution.

Begin heating the sodium chloride solution as directed by the instructor, and continue distillation until approximately 20 mL of water has been collected. Shut off the heat to prevent the distillation flask from heating to dryness.

Transfer approximately 1 mL of the distilled water to a clean small test tube.

To demonstrate that the distilled water is now free of sodium chloride, test the sample of original 1% sodium chloride solution that was reserved before the distillation, as well as the 1-mL sample of water that has been distilled, with a few drops of 0.1 *M* silver nitrate solution (*Caution!*). Silver ion forms a *precipitate* of insoluble AgCl when added to a chloride ion solution. No precipitate should form in the water that has been distilled.

3. Fractional Distillation

Fractional distillation may be used to separate mixtures of volatile substances that differ by at least several degrees in their boiling points. The vapor of the liquid being boiled passes into a *fractionating column*, which provides a great deal of surface area and the equivalent of many separate simple distillations.

Your instructor has set up a fractional distillation apparatus in one of the exhaust hoods. (See Figure 8-3.)

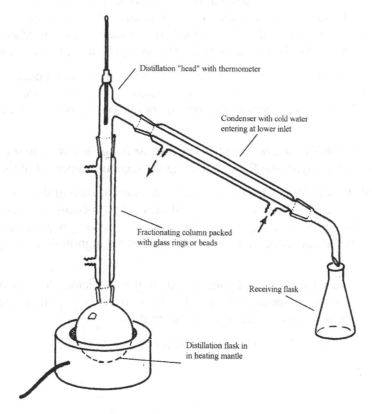

Distillation "head" with thermometer

Condenser with cold water entering at lower inlet

Fractionating column packed with glass rings or beads

Receiving flask

Distillation flask in in heating mantle

Figure 8-3. Fractional distillation. The tall vertical fractionating column is packed with bits of glass that provide a large surface area.

Compare the fractional distillation apparatus with the simple distillation apparatus used in Part 2, and note the differences. Your instructor will explain the operation of the fractional distillation apparatus. The apparatus is set up in the hood, because the mixture you will distill is very volatile and may be flammable.

Obtain an unknown mixture for fractional distillation and record its identification number. Use a graduated cylinder to transfer 40 mL of the unknown mixture to the distillation flask. Record the exact volume of the mixture used. During the distillation, carefully watch the thermometer that is part of the apparatus. The temperature is used to monitor the distillation, because the temperature will *increase very suddenly* as one component finishes distilling and another component begins to distill.

Place a clean, dry flask under the mouth of the condenser to collect the first component of the mixture as it distills. Have ready a second clean, dry flask for collection of the second component. Have ready corks or rubber stoppers that fit snugly in the two collection flasks.

Have your instructor approve the apparatus, and then begin heating the distillation flask with very low heat until vapor begins to rise into the fractionating column.

Allow the vapor to rise to the level of the thermometer bulb, and adjust the heat so that the thermometer will remain bathed in droplets of liquid as the mixture distills. Record the temperature indicated by the thermometer as the first component of the mixture begins to distill. Collect the distillate coming from the condenser.

Continue heating the distillation flask, using the smallest amount of heat that will maintain distillation. Monitor the temperature constantly. At the point at which the first component of the mixture has finished distilling, the temperature will rise *suddenly* and abruptly by several degrees. At this point, remove the flask used to collect the first component of the mixture, and replace it with the second flask. Stopper the flask containing the first component to prevent its evaporation.

Record the temperature indicated by the thermometer as the second component of the mixture begins to distill. Continue the distillation *until approximately 10 mL of liquid remains in the distillation flask.* Remove the source of heat, but do not remove the collection flask until distillation stops.

Do not heat the distillation flask to complete dryness, or it may break from the heat. When distillation is complete, stopper the flask containing the second component of the mixture.

With a graduated cylinder, determine the respective volumes of each of the two components of the mixture. Calculate the *approximate composition* of the original mixture, in terms of the percentages of low-boiling and high-boiling components. This percentage is only approximate, because some of the vapor being distilled may have been lost, and not all of the high-boiling component may have been isolated.

Report the approximate composition of your mixture to the instructor, along with the boiling temperatures of the two components. (Assume that the 5 mL remaining in the distillation flask is predominantly the higher-boiling component of the mixture).

Turn in the two flasks of distillate to the instructor for proper disposal.

Name: _____ Section: _____

Lab Instructor: _____ Date: _____

EXPERIMENT 8

Resolution of Mixtures 1: Filtration and Distillation

Pre-Laboratory Questions

1. Given that sodium chloride is fairly soluble in cold water, that benzoic acid is soluble in hot water but not in cold water, and that charcoal is not soluble in either hot or cold water, devise a procedure that could separate a mixture of these three materials.

2. Why should the flask containing a mixture to be distilled never be heated to complete *dryness*?

3. When separating a mixture of two volatile substances by fractional distillation, what indication is seen when the lower-boiling component of the mixture has finished distilling?

4. In the simple distillation of a sodium chloride solution to be performed in this experiment, a simple test using silver nitrate, $AgNO_3$, will be performed on the original solution and on the distillate separated from the original solution.

   a. Write the equation for the reaction that occurs between silver nitrate and sodium chloride.

   b. How does this test demonstrate that the distillation has been effective in resolving the mixture of water and sodium chloride?

5. Some mixtures of volatile liquids cannot be completely separated by fractional distillation because of the formation of what are called *azeotropes* among the liquids. Use your textbook, a chemical dictionary, or an online source to find a definition and an example of an *azeotrope*.

Name: ______________________________ **Section:** __________

Lab Instructor: ______________________________ **Date:** __________

EXPERIMENT 8

Resolution of Mixtures 1: Filtration and Distillation

Results/Observations

1. **Resolution of a Solid Mixture**

   Observation on dissolving sample in hot water

   Observation on material remaining on filter paper

   Observation of filtrate as it cools

   Appearance of crystals collected

   Appearance of second crop of crystals

   Melting point of dry benzoic acid ______________________

   Literature value for melting point ______________________

   Error in melting-point determination ______________________

2. **Simple Distillation**

   Observation of distillation

Silver nitrate test:

On original sample _____

On distilled sample _____

3. **Fractional Distillation**

ID number of mixture for fractional distillation _____

Observation of fractional distillation

Boiling point of first (low-boiling) component _____

Volume of first component collected _____

Boiling point of second (high-boiling) component _____

Approximate volume of second component _____

Approximate percentage composition of mixture _____

Questions

1. In this experiment you separated two solids by making use of the fact that one solid was soluble in hot water while the other was not soluble, allowing the two substances to be separated by filtration. Use a chemical encyclopedia or the internet to describe the technique of *fractional crystallization* by which several solid substances may be separated from one another based on differences in the solubilities of the substances in a solvent.

2. How do simple and fractional distillation *differ*? Under what circumstances is one method likely to be used in preference to the other method?

EXPERIMENT 9

Resolution of Mixtures 2: Chromatography

Objective

The field of separation science is one of the most important in chemistry today. The particular branch of chemistry called **analytical chemistry** is concerned with the separation of mixtures and the analysis of the amount of each component in the mixture. In this experiment, you will study the effect of the solvent mixture on a thin-layer chromatographic analysis.

General Introduction to Chromatography

The word **chromatography** means "color-writing." The name was chosen at the beginning of this century when the method was first used to separate colored components from plant leaves. Chromatography in its various forms is perhaps the most important known method of chemical analysis of mixtures.

Paper chromatography (Part A of this experiment) and thin-layer chromatography (Part B) are simple techniques that can be used to separate mixtures into their individual *components*. The methods are very similar in operation and principle, differing primarily in the medium used for the analysis. There are two parts to any chromatographic system: the stationary phase (the surface) and the mobile phase (the solvent).

Paper chromatography uses ordinary filter paper as the stationary phase. It consists primarily of the polymeric carbohydrate, cellulose, as the medium on which one applies the sample mixture for separation. A single drop or 'spot' of the unknown mixture is applied about half an inch from the end of a strip of filter paper. The filter strip is then placed in a shallow layer of solvent (or a mixture of solvents) in a jar or beaker. Because filter paper is permeable to liquids, the solvent begins rising by capillary action.

Thin-layer chromatography (universally abbreviated as **TLC**) uses a thin coating of aluminum oxide (alumina) or silica gel on a glass microscope slide or plastic sheet to which the mixture to be resolved is applied.

As with paper chromatography, a single drop or spot of the unknown mixture to be analyzed is applied about half an inch from the end of a strip of a TLC slide. The TLC slide is then placed in a shallow layer of solvent or solvent mixture in a jar or beaker. Because the coating of the TLC slide is permeable to liquids, the solvent begins rising by capillary action.

As the solvent rises to the level at which the spot of mixture was applied, various effects can occur, depending on the constituents of the spot. Those components of the spot that are completely soluble in the solvent will be swept along with the solvent front as it continues to rise. Those components that are not at all soluble in the solvent will be left behind at the original location of the spot. Most components of the unknown spot mixture behave in an intermediate manner as the solvent front passes. Components in the spot that are *somewhat* soluble in the solvent will be swept along by the solvent front, but to *different extents,* reflecting their specific solubilities and specific affinities for the slide coating. By this means, the original spot of mixture is spread out into a series of spots or bands, with each spot or band representing one single component of the original mixture.

The separation of a mixture by chromatography is not solely a function of the solubility of the components in the solvent used, however. The f TLC slide coating used in chromatography is not inert but rather, consists of molecules that may *interact* with the molecules of the components of the mixture being separated. Each component of the mixture is likely to have a different extent of interaction with the filter paper or slide coating. This differing extent of interaction between the components of a mixture and the molecules of the support forms an equally important basis for the separation. The TLC slide coating adsorbs molecules on its surface to differing extents, depending on the structure and properties of the molecules involved.

To quantify a chromatographic separation, a mathematical function called the **retention factor**, R_f, is defined:

$$R_f = \frac{\text{distance traveled by spot}}{\text{distance traveled by solvent front}}$$

The retention factor depends on what solvent is used for the separation and on the specific composition of the slide coating used for a particular analysis. Because the retention factors for particular components of a mixture may vary if an analysis is repeated under different conditions, a *known* sample is generally analyzed at the *same time* as an *unknown* mixture on the same sheet of filter paper or slide. If the unknown mixture produces spots having the same R_f values as spots from the known sample, then an identification of the unknown components has been achieved.

Paper chromatography and thin-layer chromatography are only two examples of many different chromatographic methods. Mixtures of volatile liquids are commonly separated by **gas chromatography**, in which a mixture of liquids is vaporized and passed through a column of a solid adsorbent material that has been coated with an appropriate liquid. The vaporized mixture is pushed along by the action of a carrier gas (usually helium). As in all varieties of chromatography, the components of the mixture have different solubilities in the mobile phase and different attractions for the stationary phase. Separation of the mixture components thus occurs as the mixture progresses through the tube. The individual components emerge one by one and are detected by electronic means. A final very important technique is called high-performance liquid chromatography (HPLC). In HPLC, liquid mixtures to be analyzed are blown through a column of adsorbent material under high pressure from a pump, resulting in a rapid passage through the column. HPLC is routinely used in medical and forensic laboratories to analyze biological samples. Blood samples, for example, can be quickly analyzed for the presence of alcohol or illicit drugs in just a few minutes using HPLC.

Although paper chromatography is a very simple technique, it is still frequently used for analysis of colored substances (or substances that can be rendered colored by treatment with an appropriate reagent). Biologists often use paper chromatography for quick analysis of plant pigments. The method can also be used for simple analyses of protein abstracts (amino acids).

In the first portion of this experiment, you will do some very simple paper chromatographic analysis of felt-tip pen inks. As you know, inks for such pens come in many different, bright colors – particularly in pen sets used by small children for working on coloring books. Such brightly colored inks are often mixtures of primary-color inks. For example, a felt-tip pen with what appears to be bright purple in may actually contain a mixture of blue and red inks. Similarly, an orange ink might be a mixture of red and yellow inks. This choice clearly demonstrates the basis for chromatographic separation methods and will give you some insight into the importance of the various chromatographic methods.

In Part B of the Procedure, you will investigate the effects of various solvent mixtures on some organic indicator dyes. Indicators are organic compounds that are typically used to signal a change in pH in acid/base titration analyses. Such indicators are dyes that exist in different colored forms at different pH levels, and the change in color of the indicator is the signal that the titration analysis is complete. In most cases, indicator dyes are very *intensely* colored, and only a very tiny quantity of the indicator is needed.

You will perform a thin-layer chromatographic analysis of a mixture of the dyes bromcresol green, methyl red, and xylenol orange. These dyes have been chosen because they have significantly different retention factors, and a nearly complete separation should be possible in the appropriate solvent system. You will also investigate the effect of the solvent on TLC analyses, by attempting the separation in several different solvent systems.

In real practice, thin-layer chromatography has several uses. When a new compound is synthesized, for example, a TLC of the new compound is routinely done to make certain that the new compound is pure (a completely pure compound should only give a single TLC spot; impurities would result in additional spots). TLC is also used to separate the components of natural mixtures isolated from biological systems. For example, the various pigments in plants can be separated by TLC of an extract made by boiling the plant leaves in a solvent. Once the components of a mixture have been separated by TLC, it is even possible to isolate small quantities of each component by scraping its spot from the TLC slide and redissolving the spot in some suitable solvent.

Safety Precautions	• **Protective eyewear approved by your institution must be worn at all time while you are in the laboratory.** • **Acetone is highly flammable. No flames are permitted in the laboratory. Acetone may be toxic if inhaled or absorbed through the skin. Dispose of the acetone as directed by the instructor. Do not pour it down the drain.** • **The organic indicator dyes used in this experiment will stain skin and clothing; many such dyes are toxic or mutagenic. Wash after handling.** • **The solvents used for the chromatographic separations are highly flammable and their vapors are toxic. No flames are permitted in the room during the use of these solvents. Use the solvents only in the fume exhaust hood.**

Apparatus/Reagents Required

Part A: Filter paper for chromatography (5×10 cm), latex or vinyl surgical gloves, ruler, pencil, heat gun, acetone-water mixture (50% v/v), felt-tip pens (water soluble and permanent), 600 mL beaker, plastic wrap, rubber bands.

Part B: Bakerflex® plastic TLC slides (1 × 4 inch), latex or vinyl surgical gloves, ruler, pencil, plastic wrap or Parafilm®, 10-λ micropipets, ethanolic solutions of the indicator dyes (methyl red, xylenol orange, bromcresol green), acetone, ethyl acetate, hexane, ethanol, six 400-ml beakers

Procedure – Part A: Paper Chromatography

Record all observations directly in your notebook in ink.

Because the skin contains oils that can interfere with the chromatogram, surgical gloves should be worn from this point onward in the procedure, to prevent contamination of the chromatogram. A pencil is used for marking the chromatogram because ink from a ballpoint pen would also undergo separation.

Obtain a sheet of filter paper prepared for the chromatographic analysis. With the paper positioned so that the 10" dimension is the vertical, and using a ruler, draw a light pencil line across the paper about ½ inch from each end. On the lower pencil line, lightly mark three or four small circles (free-hand). These circles will be where the ink samples are to be applied to the paper.

From your instructor, obtain several felt-tip pens containing *water-soluble* inks. Apply a single small spot of different ink to each of the pencil circles. Allow the spots to dry completely. Record in your notebook the original color of each ink applied to the filter paper, as well as any identifying code numbers marked on the pen barrels.

When the spots are completely dry, apply a second spot of each ink directly on top of the previous spot for that ink. This builds up a larger sample of the dye on the paper.

Clean a 600-mL beaker (or special chromatography jar) for use in developing the chromatogram. Cut a square of plastic wrap for the beaker.

Add distilled water to the beaker to a depth of approximately ¼ inch. Avoid splashing water on the beaker walls and be sure the depth is less than ½ inch.

Make certain that the ink spots on the filter paper are completely dry. When they are, fold the filter in half lengthwise.

Carefully lower the filter paper – with the spots on the bottom! – into the water in the beaker. Make certain not to wet the spots as you lower the paper, and do not move the beaker, so as to avoid sloshing water onto the spots. Cover the beaker with the plastic wrap, using a rubber band to secure the plastic wrap. See Figure 9-1.

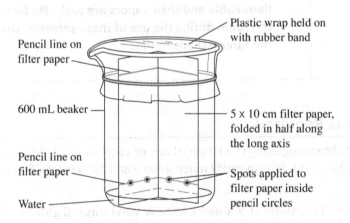

Figure 9-1. Set-up for Paper Chromatography

Allow the water to rise along the filter paper until it reaches the upper pencil line, then carefully remove the filter paper and set it on a clean paper towel. Quickly dry the chromatogram using a heat gun or hair dryer.

Use a ruler to determine R_f for each of the ink spots and record your results.

Your instructor will provide you with several *permanent* ink markers (which are *not* water soluble). Repeat the chromatographic procedure, using a new strip of filter paper, and replacing the water with a 50% acetone-water mixture as the solvent (mobile phase). Run the chromatographic separation as before and determine the R_f values for the permanent inks.

Save your two chromatograms, properly labeled, for submitting with your report.

Procedure – Part B: Thin-layer Chromatography

Clean and dry six 400-mL beakers to be used as the chambers for the chromatography. Obtain several squares of plastic wrap or Parafilm® to be used as covers for the beakers.

The chromatographic separation will be attempted in several solvent mixtures to investigate which gives the most nearly complete resolution of the three dyes. All of these solvents are highly flammable: no flames should be open in the lab. A total of only 10–15 mL of each solvent mixture is necessary. Prepare mixtures of the solvents below, in the proportions indicated by volume, and transfer each to a separate 400-mL beaker. Cover the flasks after adding the solvent mixture, and label the flask with the identity of the mixture it contains.

> Acetone 60% / hexane 40%
>
> Ethyl acetate 60% / hexane 40%
>
> Acetone 50% / ethyl acetate 50%
>
> Acetone 50% / ethanol 50%
>
> Ethyl acetate 50% / ethanol 50%
>
> Hexane 50% / ethanol 50%

Wearing plastic surgical gloves to avoid oils from the fingers, prepare six plastic TLC slides by marking *lightly* with pencil (not ink) a line across both the top and the bottom of the slide. Do not mark the line too deeply or you will remove the coating of the slide. See Figure 9-2.

On one of the lines you have drawn on each slide, mark four small pencil dots (to represent where the spots are to be applied). *Above* the other line on each slide, mark the following letters: *R* (methyl red), *X* (xylenol orange), *G* (bromocresol green), and *M* (mixture). See Figure 10-1.

Figure 9-2. Plastic TLC slide with spots of the three dyes and the mixture applied.
Keep the spots applied *small*.

Obtain small samples of the ethanol solutions of the three dyes (methyl red, xylenol orange, bromcresol green). Also obtain several micropipets: use a separate micropipet for each dye, and be careful not to mix up the pipets during the subsequent application of the dyes.

Apply a single small droplet of the appropriate dye to its pencil spot on each of the TLC slides you have prepared (wipe the outside of the micropipet if necessary before applying the drop to remove any excess dye solution). Keep the spots of dye as small as possible.

Apply one droplet of each dye to the spot labeled *M* (mixture) on each slide, being sure to allow each previous spot to dry before applying the next dye. Allow the dye spots on the TLC slides to dry completely before proceeding.

Gently lower one of the TLC slides, spots downward, into one of the solvent systems. Be careful not to wet the spots or to slosh the solvent in the beaker. Do not move or otherwise disturb the beaker after adding the TLC slide. Carefully cover the beaker with plastic wrap.

Allow the solvent to rise on the TLC slide until it reaches the upper pencil line (this will not take very long).

When the solvent has risen to the upper pencil mark, remove the TLC slide and quickly mark the exact solvent front before it evaporates. Mark the TLC slide with the identity of the solvent system used for development. Set the TLC slide aside to dry completely.

Repeat the process using the additional TLC slides and solvent systems. Be certain to mark each slide with the solvent system used.

Use a ruler to determine R_f for each dye in each solvent system, and record your results. Which solvent system led to the most nearly complete resolution of the dye mixture? If no mixture gave a complete resolution, your instructor may suggest other solvents for you to try or other proportions of the solvents already used.

Save your TLC slides and staple them to the lab report page for this experiment.

EXPERIMENT 9

Resolution of Mixtures 3:
Thin-Layer Chromatography

Pre-Laboratory Questions

1. On a paper chromatogram of inks, a spot produced by a red ink traveled 3.1 cm. In the same length of time, the solvent front traveled 6.8 cm. Calculate the retention factor, R_f, for this red ink.

2. In the second segment of Part A, you were told to switch from using just water as the mobile phase to a mixture of acetone and water. Suggest a reason for this change.

3. You are advised to keep the spots of dye being applied to the TLC slides as small as possible. Suggest some reasons why small spots might be preferable to larger spots.

4. The indicator dyes used in this experiment are also used in acid/base titration analyses because they change color at particular values of pH. Use a handbook of chemical data or online source to find the colors of each of these dyes at low- and high-pH, and the pH range at which each changes color.

Indicator	Color at Low pH	Color at High pH	pH range for change
Methyl red			
Xylenol orange			
Bromcresol green			

5. TLC slides are most commonly coated with alumina, less commonly with silica gel. Use an encyclopedia of chemistry or a handbook to find out the *composition* of each of these materials.

Name: _____ Section: _____

Lab Instructor: _____ Date: _____

EXPERIMENT 9

Resolution of Mixtures 2: Chromatography

Results/Observations – Part A

Water-soluble inks

Pen	Pen color	Identification code
1	_____	_____
2	_____	_____
3	_____	_____
4	_____	_____

Distance traveled by solvent front   _____

Pen	Spot 1 Distance	Spot 1 R_f	*Spot 2 Distance	*Spot 2 R_f
1				
2				
3				
4				

*For those inks which resolved into more than one spot

Acetone-soluble inks

Pen	Pen color	Identification code
1	_____	_____
2	_____	_____
3	_____	_____
4	_____	_____

Distance traveled by solvent front   _____

Pen	Spot 1 Distance	Spot 1 R_f	*Spot 2 Distance	*Spot 2 R_f
1				
2				
3				
4				

*For those inks which resolved into more than one spot

Questions

1. Which pens appear to contain single-component inks and which contain mixtures? In the table, list the identification code numbers of the pens you used.

Single Dye	Mixture of Dyes

2. Why was it necessary for you to wear plastic surgical gloves while preparing and developing your chromatograms? What effect might oils from the skin cause if you had handled the chromatography paper with your bare fingers?

3. Why was the mixture of acetone and water used to chromatograph the permanent inks?

Results/Observations – Part B

For each of the solvent mixtures studied, calculate R_f for each of the spots:

acetone/hexane   Distance traveled by solvent front   _____

Distance traveled by spot   Calculated R_f

xylenol orange   _____   _____

bromcresol green   _____   _____

methyl red   _____   _____

ethyl acetate/hexane   Distance traveled by solvent front   _____

Distance traveled by spot   Calculated R_f

xylenol orange   _____   _____

bromcresol green   _____   _____

methyl red   _____   _____

acetone/ethyl acetate   Distance traveled by solvent front   _____

Distance traveled by spot   Calculated R_f

xylenol orange   _____   _____

bromcresol green   _____   _____

methyl red   _____   _____

acetone/ethanol   Distance traveled by solvent front   _____

Distance traveled by spot   Calculated R_f

xylenol orange   _____   _____

bromcresol green   _____   _____

methyl red   _____   _____

ethyl acetate/ethanol	Distance traveled by solvent front	_____
	Distance traveled by spot	Calculated R_f
xylenol orange	_____	_____
bromcresol green	_____	_____
methyl red	_____	_____
hexane/ethanol	Distance traveled by solvent front	_____
	Distance traveled by spot	Calculated R_f
xylenol orange	_____	_____
bromcresol green	_____	_____
methyl red	_____	_____

Questions

1. Why was it necessary to keep the beaker used for chromatography tightly covered with plastic wrap while the solvent was rising through the TLC slide?

2. Which solvent mixture gave the most complete resolution of the three indicator dyes? Which solvent mixture gave the poorest resolution?

3. Of the solvents used, some were very polar (e.g., acetone, ethanol) and others were very nonpolar (e.g., hexane). Did the polarity of the various solvent mixtures seem to affect the completeness of the separation of dyes? Why might this be so?

EXPERIMENT 10

Stoichiometry 1: Limiting Reactant

Objective

Stoichiometric measurements are among the most important in chemistry; they indicate the *proportions* in which various substances react. In this experiment, you will use the temperature change produced by a reaction as an index of the extent of reaction.

Introduction

The concept of a limiting reactant is very important in the study of the stoichiometry of chemical reactions. The **limiting reactant** is the reactant that controls the amount of product possible for a process, because once the limiting reactant has been consumed, no further reaction can occur.

Consider the following balanced chemical equation:

$$A + B \longrightarrow C$$

Suppose we set up a reaction in which we combine 1 mol of substance A and 1 mol of substance B. From the stoichiometry of the reaction, and from the amounts of A and B used, we would predict that exactly 1 mol of product C should form.

Suppose we set up a second experiment involving the same reaction, in which we again use 1 mol of A, but this time we only use 0.5 mol of B. Clearly, 1 mol of product C will *not* form. According to the balanced chemical equation, substances A and B react in a 1:1 ratio, and with only 0.5 mol of substance B, there is not enough substance B present to react with the entire 1 mol of substance A. Substance B would be the *limiting reactant* in this experiment. There would be 0.5 mol of unreacted substance A remaining once the reaction had been completed, and only 0.5 mol of product C would be formed.

Suppose we set up a third experiment, in which we combine 1 mol of substance A with 2 mols of substance B. In this case, substance A would be the limiting reactant, and it would control how much product C is formed. There would be an excess of substance B present once the reaction was complete.

One way of determining the extent of reaction for a chemical process is to monitor the *temperature* of the reaction system. Chemical reactions nearly always either absorb or liberate thermal energy as they occur, and the amount of heat transferred is *directly proportional* to how much product has formed. In this experiment, you will use a thermometer or temperature probe to monitor temperature changes as an indication of the extent of reaction.

In this experiment, you will prepare solutions of hydrochloric acid and sulfuric acid and then determine the stoichiometric ratio in which these acids react with the base, sodium hydroxide. You will perform several trials of these acid/base reactions, in which the amount of each reagent used is systematically varied between the trials. By monitoring the temperature changes that take place as the reaction occurs, you will have an index of the extent of reaction. The maximum extent of reaction will occur when the reactants have been mixed together in the correct stoichiometric ratio for reaction. If the reactants for a particular trial are *not* in the correct stoichiometric ratio, then one of these reactants will *limit* the extent of reaction and will also limit the temperature increase observed during the experiment.

Safety Precautions	• Protective eyewear approved by your institution must be worn at all time while you are in the laboratory.
	• The acid and base solutions used in the experiment are damaging to skin, eyes, and clothing. Wash after handling. If the solutions are spilled on the bench, consult with the instructor about clean up. Sulfuric acid solutions become more concentrated, and consequently more dangerous, as the water evaporates from them

Reagents/Apparatus Required

3.0 M solutions of HCl, H_2SO_4, and NaOH; plastic foam cups (6 oz); thermometer; temperature probe (if using); 600-mL or larger beakers for storage

Procedure

Record all data and observations directly on the report pages in ink.

Check your pre-laboratory calculations with the instructor before preparing the solutions needed for this experiment.

1. **Reaction Between HCl and NaOH**

   When preparing dilute acid/base solutions, slowly add the more-concentrated stock solution to the appropriate amount of water, with stirring.

   Using 600-mL (or larger) beakers for storage, prepare 500. mL each of 1.0 M NaOH and 1.0 M HCl, using the stock 3.0 M solutions. Use a clean graduated cylinder to measure the volumes of concentrated acid or base and water required.

   Stir the solutions vigorously with a stirring rod for 1 minute to mix. Keep the solutions covered with a watch glass when not in use.

   Frequently, diluting a concentrated solution with water will result in a temperature increase. *Allow the solutions to stand for 5–10 minutes until all have come to the same temperature* (within $\pm 0.2°C$). Be sure to rinse and wipe the thermometer before switching between solutions.

   Obtain a plastic foam cup for use as a reaction vessel: an insulated cup is used so that the heat liberated by the reactions will not be lost to the room during the timeframe of the experiment. To minimize risk of the cup tipping over, support it in a beaker. (A 250-mL beaker works well for a 6 oz cup.)

   Using a graduated cylinder, measure 45 mL of 1.0 M HCl. Pour it into the plastic cup, and determine the temperature of the HCl (to the nearest 0.2°C). Record this temperature on your report sheet.

   Measure 5.0 mL of 1.0 M NaOH in a 10-mL graduated cylinder.

   Add the NaOH to the HCl in the plastic cup all at once, and *carefully* stir the mixture with the probe or thermometer. Determine and record the *highest temperature reached* as the reaction occurs.

   Rinse and dry the plastic cup.

   Perform the additional reaction trials indicated in Table I on the report sheet. In each case, measure the amounts of each solution carefully with a graduated cylinder. For each trial, record the initial temperature of the solution in the plastic cup, as well as the highest temperature reached during the reaction.

To ensure meaningful temperature readings, always start with the reagent with the greatest volume in the foam cup. Thus, HCl will be in the cup for the first five trials, then NaOH for the last four.

2. **Reaction Between H$_2$SO$_4$ and NaOH**

When preparing dilute acid/base solutions, slowly add the more-concentrated stock solution to the appropriate amount of water, with stirring.

Using 600-mL (or larger) beakers for storage, prepare 500. mL each of 1.0 *M* NaOH and 1.0 *M* H$_2$SO$_4$, using the stock 3.0 *M* solutions. Be sure to measure the amounts of concentrated acid or base and water required with a graduated cylinder.

Stir the solutions vigorously with a stirring rod for 1 minute to mix. Keep the solutions covered with a watch glass when not in use.

Allow the solutions to stand for 5–10 minutes until they have come to the same temperature (within ± 0.2°C). Be sure to rinse and wipe the thermometer before switching between solutions.

Using the procedure described earlier, perform the reaction trials indicated in Table II on the report sheet for H$_2$SO$_4$ and NaOH. In each case, measure the amounts of each solution carefully with a clean graduated cylinder. For each trial, record the initial temperature of the solution in the plastic cup, as well as the highest temperature reached during the reaction.

Interpretation of Results

From the volume and concentration of the solutions used in each experiment, calculate the number of moles of each reactant used in each trial. For example, if you used 25.0 mL of 3.0 *M* HCl solution, the number of moles of HCl present would be

$$25.0 \ mL \times \frac{1 \ L}{10^3 mL} \times 3.0 \ \frac{mol \ HCl}{L} = 0.075 \ mol \ HCl$$

Record these values in Tables I and II on the report sheet, as appropriate.

To demonstrate most clearly how the extent of reaction in each of the experiments is determined by the limiting reactant, prepare two **graphs** of your experimental data. One graph should be for the reaction between HCl and NaOH; the second graph should be for the reaction between H$_2$SO$_4$ and NaOH.

Set up your graphs so that the vertical axis represents the *temperature change* measured for a given trial. Set up the horizontal axis to represent the *number of moles of NaOH* used in the trial. Plot each of the data points with a sharp pencil.

You should notice that each of your graphs consists of a set of *ascending* points and a set of *descending* points. For each of these sets of points, use a ruler to draw the best straight line through the points.

The **intersection** of the lines through the ascending points and the descending points in each graph represents the maximum extent of reaction for the particular experiment. From the maximum point on your graph, draw a straight line down to the horizontal axis. Read off the horizontal axis the number of moles of NaOH (and acid) that has reacted at this point.

According to your graph, in what stoichiometric ratio do HCl and NaOH react?

According to your graph, in what stoichiometric ratio do H$_2$SO$_4$ and NaOH react?

For each of your graphs, indicate which points represent the acid as limiting reactant and which points represent NaOH as limiting reactant. Staple your two graphs to the report pages.

Example: Suppose the reaction $A + B \rightarrow C$ was studied by measuring the temperature changes associated with several combinations of the two reactants, giving the data below:

mol A	mol B	ΔT (°C)
0.1	0.9	18.1
0.2	0.8	19.9
0.3	0.7	21.9
0.4	0.6	24.2
0.5	0.5	24.9
0.6	0.4	24.1
0.7	0.3	21.9
0.8	0.2	20.1
0.9	0.1	17.9

Construct a graph of the data, plotting the temperature *versus* the number of moles of B used in each experiment. Show graphically how the maximum extent of reaction is determined.

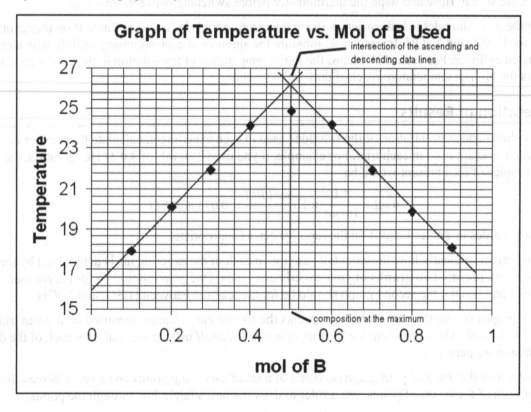

Name: ______________________________ Section: __________

Lab Instructor: ______________________________ Date: __________

EXPERIMENT 10

Stoichiometry 1: Limiting Reactant

Pre-Laboratory Questions

1. Define the following terms:

   a. Limiting reactant

   b. Stoichiometry

2. Calcium metal reacts with chlorine gas to produce calcium chloride, $CaCl_2$.

   a. Write the balanced chemical equation for the reaction.

   b. If 5.00 g calcium is combined with 10.0 g of chlorine, show by calculation which substance is the limiting reactant, and calculate the theoretical yield of calcium chloride for the reaction.

3. Describe how you would prepare exactly 250. mL of 1.0 M NaOH solution from a 3.0 M stock NaOH solution. Show all calculations.

4. Consider the following data table, which is similar to the data you will collect in this experiment.

    Complete the entries in the table. On graph paper from the end of this manual, construct a graph of the data, plotting the *temperature change* measured for each run versus the *number of moles* of substance A used for the run. Use your graph to determine the stoichiometric ratio in which substances A and B react. Explain your reasoning. Space is left for calculations below the table.

1.0 M NaOH (mL)	Mol NaOH	1.0 M HCl (mL)	Mol HCl	ΔT (°C)
5.0		45.0		1.52
10.0		40.0		3.01
15.0		35.0		4.52
20.0		30.0		6.01
25.0		25.0		7.52
30.0		20.0		5.99
35.0		15.0		4.49
40.0		10.0		2.99
45.0		5.0		1.49

EXPERIMENT 10

Stoichiometry 1: Limiting Reactant

Results/Observations

Table 1

1. **Reaction Between Hydrochloric Acid and Sodium Hydroxide**

1.0 M HCl (mL)	Mol HCl	1.0 M NaOH	Mol NaOH	ΔT ($^{\circ}$C)
45.0		5.0		
40.0		10.0		
35.0		15.0		
30.0		20.0		
25.0		25.0		
20.0		30.0		
15.0		35.0		
10.0		40.0		
5.0		45.0		

Intersection of ascending and descending portions of graph, mol NaOH   _____

Stoichiometric ratio, mol HCl/mol NaOH, at intersection   _____

Balanced chemical equation for reaction between HCl and NaOH

In which of the reactions was HCl the limiting reactant? For any one of the reactions in which HCl was the limiting reactant, give an example of how you would demonstrate this.

2.   Reaction Between Sulfuric Acid and Sodium Hydroxide

Table II

1.0 M H_2SO_4 (mL)	Mol H_2SO_4	1.0 M NaOH (mL)	Mol NaOH	ΔT (°C)

Intersection of ascending and descending portions of graph, mol NaOH _____

Stoichiometric ratio, mol H_2SO_4/mol NaOH, at intersection _____

Balanced chemical equation for reaction between H_2SO_4 and NaOH

Questions

1.   Why was a plastic coffee cup used as the reaction vessel? What error might have been introduced into the experiment if heat had been lost from the reaction vessel?

2.   Suppose a similar experiment had been done using 1.0 M H_3PO_4 solution as the acid. What would be the stoichiometric ratio, mol H_3PO_4/mol NaOH, at the intersection of the ascending and descending portions of the graph for such an experiment? Explain.

EXPERIMENT 11

Stoichiometry 2:
Stoichiometry of an Iron(III)-Phenol Reaction

Objective

Stoichiometric measurements are among the most important in chemistry; they indicate the *proportions* in which various substances react. In this experiment, you will use a spectrophotometer to determine the amount of colored product produced when a solution of iron(III) nitrate reacts with a solution of a phenol derivative.

Introduction

Solutions of iron(III) react with many of the members of the family of organic chemical compounds called *phenols* to produce intensely colored products. The interaction is between iron(III) and a hydroxyl group (–OH) bonded to a benzene ring. The reaction between iron(III) and salicylic acid, for example, is used as a qualitative and quantitative test for the purity of aspirin. Aspirin is synthesized from salicylic acid, a phenol. Aspirin itself is *not* a phenol, however. When treated with iron(III), pure aspirin should *not* produce a color. If unreacted salicylic acid starting material is present, however, then the aspirin sample will produce a purple color when treated with iron(III), with the *intensity* of the purple color an index of the *amount* of unreacted salicylic acid present.

phenol   salicylic acid   acetylsalicylic acid (aspirin)

In this experiment, you will determine the stoichiometry of the reaction between iron(III) and the phenol known as 5-sulfosalicylic acid. This phenol is chosen because it is more soluble in water than phenol itself or salicylic acid.

5-sulfosalicylic acid

You will prepare a series of solutions in which you *systematically vary* the relative amounts of iron(III) and 5-sulfosalicylic acid (SSA). The intensity of the color of the resulting mixtures will be used to identify the stoichiometric ratio in which iron(III) and 5-sulfosalicylic acid react: the most intense color will be found in the solution in which the two reactants have been combined in the correct stoichiometric ratio for reaction.

Although you should be able to identify the most intensely colored solution with the eye, a colorimeter or spectrophotometer will be used to quantify the color intensities. Solutions appear colored because components in the solution absorb some wavelengths of visible light, leaving other wavelengths to be seen by the eye. The spectrophotometer measures what percentage of light entering a solution is able to pass through the solution without being absorbed.

Safety Precautions	• Protective eyewear approved by your institution must be worn at all time while you are in the laboratory. • The iron(III) and 5-sulfosalicylic acid solutions also have nitric acid present as a preservative. This makes the solutions corrosive to eyes, skin, and clothing. Wash immediately if the solutions are spilled on the skin, and clean up all spills on the benchtop. • Always use a rubber safety bulb to apply the suction force when using the pipet. Pipetting by mouth is dangerous and is not permitted.

Apparatus/Reagents Required

1.7×10^{-3} M iron(III) nitrate (in 0.1 M HNO$_3$), 1.7×10^{-3} M 5-sulfosalicylic acid (in 0.1 M HNO$_3$), $18 \times$ 150-mm test tubes, marking pen or labels, 10-mL Mohr pipet and rubber safety bulb, spectrophotometer.

Procedure

Label nineteen 18×150-mm test tubes with the numbers 1-19 inclusive, and set them up for convenience, in numerical order, in a test tube rack. Obtain about 100 mL each of the available iron(III) nitrate and 5-sulfosalicylic acid solutions. Record the exact concentrations of these solutions on the report page and in your notebook.

Using a rubber safety bulb to provide the suction force, use the 10-mL Mohr pipet to add the required quantities of iron(III) nitrate and 5-sulfosalicylic acid solutions to the appropriate test tubes as indicated in the table below. It is necessary to prepare all the solutions before making any measurements of the solutions with the spectrophotometer.

Test tube	mL Fe(III)	mL 5-SSA		Test tube	mL Fe(III)	mL 5-SSA
1	0.50	9.50		11	5.50	4.50
2	1.00	9.00		12	6.00	4.00
3	1.50	8.50		13	6.50	3.50
4	2.00	8.00		14	7.00	3.00
5	2.50	7.50		15	7.50	2.50
6	3.00	7.00		16	8.00	2.00
7	3.50	6.50		17	8.50	1.50
8	4.00	6.00		18	9.00	1.00
9	4.50	5.50		19	9.50	0.50
10	5.00	5.00				

Use a clean stirring rod to mix the contents of each test tube, being sure to rinse and wipe dry the stirring rod before stirring the next solution.

Let the test tubes stand for 5 minutes to allow the color of the product of reaction to develop fully. Examine the test tubes. In which test tube does the purple color seem most intense (darkest)?

Several visible spectrophotometers or colorimeters have been set up for your use in the laboratory. Your instructor will provide you with detailed instructions on how these instruments should be used. The typical spectrophotometer used in the general chemistry lab measures the % transmittance of a solution: that is, it measures the fraction of the incident light that is passed through the solution without being absorbed, expressed in terms of a percentage. Modern instruments are capable of converting the percent transmittance values (%T) to units of absorbance (*Abs*). Although the spectrophotometer will almost certainly make the conversion for you, the relationship between %T and *Abs* is given by:

$$Abs = -\log T = -\log (\%T/100) = -\log \%T - \log (1/100) = -\log \%T - (-2) = 2 - \log \%T$$

It is these absorbance values that you will use to determine the stoichiometric ratio in the reaction between Iron(III) ions and SSA. Since absorbance will vary somewhat with temperature, it is important to have all solution mixtures prepared before any measurements are made. Measurements should be made quickly, over as short a time span as possible. Since we are making measurements at the wavelength of maximum of absorption of the colored complex (known as λ_{max} – "lambda-max"), it is also important that the wavelength setting not be changed, even slightly, during the measurements.

In general, the steps in using a visible spectrophotometer are as follows:

1. The spectrophotometer must be warmed up for at least 15 minutes before being used for measurements.

2. The spectrophotometer must be set to the wavelength absorbed most strongly by the colored component in the solution. For this analysis, the wavelength to be used is 505 nm.

3. The spectrophotometer must be calibrated to read 0% transmittance when there is no sample present in the sample chamber, and to read 100% transmittance when only the solvent used for the system is present. With most modern devices, this can be accomplished in a single calibration step.

4. After calibration, the solutions to be measured must be measured quickly in sequence before the spectrophotometer has a chance to warm up appreciably, as the response of such spectrophotometers can change with time.

5. Special cells (called *cuvettes*), made of optical-quality glass or plastic, serve as sample holders with all spectrophotometers. The cuvette must be handled carefully and should be wiped with a lint-free tissue to remove finger marks before measurements are made. The cell should be aligned with the same orientation each time it is inserted into the spectrophotometer chamber. Many plastic cuvettes are square, with two ribbed sides and two smooth sides. Handle only the ribbed sides.

Following the instructions provided by your instructor, determine the absorbance of each of your 19 solutions with the spectrophotometer. Record your data to the nearest 0.1 absorbance unit.

Make certain that the wavelength selected on the spectrophotometer is set correctly (505 nm) and that the spectrophotometer has been calibrated. Do not adjust the wavelength knob once it has been set to the correct wavelength.

Rinse the spectrophotometer cell with small portions of the solution in test tube 1. Then half-fill the cell with the solution from test tube 1, wipe the bottom half of the cell with a tissue to remove finger marks, and place the cell into the sample chamber of the spectrophotometer. Record the absorbance reading on the report sheet.

Repeat the process for each of the remaining 18 solutions. Be sure to rinse the spectrophotometer cell with small portions of each solution before filling the cell, and make certain you wipe the cell with a tissue to remove finger marks before each measurement. Record the absorbance for each solution to the nearest 0.1 absorbance unit.

Calculations

For each solution, calculate the *mole fractions* of iron(III) and 5-sulfosalicylic acid (SSA) present:

$$\chi_{Fe} = mole\ fraction\ Fe^{3+} = \frac{moles\ of\ Fe^{3+}}{(moles\ of\ Fe^{3+} + moles\ of\ SSA)}$$

$$\chi_{SSA} = mole\ fraction\ SSA = \frac{moles\ of\ SSA}{(moles\ of\ Fe^{3+} + moles\ of\ SSA)} = 1.000 - \chi_{Fe}$$

Note that, if the concentrations of the SSA and Fe^{3+} solutions are equal, then the volumes of solution can be substituted for the numbers of moles, since each solution has the same number of moles of solute per milliliter of solution. Of course, if the solution molarities are not the same, this substitution is not possible. (See Prelaboratory Question 2.)

Construct a graph of the *absorbance* of the solutions (vertical axis) versus the *mole fraction of iron* in the solutions (horizontal axis). The graph typically will consist of a set of *ascending* data points and a set of *descending* data points.

The graph may "round off" near the maximum because the concentration of the colored product is largest near the maximum, which may make the graph appear to be an inverted parabola. At very high concentrations, especially for intensely colored materials, solutions may not follow Beer's Law.

Draw a straight line through the ascending data points, and a second straight line through the descending data points. You may exclude the data points closest to the maximum which do not appear to lie on the straight lines formed by the first several (and last several) data points.

Drop a vertical line from the intersection of the two straight lines to the mole fraction axis. At this point, the maximum extent of reaction took place, which resulted in the maximum possible formation of colored product and, therefore, the maximum absorbance by the solution.

Determine the relative mole fractions of iron(III) and 5-sulfosalicylic acid at the maximum in your graph. In what stoichiometric ratio do iron(III) and 5-sulfosalicylic acid react? Attach your graph to your laboratory report.

Name: _____ Section: _____

Lab Instructor: _____ Date: _____

EXPERIMENT 11

Stoichiometry 2:
Stoichiometry of an Iron(III)–Phenol Reaction

Pre-Laboratory Questions

1. Briefly describe the procedure for this experiment, including what you will measure, how you will measure it, and what you hope to determine from your measurements.

2. Calculate how many moles of 5-sulfosalicylic acid are contained in 3.50 mL of 1.71×10^{-3} M 5-sulfosalicylic acid solution. Calculate how many moles of iron(III) are contained in 6.50 mL of 1.69×10^{-3} M iron(III) nitrate solution.

3.   Spectrophotometers, such as the one you will use in this experiment, directly measure the percent transmittance of a colored solution. The percent transmittance represents the fraction of light that passes through the solution without being absorbed, expressed on a percentage basis. Most commonly, however, we need the *absorbance* of the solution: the absorbance of a colored solution is directly proportional to the concentration of the color species in the solution. The absorbance (A) of a solution is related to its % transmittance ($\%T$) by the relationship $A = 2 - \log(\%T)$

Calculate the absorbance of solutions with the following values for the $\%T$:

   19.2%T _____

   37.9%T _____

   57.1% _____

4.   As mentioned in the *Introduction* to this experiment, iron(III) reacts with most water-soluble phenols to produce a brightly colored complex. In the reaction by which aspirin is synthesized from salicylic acid,

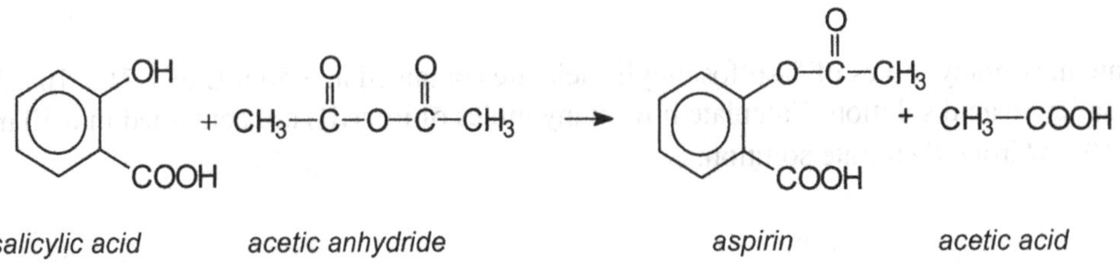

   salicylic acid   *acetic anhydride*   *aspirin*   *acetic acid*

iron(III) is used to determine how successful the conversion of salicylic acid to aspirin has been. Use a scientific encyclopedia or online reference to explain how adding iron(III) to the products of this reaction can tell you whether or not the conversion has been successful.

Name: _____ Section: _____

Lab Instructor: _____ Date: _____

EXPERIMENT 11

Stoichiometry 2:
Stoichiometry of an Iron(III)–Phenol Reaction

Results/Observations

Tube	mL Fe^{3+}	mL SSA	χ_{Fe}	χ_{SSA}	Abs
1					
2					
3					
4					
5					
6					
7					
8					
9					
10					
11					
12					
13					
14					
15					
16					
17					
18					
19					

1. Using your data for Tube 10 in the table on the previous page, show how you calculated each of the following quantities:

    a.   Mole fraction Fe^{3+}

    b.   Mole fraction SSA

    c.   Absorbance

2. Based on your graph, in what molar ratio do Fe^{3+} and 5-sulfosalicylic react? Explain.

Questions

1. In your experimental data, which reagent [iron(III) or 5-sulfosalicylic acid] is the limiting reactant in test tube 3? In test tube 17? Explain your reasoning, based on the mole ratio you determined above.

2. Use a chemical dictionary, chemical encyclopedia, or online reference to find specific definitions for the following terms.

    a.   Percent transmittance, %T

    b.   Absorbance, *Abs*

EXPERIMENT 12

Composition 1:   Percentage Composition – The Empirical Formula of Magnesium Oxide

Objective

The percentage composition and empirical formula of magnesium oxide will be determined by heating a small sample of magnesium in air.

Introduction

Magnesium metal is a moderately reactive element. At room temperature, magnesium reacts only very slowly with oxygen and can be kept for long periods of time without appreciable oxide buildup. At elevated temperatures, however, magnesium will ignite in an excess of oxygen gas, burning with an intensely white flame and producing magnesium oxide. Because of the brightness of its flame, magnesium is used in flares and photographic flashbulbs.

In this experiment, however, you will be heating magnesium in a closed container called a crucible, exposing it only gradually to the oxygen of the air. Under these conditions, the magnesium will undergo a more controlled oxidation, gradually turning from shiny metal to grayish-white powdered oxide.

$$2Mg(s) + O_2(g) \longrightarrow 2MgO(s)$$

Because the air also contains a great deal of nitrogen gas, a portion of the magnesium being heated may be converted to magnesium nitride, Mg_3N_2, rather than to magnesium oxide. Magnesium nitride will react with water and, on careful heating, is converted into magnesium oxide.

$$Mg_3N_2(s) + 3H_2O(l) \longrightarrow 3MgO(s) + 2NH_3(g)$$

The ammonia produced by this reaction can be detected by its odor, which is released on heating of the mixture.

Magnesium is a Group IIA metal, and its oxide should have the formula MgO. Judging on the basis of on this formula, magnesium oxide should consist of approximately 60% magnesium by mass. Comparing the mass of magnesium reacted and the mass of magnesium oxide that results from the reaction, will confirm this.

The empirical formula of a compound represents the smallest whole number ratio in which atoms combine to produce the compound. The empirical formula is calculated from the masses of elements that combine to form the compound, and does not take into account any molecular structure the compound may possess.

Safety Precautions	• Protective eyewear approved by your institution must be worn at all time while you are in the laboratory.
	• Magnesium produces an intensely bright flame if ignited, and looking directly at this flame may be damaging to the eyes. If the magnesium ignites, immediately cover the crucible and stop heating. Do not look directly at magnesium while it is burning.
	• The crucible must be cold before water is added to it to convert magnesium nitride to magnesium oxide or the crucible will shatter. After water is added to the crucible, the contents of the crucible may spatter when heated. Use only gentle heating in evaporating the water. Do not heat the crucible strongly until nearly all the water has been removed.
	• Do not attempt to adjust the iron ring, clay triangle, or crucible unless the heat has been shut off for at least 5 minutes so that the system will have cooled. Crucible tongs should be used only to open/close the cover of the crucible and not to pick up or move the crucible itself when the crucible is hot. The tongs are metal and heat will be transferred through the tongs to your fingers.
	• Hydrochloric acid is damaging to skin, eyes, and clothing. If it is spilled, wash immediately and inform the instructor.

Apparatus/Reagents Required

Porcelain crucible and cover, crucible tongs, clay triangle, magnesium turnings (or ribbon), pH paper, 6 M HCl

Procedure

Record all data and observations directly in your notebook in ink.

Obtain a crucible and cover, and examine them for any cracks. The crucible and cover are extremely fragile and expensive. Use caution in handling them.

If there is any loose dirt in the crucible, moisten and rub it gently with a paper towel to remove the dirt. If dirt remains in the crucible, bring it to the hood, add 5–10 mL of 6 M HCl, and allow the crucible to stand for 5 minutes. Discard the HCl in an appropriate waste container and rinse the crucible with water. If the crucible is not clean at this point, consult with the instructor about other cleaning techniques, or replace the crucible. After the crucible has been cleaned, use tongs or forceps to handle the crucible and cover.

Set up a clay triangle on a ringstand. Transfer the crucible and cover to the triangle. The crucible should sit firmly in the triangle (the triangle's arms can be bent slightly if necessary). See Figure 12-1.

Begin heating the crucible and cover with a small flame to dry them. When the crucible and cover show no visible droplets of moisture, increase the flame to full intensity by opening the air vent of the burner fully, and heat the crucible and cover for 5 minutes.

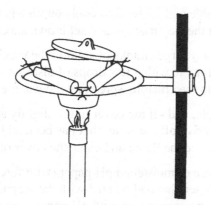

Figure 12-1. Set-up for oxidation of magnesium.

Position the ring so that the bottom of the crucible will be approximately 2 inches above the top of the burner.

Remove the flame, and allow the crucible and cover to cool to room temperature (~ 10 minutes).

When the crucible and cover are completely cool, use tongs or a paper towel to move them to a clean, dry watch glass or flat glass plate. Do not place the crucible directly on the lab bench. Weigh the crucible and cover to the nearest milligram (0.001 g).

Return the crucible and cover to the clay triangle. Reheat in the full heat of the burner flame for 5 minutes. Allow the crucible and cover to cool completely to room temperature.

Reweigh the crucible after it has cooled. If the mass this time differs from the earlier mass by more than 5 mg (0.005 g), reheat the crucible for an additional 5 minutes and reweigh when cool. Continue the heating/weighing until the mass of the crucible and cover is constant to within 5 mg. At this point, any impurities that might affect your experimental results will have been burned off. Any discoloration of the porcelain is normal and not a concern.

Add to the crucible approximately half a teaspoon of magnesium turnings (or about 8 inches of magnesium ribbon coiled into a spiral).

Using tongs or forceps, transfer the crucible, cover and magnesium to the balance and weigh to the nearest milligram (0.001 g).

Set up the crucible on the clay triangle with the cover very slightly ajar. (See Figure 12-1.) With a very small flame, begin heating the crucible gently, holding the base of the burner and 'brushing' the crucible with the flame. After a minute or two, increase the intensity of the flame by opening the air vent on the burner.

If the crucible begins to smoke when heated, immediately cover the magnesium completely and remove the heat for 2–3 minutes. The smoke consists of the magnesium oxide product and must not be lost from the crucible.

Continue to heat for 5–10 minutes with the cover of the crucible slightly ajar. Remove the heat and allow the crucible to cool for 1–2 minutes.

Remove the cover and examine the contents of the crucible. If the shiny appearance of the free metal is still evident, replace the cover and heat with a hot flame for an additional 5 minutes; then re-examine the contents. Continue heating with a small flame until no shiny metallic pieces are visible.

When the shiny magnesium metal appears to have been converted fully to the dull gray/white oxide, return the cover to its slightly ajar position, and heat the crucible with the full heat of the burner flame for 5 minutes. Then slide the cover to about the half-open position, and heat the crucible in the full heat of the burner flame for an additional 5 minutes.

Remove the heat and allow the crucible and contents to cool completely to room temperature (at least 10 minutes). Remove the crucible from the clay triangle and set it on a sheet of clean paper on the lab bench.

With a stirring rod, gently break up any large chunks of solid in the crucible. Rinse any material that adheres to the stirring rod into the crucible with a few drops of distilled water. With a dropper, add a total of about 10 drops of distilled water to the crucible, spreading the water evenly throughout the solid.

Return the crucible to the clay triangle, and set the cover in the slightly ajar position. With a very small flame, begin heating the crucible to drive off the water that has been added. Beware of spattering during the heating. If spattering occurs, remove the flame and close the cover of the crucible.

As the water is driven off, hold a piece of moistened pH paper (with forceps) in the stream of steam being expelled from the crucible. Any nitrogen that had reacted with the magnesium is driven off as ammonia during the heating and should give a basic response with pH paper (you may also note the odor of ammonia).

When it is certain that all the water has been driven off, slide the cover so that it is in approximately the half-open position, and increase the size of the flame. Heat the crucible and contents in the full heat of the burner for 5 minutes.

Allow the crucible and contents to cool completely to room temperature. When they are completely cool, weigh the crucible and contents to the nearest milligram (0.001 g).

Return the crucible to the triangle and heat for another 5 minutes in the full heat of the burner flame. Allow the crucible to cool completely to room temperature and reweigh.

The two measurements of the crucible and contents should give masses that agree within 5 mg (0.005 g). If this agreement is not obtained, heat the crucible for additional 5-minute periods until two successive mass determinations agree within 5 mg.

Clean out the crucible, and repeat the determination.

Calculations

Calculate the mass of magnesium used, as well as the mass of magnesium oxide that was present after the completion of the reaction for each of your determinations.

Calculate the mass of oxygen that combined with each of the magnesium samples.

Based on the masses of magnesium and oxygen for each of your determinations, calculate the empirical formula of magnesium oxide based on each determination.

Calculate the percentage of magnesium in the magnesium oxide from your experimental data.

> **Example:** Suppose 0.532 g of magnesium is heated, resulting in the production of 0.875 g of magnesium oxide. Calculate the experimental percentage of magnesium.

$$\% \text{ Mg} = \frac{\text{experimental mass of Mg}}{\text{experimental mass of magnesium oxide}} \times 100\% = \frac{0.532 \ g}{0.875 \ g} \times 100\% = 60.8\%$$

Calculate the mean percentage of magnesium in magnesium oxide for your two determinations.

Calculate the theoretical percentage of magnesium (by mass) in magnesium oxide, and compare this to the mean experimental value. Calculate the percent error in your determination.

Name: _____ Section: _____

Lab Instructor: _____ Date: _____

EXPERIMENT 12

Composition 1:
Percentage Composition – The Empirical Formula
of Magnesium Oxide

Pre-Laboratory Questions

1. Calculate the percentage by mass of each element present in the following compounds.

   a. Magnesium oxide, MgO

   b. Magnesium nitride, Mg_3N_2

2. Suppose 1.043 g of magnesium is heated in air. What is the theoretical amount of magnesium oxide that should be produced?

3. A compound containing only carbon and hydrogen is analyzed, and is found to contain 20.11% hydrogen on a mass basis. Calculate the *empirical formula* of the compound. Show work below.

4. If 1.000 g of elemental carbon is heated in a stream of chlorine gas, $Cl_2(g)$, 12.810 g of a compound containing only carbon and chlorine is produced. Calculate the *empirical formula* of the compound.

EXPERIMENT 12

Composition 1:
Percentage Composition – The Empirical Formula
of Magnesium Oxide

Results/Observations

	Trial 1	Trial 2
Mass of empty crucible after first heating		
Mass of empty crucible after second heating		
Mass of crucible with Mg		
Mass of Mg used		
Mass of crucible and oxide after first heating		
Mass of crucible and oxide after second heating		
Mass of magnesium oxide produced		
Mass percent of magnesium in magnesium oxide		

Mean value for the mass percent of magnesium in magnesium oxide _____ %

Calculated value for the mass percent of magnesium in magnesium oxide _____ %

Percent error between calculated and experimental mean value for % Mg _____ %

Questions

1. If you had not added water to remove any Mg_3N_2 that was present in your sample, would your experimental % Mg have been *too high* or *too low*? Show a calculation to support your answer.

2. Suppose a similar experiment is to be performed, in which you will heat a weighed sample of calcium metal in a crucible (in place of magnesium).

   a. Calculate the theoretical percentage of calcium in calcium oxide.

   b. If a 0.9358 g sample of calcium metal were heated in oxygen, what mass of calcium oxide would be produced?

Experiment 13

Composition 2: Hydrates and Thermal Decomposition

Objectives

Some characteristics of hydrates will be investigated, the number of waters of hydration in a hydrated salt will be determined, and the formula for a compound will be established by heating to constant mass and identifying the products.

Introduction

Hydrates are chemical substances of definite composition formed from ionic compounds and integer or half-integer numbers of water molecules. In this experiment you will investigate some of the properties of hydrates and you will verify the number of water molecules present in a formula unit of one particular hydrate compound. A common hydrate is bright blue copper sulfate pentahydrate, formula $CuSO_4 \cdot 5H_2O$. The dot ($\cdot$) indicates that five water molecules are bound to the copper and sulfate ions. Water bound in this manner is called a *ligand*. Note that the dot is *not* a multiplication symbol. When you calculate the molar mass of a hydrate, the dot actually functions as a "plus" sign: the mass of five moles of water is added to the mass of one mole of copper(II) sulfate.

One of the most important hydrates is calcium sulfate dihydrate, $CaSO_4 \cdot 2H_2O$. The mineral gypsum is a hard, dense form of this compound. When it is heated and the two molecules of water are driven off, the *anhydrous* (water-free) calcium sulfate that remains is known as plaster of Paris. The addition of water to plaster of Paris to form the dihydrate once again is quite exothermic, producing plaster. The plaster 'sets' rapidly to a highly interlocked crystalline form that has water molecules arrayed between layers of calcium and sulfate ions.

Anhydrous calcium chloride, $CaCl_2$, aggressively picks up water, ultimately forming a hexahydrate, $CaCl_2 \cdot 6H_2O$. The reaction is exothermic and the resulting solution has a very low freezing point, so the primary use of the compound is for deicing roads, which it is capable of doing down to temperatures as low as $-51°C$! Anhydrous calcium chloride is commonly used in the laboratory as a *desiccant*; it will remove moisture from air (in a closed vessel called a desiccator) or from gases that flow through particles of the solid. It also has the property of removing water from organic liquids such as diethyl ether, which can dissolve about 8% of its weight in water. It can be useful in the summer, controlling dust in highway construction by removing moisture from the air to form a damp solid.

Copper sulfate pentahydrate is blue, while the anhydrous salt is white. Cobalt chloride in moist air is a hexahydrate, $CoCl_2 \cdot 6H_2O$, which is maroon-red, while in dry, cool air it is the blue anhydrous salt. At moderate humidity levels it is present as the purple dihydrate, $CoCl_2 \cdot 2H_2O$, so it can serve as an indicator of relative humidity. It is incorporated, for example, with the desiccant silica gel found in the packaging of electronic equipment; when the desiccant turns pink it needs to be replaced or reactivated by heating.

Some hydrates lose water at room temperature and normal humidity; they are said to *effloresce* (or, they are *efflorescent*). Other compounds, such as calcium chloride and sodium hydroxide, pick up so much moisture from the air that they dissolve in the resulting liquid. These substances *deliquesce*; they are *deliquescent*. Sodium hydroxide is deliquescent.[1]

---

[1] A similar term, also referring to the tendency to absorb and hold moisture from the air is "hygroscopic."

A very large class of organic compounds is called carbohydrates. This might seem to imply that they are hydrates of carbon; that misconception arose in the middle of the 18th century when only the empirical formulas, [C(H$_2$O)$_n$], of many sugars were known. The empirical formula for glucose, C$_6$H$_{12}$O$_6$, seems as if it could be rewritten as C$_6$(H$_2$O)$_6$, but this sugar is not carbon hexahydrate. True hydrate formation is completely reversible; this is not the case with carbohydrates. While they can be partially dehydrated; that is, some of the material can be driven off as water as a result of heating, when a carbohydrate is heated it will lose the elements of water, but be decomposed irreversibly in the process. Partial dehydration of sucrose, C$_{12}$H$_{22}$O$_{11}$, produces taffy, caramel, and finally the dark brown substance from which peanut brittle is made.

As you may be aware, hydrates are not the only compounds that can decompose when heated strongly. The Inquiry portion of this experiment addresses another, similar situation.

Safety Precautions	<ul><li>Protective eyewear approved by your institution must be worn at all times while you are in the laboratory.</li><li>Cobalt compounds are moderately toxic. Avoid getting your hands near your face. Wash your hands thoroughly with soap and water before leaving the laboratory.</li><li>Hot and cold glassware look alike. Be careful handling the test tubes, watch glasses, and crucibles in this experiment.</li><li>Use tongs or a towel to protect your hands from hot glassware.</li></ul>

Apparatus/Reagents Required

Crucible with cover, hotplate, ring stand set-up with wire triangle and wire screen with ceramic center. Sample of hydrates to be examined: calcium sulfate, cobalt(II) sulfate, copper(II) chloride, nickel(II) chloride, potassium carbonate, potassium chloride, sucrose, copper(II) sulfate pentahydrate, and copper(II) carbonate (formula to be determined)

Procedure

Part A

Turn on the hotplate, if using, to a low setting. Place a small amount of each of the eight known solids into separate 10 x 75-mm culture tubes (use only enough to cover the bottom of the tube). If the solid crystals are large, *gently* crush the solids to a fine powder with a glass stirring rod. Using a suitable holder, and keeping the tube at about a 45° angle, heat the bottom of each tube on a hot plate or *very gently* in a burner flame, and observe any changes. If the compound loses water on heating, moisture should collect on the upper portion of the tube. This is an indication the compound you are testing is a hydrate. Note the appearance (especially color) of the solid before and after heating. Record which tubes showed evidence of dehydration and specify the evidence that you saw.

After the tubes cool, add 10-15 drops of distilled water to each solid using a pipet or eyedropper, stir or agitate the tube and contents, and observe what happens. If the substance is a hydrate it should dissolve and give a solution close in color to the original hydrated salt.

Part B.

Set up an empty crucible as shown in **Figure 13.1**, but without the lid. Heat it strongly with a hot burner flame for 2-3 minutes, to drive off moisture and any impurities present. Allow the crucible to cool completely before you proceed.

Obtain a sample of copper(II) sulfate pentahydrate, $CuSO_4 \cdot 5H_2O$. If necessary crush the material to a relatively fine granular or powdered state, using a mortar and pestle or a glass stirring rod in a beaker. Carefully weigh ($\pm$ 0.001g) the cooled crucible with its cover, then add about 1.5 – 2.0 g ($\pm$ 0.001 g) of the hydrate to the crucible. Weigh crucible, cover, and contents ("the system") again, to the same milligram precision.

Place the cover on the crucible and transfer the system to a wire triangle supported, by an iron ring. The bottom of the crucible should be just high enough that it will be in the hottest part of a strong burner flame. Set the cover very slightly ajar, just enough so that steam can escape. See **Figure 13.1**.

Heat the crucible *gently* by holding the burner in your hand and moving it back and forth for about 5 or 6 minutes. Much of the water of hydration will be driven off during this time. Gentle heating is necessary at first to avoid loss of material due to spattering.

Now increase the intensity the flame by opening the air vents. Heat for 10 minutes or longer, but keep an eye on the bottom surface of the crucible – if it starts to glow orange or red, back off on the heating for a moment. Copper(II) sulfate is stable up to about 650 °C, but will start to decompose above that temperature. If you notice even a hint of a foul, choking odor, reduce the heating immediately.

At the conclusion of heating, remove and turn off your burner. Allow the crucible and contents to cool for at least 10 minutes, then weigh the system. Determine the mass of water lost

Figure 13.1: Heating the Crucible
Note the slightly ajar lid position.

Return the crucible to the wire triangle and heat strongly for an additional 5 minutes. Allow the crucible and contents to cool fully (~10 minutes) and weigh the system again. Determine the mass change from the previous weighing. The change in mass should be less than 10 mg; this is referred to as *heating to constant mass*, and is one way to be certain that the dehydration was complete. If dehydration is not complete, heat for an additional 5-minute period, cool, and reweigh. The final mass will be used in the calculations.

Your instructor will direct you as to what to do with the anhydrous copper(II) sulfate product. After disposal, clean the crucible with soap and water and dry it thoroughly. For stubborn residues, a rinse with dilute (6 *M*) hydrochloric acid may help.

Part C.

When metal carbonates are heated, the carbonate ion generally undergoes thermal decomposition. Metal carbonates typically decompose to the solid metal oxide and gaseous carbon dioxide. The decomposition of calcium carbonate, for example, is shown by:

$$CaCO_3(s) \xrightarrow{\text{(Strong heating)}} CaO(s) + CO_2(g)$$

Copper(II) carbonate will likewise undergo thermal decomposition, but the process is not quite so straightforward as that shown above for calcium carbonate. You are going to investigate the decomposition of solid copper(II) carbonate in an effort to determine what the products are, and ultimately the formula for the compound.

As you did in Part B, heat the uncovered crucible strongly with a hot burner flame for 2-3 minutes to drive off any impurities that may be present. Allow the crucible to cool for 8-10 minutes before you proceed. *It is important that all weighings be done at or near room temperature.*

Obtain a sample of solid copper(II) carbonate. Carefully weigh (±0.001g) the cooled crucible with its cover, and then add about 2.0 – 2.5 g (± 0.001 g) of the hydrate to the crucible. Weigh crucible, cover, and contents ("the system") again, to the same precision.

Place the cover on the crucible and transfer the system to a wire triangle supported, by an iron ring. The bottom of the crucible should be just high enough that it will be in the hottest part of a strong burner flame. Set the cover very slightly ajar, just enough so that steam can escape. See **Figure 13.1**.

Heat the crucible *gently* for about 5-6 minutes. Gentle heating is necessary to avoid loss of material due to spattering. Much, but not all of the vaporizable material will be driven off during this time.

Now adjust the flame so that the hottest part is in contact with the bottom of your crucible. The expected product, copper(II) oxide is thermally stable, so there is no concern about it decomposing. The bottom of the crucible will take on a bright orange glow – this is expected. Heat for 10 minutes or longer. At the conclusion of heating, remove and turn off your burner. Allow the crucible and contents to cool for at least 10 minutes then weigh the system (crucible, cover, and contents). Determine the mass lost due to heating and the mass of solid residue remaining in the crucible.

Return the crucible to the wire triangle and heat strongly for an additional 5 minutes – start timing from when the bottom of the crucible begins to show the orange glow. Allow the crucible and contents to cool for 10 minutes and weigh the system again. Determine the mass change from the previous weighing. The change in mass should not exceed 0.04 g; this is referred to as *heating to constant mass*, and is one way to be certain that the decomposition is complete. If the mass has changed by more than 0.04 g, heat for an additional 5-minute period, cool for 10 minutes, and reweigh. The final mass will be used in the calculations.

The copper(II) oxide should come out of the crucible easily. Dispose of it as your instructor directs.

EXPERIMENT 13

Hydrates and Thermal Decomposition

Pre-Laboratory Questions

1. Define the term *hydrate* (without copying words from the Introduction) and explain the function of the dot (·) in the formulas of hydrates. How is the dot treated when you are calculating the molar mass of a hydrate?

2. What is the meaning of the term, *anhydrous*?

3. In Part B of this experiment you will carry out the dehydration of copper(II) sulfate pentahydrate, $CuSO_4 \cdot 5H_2O$. Determine the theoretical mass percent of water in this compound.

4. In Part C of this experiment you will thermally decompose a compound labeled as copper(II) carbonate.

    (a) Based on its name alone, what is the most likely formula for copper(II) carbonate?

    (b) What is the theoretical percent by mass of copper in a compound with the formula you wrote in Part (a)?

EXPERIMENT 13

Hydrates and Thermal Decomposition

Pre-Laboratory Questions

1. Define the term *hydrate* without copying the words from the Introduction and explain the function of the dot (·) in the formulas of hydrates. How is the dot treated when you are calculating the molar mass of a hydrate?

2. What is the meaning of the term *anhydrous*?

In Part (b) of this experiment you will carry out the dehydration of copper(II) sulfate pentahydrate, $CuSO_4 \cdot 5H_2O$. Determine the theoretical mass percent of water for this compound.

3. In Part C of this experiment you will thermally decompose the compound labeled as copper(II) carbonate.

   (a) Based on its composition, what is the most likely formula for copper(II) carbonate?

   (b) What is the theoretical percent by mass of copper in a compound with the formula you wrote in Part (a)?

Name: _____ Section: _____

Lab Instructor: _____ Date: _____

EXPERIMENT 13

Hydrates and Thermal Decomposition

Results/Observations

Part A – Hydrate Testing

Compound	Original Color	Color After Heating	Effect of Adding Water	Hydrate? (Y/N)
Calcium sulfate				
Cobalt(II) sulfate				
Copper(II) chloride				
Nickel(II) chloride				
Potassium sulfate				
Potassium chloride				
Sucrose				

Part B – Determination of the Number of Waters of Hydration

Mass of empty crucible and cover, after preliminary heating	
Mass of crucible, cover and copper(II) sulfate sample before heating	
Mass of crucible, cover, and sample after first 10-minute heating	
Mass of crucible, cover, and sample after 2nd heating (5 minutes)	
Mass of crucible, cover, and sample after 3rd heating (if needed)	

Part C – Thermal Decomposition of Copper(II) Carbonate

Mass of crucible and cover only, after preliminary heating	
Mass of crucible, cover, and original sample before heating	
Mass of crucible, cover, and sample after 1st (10-minute) heating	
Mass of crucible, cover, and sample after 2nd heating (5 minutes)	
Mass of crucible, cover, and sample after 3rd 5 heating (if needed)	

Questions

Part A

1. Of the seven solids tested in *Part A*, identify the ones that were hydrates, citing evidence for your choices. Then explain how you know that the others were not hydrates.

2. When a white solid dissolves in water, what is the color of the solution? How do you account for this observation? (Hint: solids reflect color, solutions transmit color.)

3. In this experiment, you used hydrates that included two compounds, both containing copper(II) ions: copper(II) chloride, $CuCl_2$, and copper(II) sulfate, $CuSO_4$.

   (a) Potassium chloride contains chloride ions, and calcium sulfate contains sulfate ions; what color are these solids?

   (b) If both copper(II) compounds were blue (although of different shades), and if at least some compounds containing chloride and sulfate ions are white, speculate as to whether the water molecules in hydrated compounds are more likely bonded to the cation or the anion in each case, and defend your answer.

Part B

4. Based on your data for Part B of this experiment, carry out the following calculations.

    (a) Determine the mass and number of moles of anhydrous copper(II) sulfate present in your crucible at the conclusion of the procedure.

    (b) Determine the mass and number of moles of water that were lost during heating.

    (c) Determine the ratio of moles of water per mole of anhydrous solid.

5. Assume that the number of waters of hydration must be a whole number.

    (a) What is the nearest whole-number value for the ratio of moles of water to moles of anhydrous salt?

    (b) Write the formula of the hydrate in the form, $CuSO_4 \cdot Y\,H_2O$, where Y is the number of moles you calculated in **6b**.

Part C

6. The black solid that remained in your crucible at the completion of Part C was copper(II) oxide.

    (a) Write the formula for copper(II) oxide in the blank: _____ .

    (b) Determine the number of moles of copper(II) oxide recovered in your experiment. Show your work.

    (c) Determine the number of moles and the mass of elemental copper present in your copper(II) oxide. Show your work. This is also the mass and number of moles of copper that must have been in your original copper(II) carbonate sample. Why?

7. Calculate the percent by mass of copper in copper(II) carbonate, based on your answers to question 6. Show your calculations.

8. In all likelihood, your answer to question 7 does not match the predicted mass percent of copper from Prelaboratory Question 4(b). The reason for this is that copper(II) carbonate does not have the formula that would be expected, based on its name. Consider the following possible formulas:

   $CuCO_3$ $\qquad$ $CuCO_3 \cdot H_2O$ $\qquad$ $CuCO_3 \cdot 2H_2O$ $\qquad$ $Cu_2(OH)_2CO_3$ $\qquad$ $Cu_2(OH)_2CO_3 \cdot 2H_2O$

   One of these is the correct formula for copper(II) carbonate. Based on your experimental results, which one is it? Show calculations to support your answer.

9. Write the balanced formula equation for the decomposition of copper(II) carbonate, *based on the formula you chose in question 8*, above. In addition to copper(II) oxide, the products are carbon dioxide and water, both driven off as gases. (Hint: Which of the five options in question 4 is eliminated by the fact that the products include water?)

10. If the copper(II) carbonate system is heated for too short a time, will your answer to question 7 be too high or too low? Justify your answer.

11. Write the balanced formula equation for the decomposition of copper(II) carbonate, *based on the formula you chose in question 4*, above. In addition to copper(II) oxide, the products are carbon dioxide and water, both driven off as gases. (Hint: Which of the five options in question 4 is eliminated by the fact that the products include water?)

12. If the copper(II) carbonate system is heated for too short a time, will your answer to question 3c. be too high or too low? Justify your answer.

13. Under Cleanup and Disposal, the reaction between solid copper(II) oxide and hydrochloric acid is mentioned as a means to clean your crucible. Write the balanced formula equation for this reaction.

EXPERIMENT 14

Preparation and Properties of Hydrogen and Oxygen Gases

Objective

Two of the most common gaseous elemental substances, hydrogen and oxygen gases, will be prepared by chemical reaction.

General Technique for Generation of the Gases

In each of the sections of this experiment, the gas generated will be collected by the technique known as **water displacement**. Neither hydrogen nor oxygen is appreciably soluble in water, so if they are bubbled from a closed reaction system into an inverted bottle filled with water, the water will be pushed from the bottle. The gas thus collected in the bottle will be saturated with water vapor but is otherwise free of contaminating gases from the atmosphere.

If it has not been done for you, construct the gas-generating apparatus shown in Figure 14-1 (it will be used to generate and collect both of the gases). The tall, funnel-like vertical tube shown inserted into the Erlenmeyer flask is called a **thistle tube**. It is a convenient means of delivering a liquid to the flask. If a thistle tube is not available, a long-stemmed gravity funnel may be used.

When inserting glass tubing or the stem of your thistle tube/funnel through the two-hole stopper, use plenty of glycerine to lubricate the stopper, and protect your hands with a towel in case the glass breaks. Do not force the glass: if you cannot easily insert the glass into the stopper ask your instructor for help. Make certain that the two-hole stopper fits snugly in the mouth of the Erlenmeyer flask. Rinse all glycerine from the glassware before proceeding.

Set up a pneumatic trough filled with water. Fill three or four gas bottles or flasks to the rim with water. Cover the mouth of the bottles/flasks with the palm of your hand, and invert into the water trough. Set aside several stoppers or glass plates; use them to cover the mouths of the bottles/flasks once they have been filled with gas. In the following procedure, be sure to keep the gas collection bottles or flasks under the surface of the water in the trough until a stopper or glass plate has been placed over the mouth of the bottle/flask to contain the gas.

When the trough and gas collection equipment are ready, begin the evolution of gas by the specific method discussed for each section of the experiment. Liquids required for the reaction are added to the gas-generating flask through the thistle tube or funnel. The liquid level in the Erlenmeyer flask must *cover the bottom* of the thistle tube/funnel during the generation of the gas, or gas will escape up the stem of the funnel rather than passing through the rubber tubing. Add additional liquid as needed to ensure this.

Gas should begin to bubble from the mouth of the rubber tubing as the chemical reaction begins. Allow the gas to bubble out of the rubber tubing for 2–3 minutes to sweep air out of the system. Add additional portions of liquid through the thistle tube/funnel as needed to continue the production of gas, and do not allow gas evolution to cease, or water may be pulled from the trough into the mixture in the Erlenmeyer flask.

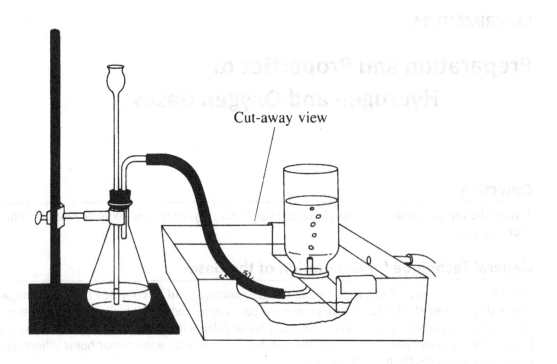

Cut-away view

Figure 14-1. Apparatus for the generation and collection of a water-insoluble gas

Be certain that the thistle tube or funnel extends almost to the bottom of the flask, and that its
end is below the liquid level in the flask.

After air has been swept from the system, collect three or four bottles full of gas by inserting the end of
the rubber tubing into the bottles of water in the trough. Gas will displace the water from the bottle.
Remember to keep the mouth of the bottle under the surface of the water in the trough. When the bottles
of gas have been collected, stopper them while they are still under the surface of the water then remove
them. If glass plates are used to contain the gas rather than stoppers, slide the plates under the mouths of
the gas bottles while they are under water, and remove the bottles from the water. Because hydrogen is
less dense than air, bottles of hydrogen must be kept in the inverted position after removing them from the
water trough. Bottles of oxygen should be stored upright, since oxygen gas is slightly more dense than air.

Introduction

Elemental hydrogen is a colorless, odorless, tasteless gas. Hydrogen can be generated by many methods,
most notably during the electrolysis of water or by replacement from acids by active metals:

$$2H_2O(l) \longrightarrow 2H_2(g) + O_2(g)$$

$$M(s) + n\,H^+(aq) \longrightarrow M^+(aq) + \left(\frac{n}{2}\right)H_2(g) \quad \text{(M represents any active metal)}$$

Pure elemental hydrogen burns quietly itself, but forms *explosive mixtures* when mixed with air prior to
ignition. Hydrogen gas is considerably less dense than air, and for this reason it was used as the buoyant
medium for airships in the earlier part of this century. Because hydrogen forms explosive mixtures with
the oxygen in air, it was replaced by helium for this purpose after the explosion of the German airship
Hindenburg. Helium is considerably more expensive than hydrogen but is not at all flammable.

Elemental hydrogen will be produced in this experiment by allowing metallic zinc to react with an excess
of hydrochloric acid. Since hydrogen gas is relatively insoluble in water, the hydrogen gas will be
collected by displacement of water from an inverted bottle.

Elemental oxygen is essential to virtually all known forms of living creatures. Oxygen is used in cellular respiration in the oxidation of carbohydrates. For example, glucose reacts with oxygen according to

$$C_6H_{12}O_6(s) + 6\ O_2(g) \longrightarrow 6\ CO_2(g) + 6\ H_2O(l)$$

This reaction is the major source of energy in the cell (the reaction is exergonic). The chief source of oxygen gas in the earth's atmosphere is the activity of green plants. Plants contain the substance chlorophyll, which permits the reverse of the preceding reaction to take place. Plants convert carbon dioxide and water vapor into glucose and oxygen gas. This reverse reaction is endergonic and requires the input of energy from sunlight (photosynthesis).

Elemental oxygen is a colorless, odorless, and practically tasteless gas. Though elemental oxygen itself does not burn, it supports the combustion of other substances. Generally, a substance will burn much more vigorously in pure oxygen than in air (which is only ~20% oxygen by volume). Hospitals, for example, have always banned smoking in patient rooms in which oxygen is in use.

Elemental oxygen is the classic oxidizing agent after which the process of oxidation was named. For example, metallic substances are oxidized by oxygen, resulting in the production of the metal oxide. Iron rusts in the presence of oxygen and moist air, resulting in the production of the rust-colored iron oxides.

$$2\ Fe(s) + O_2(g) \longrightarrow 2\ FeO(s)$$

$$4\ Fe(s) + 3\ O_2(g) \longrightarrow 2\ Fe_2O_3(s)$$

Safety Precautions	• **Protective eyewear approved by your institution must be worn at all times while you are in the laboratory.**
	• **Examine the bottles/flasks to be used to collect the gases. Replace any bottles or flasks that are cracked, chipped, or otherwise imperfect.**
	• **Elemental hydrogen is flammable and forms highly explosive mixtures with air. Although the quantity of hydrogen generated in this experiment is small, exercise caution when dealing with the gas. If the gas generator is still bubbling hydrogen after you have collected your samples, transfer it to the fume exhaust hood to vent the additional hydrogen.**
	• **Hydrochloric acid is damaging to eyes, skin, and clothing. Wash immediately if it is spilled, and inform the instructor.**
	• **Hydrogen peroxide and manganese(IV) oxide may be irritating to the skin. Wash after use.**
	• **Oxygen gas supports vigorous combustion. Be cautious of reactions involving a flame near the source of oxygen.**
	• **Sulfur dioxide gas is toxic and irritating to the respiratory tract. Generate and use sulfur dioxide only in the fume exhaust hood.**

Apparatus/Reagents Required

Gas-generating apparatus: 250-mL Erlenmeyer flask with tightly fitting two-hole stopper, glass tubing, thistle tube or long-stemmed funnel, rubber tubing, water trough

Gas collection bottles (or flasks), stoppers to fit the gas collection bottles (or glass plates to cover the mouths of the bottles), wood splints, mossy zinc, 3 *M* hydrochloric acid, glycerine, deflagrating spoon, pH paper, 6% hydrogen peroxide solution, manganese(IV) oxide, sulfur, clean iron/steel nails, 1 *M* hydrochloric acid

Procedure

Record all data and observations directly in your notebook in ink.

1. **Generation and Collection of Hydrogen Gas**

   Construct the gas generator and water trough apparatus as indicated in Figure 14-1 for use in collecting hydrogen.

   Add approximately a half-inch layer of mossy zinc chunks to the Erlenmeyer flask. Replace the two-hole stopper (with thistle tube/funnel and glass delivery tube) in the Erlenmeyer flask, and make certain that the stopper is set *tightly* in the mouth of the flask. Make certain that the stem of the funnel/thistle tube extends almost to the bottom of the flask.

When the trough and gas collection equipment are ready, begin the evolution of hydrogen by adding approximately 30 mL of 3 *M* hydrochloric acid to the zinc through the thistle tube or funnel. The liquid level must cover the bottom of the thistle tube/funnel, or hydrogen will escape up the stem of the funnel rather than passing through the rubber tubing. Add additional HCl as needed to ensure this.

After sweeping air from the system for 2-3 minutes, collect three or four bottles of hydrogen. When the hydrogen has been collected, stopper the gas bottles under the surface of the water; then remove the bottles. Since hydrogen is lighter than air, keep the bottles inverted on your lab bench to prevent loss of the gas. If the zinc is still bubbling after you have collected your hydrogen samples, transfer the flask to the fume exhaust hood to vent the additional hydrogen.

2.   **Testing Hydrogen Gas**

Bring two bottles of hydrogen to your instructor. The instructor will perform two tests on the hydrogen for you. Be sure to wear your safety glasses at all times

To demonstrate that hydrogen forms explosive mixtures with air, the instructor will hold one bottle of hydrogen, mouth downward, and will remove the stopper or glass plate. The bottle of hydrogen will be tilted *momentarily* at a 45-degree angle to allow some hydrogen to escape from the bottle and a quantity of air to enter the bottle. The instructor will then hold the bottle of hydrogen/air mixture at arm's length and will bring the mouth of the bottle near a burner flame. A loud "pop" will result as the hydrogen explodes.

The instructor will light a wooden splint. The instructor will then hold a second bottle of hydrogen in the inverted position, will remove the stopper, and will quickly insert the burning splint deeply into the bottle of hydrogen. Observe whether or not the splint continues to burn when deep in the hydrogen. The instructor will then slowly withdraw the splint to ignite the hydrogen as it escapes from the mouth of the bottle. If this is done carefully, the hydrogen should burn *quietly* with a pale blue flame (rather than exploding as in the first test).

In the exhaust hood, open the third bottle of hydrogen and turn the bottle mouth upward for 30 seconds. After the 30-second period, invert the bottle and attempt to ignite the gas in the bottle. There will be no "pop." Because hydrogen is lighter than air, the hydrogen will have already escaped from the bottle.

3.   **Generation and Collection of Oxygen Gas**

Construct the gas generator and set up the water trough as indicated in Figure 14-1 for use in collecting oxygen.

Add approximately 1 g of manganese(IV) oxide to the Erlenmeyer flask. Add also about 15 mL of water, and shake to wet the manganese(IV) oxide (it will not dissolve). Replace the two-hole stopper (with thistle tube/funnel and glass delivery tube) in the Erlenmeyer flask, and make certain that the stopper is set tightly in the mouth of the flask. Make certain that the stem of the funnel/thistle tube extends almost to the bottom of the flask.

When the trough and gas collection equipment are ready, begin the evolution of oxygen by adding approximately 30 mL of 6% hydrogen peroxide to the manganese(IV) oxide (through the thistle tube or funnel). Make certain that the liquid level in the flask covers the bottom of the stem of the thistle tube/funnel, or the oxygen gas will escape from the system. Add more hydrogen peroxide as needed to ensure this.

The hydrogen peroxide should immediately begin bubbling in the flask, and oxygen gas should begin to bubble from the mouth of the rubber tubing. After sweeping air from the system for several minutes, collect three or four bottles of oxygen. Add 6% hydrogen peroxide as needed to maintain the flow of gas. When the oxygen has been collected, stopper the gas bottles under the surface of the water; then remove the bottles.

4.   **Testing Oxygen Gas**

Ignite a wooden splint. Open the first bottle of oxygen, and slowly bring the splint near the mouth of the oxygen bottle in an attempt to ignite the oxygen gas as it diffuses from the bottle. Is oxygen itself flammable?

Ignite a wooden splint in the burner flame. Blow out the flame, but make sure that the splint still shows some glowing embers. Insert the glowing splint deeply into a second bottle of oxygen gas. Explain what happens to the splint.

Take the third stoppered bottle of oxygen to the exhaust hood. Place a tiny amount of sulfur in the bowl of a deflagrating spoon, and ignite the sulfur in a burner flame in the hood. Remove the stopper from the bottle of oxygen, and insert the spoon of burning sulfur.

Allow the sulfur to burn in the oxygen bottle for at least 30 seconds. Then add 15–20 mL of distilled water to the bottle, and stopper. Shake the bottle to dissolve the gases produced by the oxidation of the sulfur. Test the water in the bottle with pH paper. Write the equation for the reaction of sulfur with oxygen, and explain the pH of the solution measured.

Clean an iron (not steel) nail free of rust by dipping it briefly in 1 *M* hydrochloric acid; rinse with distilled water. If the surface of the nail still shows a coating of rust, repeat the cleaning process until the surface of the nail is clean.

Using tongs, heat the nail in the burner flame until it glows red. Open the fourth bottle of oxygen, and drop the nail into the bottle. After the nail has cooled to room temperature, examine the nail. What is the substance produced on the surface of the nail?

Name: _____ Section: _____

Lab Instructor: _____ Date: _____

EXPERIMENT 14

Preparation and Properties
of Hydrogen and Oxygen Gases

Pre-Laboratory Questions

1. In this experiment, simple chemical reactions are used to generate small quantities of hydrogen and oxygen gases. These methods would be too expensive and too complicated if large quantities of these gases were needed, however. Use your textbook, a chemical encyclopedia, or an online source to look up the major *industrial* methods of preparing very large quantities of these gases. Summarize your findings here.

2. The bottles of hydrogen you will generate must be kept in the *inverted* position after the gas has been produced. Why?

3. Write balanced chemical equations for the generation of hydrogen gas using hydrochloric acid and zinc metal, and for the generation of oxygen gas from the decomposition of hydrogen peroxide.

4. Write balanced chemical equations for each of the tests on the generated gases discussed in Sections 2 and 4 of the Procedure.

EXPERIMENT 14

Preparation and Properties of Hydrogen and Oxygen Gases

Results/Observations

1. **Generation and Collection of Hydrogen Gas**

   (a) Observation on adding acid to zinc

2. **Tests on the Hydrogen Gas**

   (a) Test on igniting hydrogen/air mixture

   (b) Test on igniting pure hydrogen

   (c) Attempt to ignite hydrogen after inverting bottle

3. **Generation and Collection of Oxygen Gas**

   (a) Observation on adding hydrogen peroxide to MnO_2

4. **Tests on the Oxygen Gas**

   (a) Test on attempting to ignite oxygen

   (b) Test on inserting glowing splint into oxygen

(c) Result of adding burning sulfur to oxygen

(d) pH of sulfur/oxygen product solution

(e) Equation for sulfur/oxygen reaction

(f) Test on adding glowing iron nail to oxygen

(g) Equation for iron/oxygen reaction

Questions

1. In the tests on your instructor performed on hydrogen, you should have seen that *pure* hydrogen burns quietly, but that a *mixture* of hydrogen and air is explosive. Use a handbook, scientific encyclopedia, or online source to determine how scientists describe the *explosive limits* under which a flammable substance's vapor can cause an explosion when mixed with air.

2. Manganese(IV) oxide was added as a *catalyst* to speed up the breakdown of the hydrogen peroxide used to generate oxygen. Use your textbook or a scientific encyclopedia to define the term, *catalyst*. Give an example of another reaction in which a *catalyst* is used to speed up the reaction.

EXPERIMENT 15

Gas Laws 1: Molar Volume and The Ideal Gas Constant

Objective

The molar volume of a typical gas will be determined and an experimental value for the universal gas will be calculated

Introduction

There are a number of gas laws described in Chapter 5. Some, like the laws of Avogadro, Boyle and Charles, concern how one of the four variables – pressure, volume, absolute temperature, and number of moles – varies when another is changed, with two of the four being held constant. Two other relationships are of a more general nature. As its name suggests, the Combined Gas Law, represents a combining of the simpler laws. Given that volume is inversely proportional to pressure (Boyle's Law), and directly proportional to temperature (Charles' Law), you can write an expression that incorporates both sets of observations:

$$\frac{P_1 V_1}{T_1} = \frac{P_2 V_2}{T_2}$$

Going one step further and introducing the law of Avogadro, which says that the volume occupied by a gas is directly proportional to the number of moles of gas in the system, gives the expression

$$\frac{P_1 V_1}{n_1 T_1} = \frac{P_2 V_2}{n_2 T_2}$$

In which P_1, V_1, n_1 and T_1 refer to an initial set of conditions and P_2, V_2, n_2 and T_2 refer to the final conditions, after any changes have occurred. This combined gas law allows you to do calculations involving changes in any or all of the four variables.

What the equation tells us is that the value of the expression, $\frac{PV}{nT}$, does not change; if that is the case, then we can determine a value for $\frac{PV}{nT}$ that should be valid under any set of conditions as long as the sample remains a gas. In other words we can state that

$$\frac{PV}{nT} = a\ constant, R$$

This relationship can be rearranged to the form known as the Ideal Gas Law:

$$PV = nRT$$

The constant, R, is known as the *universal gas constant*. One of your objectives in this experiment is to determine the value of R experimentally. Notice that the units of R must reflect pressure times volume, divided by number of moles times temperature. R is most often expressed in units of L·atm/mol·K ("liter-atmospheres per mole-Kelvin"). In this experiment you will calculate the value of R for each of three trials, as well as an average result that will then be used to determine your percentage error.

You will first determine the molar volume of hydrogen gas at ambient laboratory conditions. Then you will use the combined gas law determine what volume one mole of hydrogen (or any other gas) would occupy at STP (0°C and one atmosphere pressure). Once that is done, you will use your experimental values of P, V, n, and T to calculate an experimental value for R. As with the molar volume calculation, you will determine individual values for each trial, along with an average value, which you will compare with the accepted value of R: 0.0821 L atm/mol K.

Safety Precautions	<ul><li>Protective eyewear approved by your institution must be worn at all times while you are in the laboratory.</li><li>Hydrochloric acid is corrosive to skin and clothing. Clean up all spills thoroughly. If acid spills on the lab bench, use a bit of baking soda to neutralize it before wiping it up with wet paper towels.)</li><li>Do not neutralize acid spills on skin or clothing; flood the affected area with water.</li><li>Wash hands thoroughly with soap and water before leaving the laboratory.</li></ul>

Apparatus/Reagents Required

Analytical balance, 50-mL eudiometer tube, #00 1-hole rubber stopper, copper wire (10 cm), thermometer (–10 °C to 110°C), magnesium ribbon, 3 pcs, ~4-5 cm each, freshly cleaned, 6 M HCl(aq), NaHCO$_3$(s) (baking soda)

Procedure

Record all data and observations directly in your notebook in ink.

Fill a Berzelius (tall-form) 1-L beaker to within about 4-5 cm of the top, using tap water; the water should be at or near room temperature.

Obtain a short (4-5cm) piece of magnesium ribbon, clean it thoroughly with steel wool, then measure and record its mass to the nearest 0.1 mg, using the analytical balance. Use a 10-cm piece of copper wire to make a cage for your magnesium, by folding the magnesium over the wire then rolling the wire around the magnesium. Fit the wire cage into a 1-hole #00 rubber stopper. The cage should be about 5 cm from the small end of the stopper to hold the magnesium in place (Figure 15.1).

Figure 15-1. Stopper Assembly

Carefully pour about 5 mL of 6 M HCl(aq) into a 50-mL gas-measuring tube. The graduations on the tube may be used for this. Holding the tube at about a 45° angle, use a wash bottle or beaker to *carefully* add water to the tube until it is completely full. Try to direct the water down the side of the tube to minimize mixing of the water and the acid. Insert the stopper assembly (Figure 15.1, above) into the top of the tube.

Water should escape through the hole in the stopper; if it does not, remove the stopper and carefully add more water, then replace the stopper assembly. This will keep air from being trapped in the tube.

Place your finger over the hole in the stopper and invert the tube, lowering it into the beaker of water. Remove your finger when the stopper is below the level of water in the beaker. The hydrochloric acid solution is more dense than water, so it will slowly sink toward the stopper and the magnesium; observe and record evidence of reaction. (See Figure 15-2.)

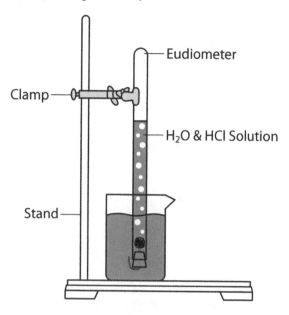

Figure 15-2. The Assembly

When the reaction is complete allow the system to stand for two or three minutes, tapping the sides of the tube to dislodge any gas bubbles that may be clinging to the glass wall. Make certain that there are no little pieces of unreacted magnesium on the wall of the cylinder. If a small piece does remain, gently shake the tube up and down to wash the metal back into the acidic solution, allowing it to finish reacting. Be careful not to lift the tube completely out of the water in the beaker.

(*Note:* Carry out this step only if the difference in water levels inside and outside the tube differ by more than 5 cm – 50 mm – or more.) Place your finger once again over the hole in the stopper, then transfer the tube and contents to a container of water that is deep enough that you can make the water levels inside and outside the tube nearly equal to each other. Record the volume of gas trapped in the tube (± 0.01 mL) and the temperature of the water near the mouth of the tube. The temperature of the escaping solution may be assumed to be the same as the temperature of the trapped gases. Record these data.

Carry out three separate determinations. After each trial allow all liquid remaining in the tube to drain into the beaker, then neutralized it with a small amount of baking soda. Clean the tube and beaker before carrying out the second and third trials.

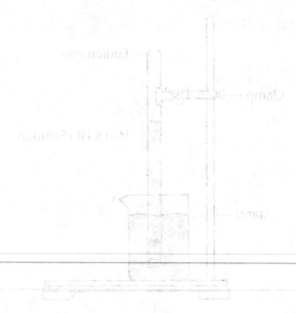

Figure 15-2. The Assembly

Name: _____ Section: _____

Lab Instructor: _____ Date: _____

EXPERIMENT 15

Gas Laws 1: Molar Volume and The Ideal Gas Constant

Pre-Laboratory Questions

1. Use Dalton's Law of Partial Pressures to explain why the pressure of hydrogen gas in the tube will be less than the observed barometric pressure in the laboratory. How will you determine the pressure of the hydrogen you produce?

2. You will collect hydrogen gas in a tube that has a maximum capacity of 50.0 mL.

   (a) Write the balanced formula equation for the reaction between hydrochloric acid and magnesium,

   (b) Calculate the mass of magnesium needed to produce 45.0 mL of gaseous hydrogen at 25 °C and 1.00 atm total pressure. Remember to allow for the vapor pressure of water, 23.8 torr at 25 °C. The mass you calculate represents the maximum mass of magnesium that you will need for each trial.

3. What two physical properties of hydrogen gas make it possible for you to collect it by displacement of water in your graduated cylinder?

4. You will start with 5 mL of 6 *M* HCl. By the time you fill the eudiometer with water and upend it into about 900 mL of water, what will be the concentration of HCl in the system? What mass of $NaHCO_3$ will be needed to neutralize this quantity of acid? Show all calculations.

5. The accepted value for the universal gas constant is 0.0821 L atm/mol·K. What would it be if the pressure was measured in torr and the volume in milliliters? Show your conversions.

EXPERIMENT 15

Gas Laws 1: Molar Volume and The Ideal Gas Constant

Results/Observations

Table 1

		Trial 1	Trial 2	Trial 3
Mass of Mg used	g			
Moles of Mg used	mol			
Moles of hydrogen collected	mol			
Volume of hydrogen collected	mL			
Temperature	°C			
Barometric pressure	torr			
Vapor pressure of water	torr			
Pressure of "dry" hydrogen[1]	torr			

1. You have measured the volume occupied by a small fraction of a mole of hydrogen, under a specific set of conditions of pressure and temperature. The volume occupied by one mole of gas is called the **molar volume** of the gas, and it is the same for all gases (behaving ideally) at a given set of pressure and temperature. For each trial:

   a. Divide the volume of hydrogen collected by the number of moles of hydrogen generated to calculate the volume that 1.00 mole of hydrogen would occupy at your experimental temperature and pressure (called "laboratory conditions"). Record your answers in the Summary Table below, in units of liters per mole, L/mol.

   b. Use the combined gas law to calculate the volume that 1.00 mole of hydrogen would occupy at 1.00 atm and 273 K. (Standard Temperature and Pressure, STP)

2. Determine the value of $\left(\frac{PV}{nT}\right)$ for each trial. These are your experimental values for R. In this expression, P is the pressure of "dry" hydrogen converted to atmosphere units, V is the volume of gas collected (L), T is the Kelvin temperature, and n is the number of moles of hydrogen generated in the reaction between $HCl(aq)$ and $Mg(s)$.

---

[1] Found by subtracting the vapor pressure of water from the ambient barometric pressure.

Summary Table

	Trial 1	Trial 2	Trial 3
Molar volume of H_2 at lab conditions			
Molar volume of H_2 at STP			
Value of $\left(\frac{PV}{nT}\right) = R$			

3. Determine the arithmetic mean of your three values for the molar volume of hydrogen at STP. Include units.

Mean value for molar volume: _____

4. Calculate the relative error (percentage error) in your determination of the molar volume of a gas. Relative error is calculated as follows:

$$\frac{|(accepted\ value) - (experimental\ value)|}{(accepted\ value)} \times 100\%$$

Percent error in determination of molar volume: _____

5. Determine the arithmetic mean of your three values for the universal gas constant, R. Include units.

Mean value of R: _____

6. Calculate the relative error (percentage error) in your determination of R.

Percent error in determination of R: _____

7. You were told to move the tube from the beaker to a deeper vessel if the water levels inside and outside the tube varied by 50 mm or more. Given the density of mercury is 13.6 g mL^{-1}, what pressure difference would this represent, expressed in torr? in atmospheres? What difference in water levels would represent a difference of 0.001 atm?

Gas Laws 2: Graham's Law

Introduction

Effusion, strictly speaking, is the passage of the molecules of a gas through a small hole (a "pinhole") into an evacuated chamber. This term is often confused with a similar word: diffusion. **Diffusion** is the spreading out of gas molecules through space when a container of gas is opened, allowing the gas to mix freely with any other gases present. Suppose a balloon were filled with hydrogen sulfide gas (which has the noxious odor of rotten eggs). If there were a tiny hole in the balloon, the hydrogen sulfide would *effuse* slowly through the hole and would then *diffuse* into the air of the room.

According to the kinetic-molecular theory of gases, the average velocity (root-mean- square velocity) of the particles in a sample of gas is inversely related to the square root of the molar mass, M, of the gas:

$$u_{rms} = \sqrt{\left(\frac{3RT}{M}\right)}$$

It is not possible experimentally to determine easily and directly the average speed of the molecules in a sample of gas, but the average speed at which gases diffuse or effuse can be measured readily. The speed of diffusion in centimeters per second can be determined by measuring how long it takes a gas to pass through a tube of known length.

Thomas Graham, a Scottish chemist, determined experimentally in the nineteenth century that the relative rate of diffusion of two different gases at the same temperature is given by the relationship

$$\frac{r_1}{r_2} = \frac{\sqrt{M_2}}{\sqrt{M_1}}$$

in which r represents the rate of diffusion of a gas and M its molar mass. Thus the relative rates of effusion of two different gases are inversely proportional to their molar masses. This equation, called Graham's law, is consistent with the postulate of the kinetic-molecular theory describing the average speed of molecules in a gas sample.

In this experiment, you will determine the relative rates of diffusion of the gases hydrogen chloride and ammonia by measuring the *distances* traveled by the two gases in the same time period. For a given period of time, a lighter gas should be able to diffuse *farther* than a heavier gas (distance traveled in a given time period is directly related to speed). Cotton balls dipped, respectively, in concentrated hydrochloric acid (HCl gas in water) and concentrated aqueous ammonia (NH_3 gas in water) will be placed in opposite ends of a glass tube. The two gases will diffuse through the tube toward each other. Hydrogen chloride and ammonia gases *react* with each other, forming the salt ammonium chloride:

$$HCl(g) + NH_3(g) \longrightarrow NH_4Cl(s)$$

As the gases meet and react, a white *ring* of $NH_4Cl(s)$ will appear in the tube. The position of this white ring along the length of the tube can be used to determine which of the two gases has diffused farther.

Safety Precautions	• Protective eyewear approved by your institution must be worn at all times while you are in the laboratory.
	• Concentrated hydrochloric acid and concentrated ammonia solutions are each damaging to the skin; wear gloves while handling them.
	• The fumes of both HCl and NH₃ are extremely irritating and are dangerous to the respiratory tract. *Use these substances only in the fume exhaust hood.*
	• When cutting glass, protect your hands with a towel and fire-polish all rough edges.

Apparatus/Reagents Required

3–4 feet of 1-cm-diameter glass tubing, concentrated HCl(*aq*), concentrated NH₃(*aq*), cotton balls, forceps, nitrile surgical gloves, rubber stoppers that fit the ends of the glass tubing, plastic wrap, meter stick, china marker

Procedure

Obtain a length of 1-cm-diameter glass tubing and make certain that the ends have been fire-polished. Obtain two rubber stoppers that will fit snugly in the ends of the tubing. Set up the tubing in the exhaust hood, using two adjustable clamps to hold the tubing in a steady horizontal position (see Figure 16-1).

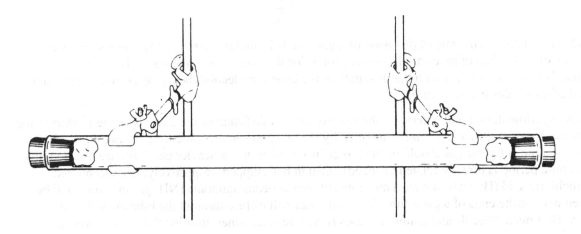

Figure 16-1. Graham's law apparatus. A white ring of ammonium chloride will appear in the tube where the two gases meet each other and react.

Wear gloves during the following procedure, and work in the exhaust hood.

In separate small beakers, obtain 3–4 mL each of concentrated HCl and concentrated NH$_3$ (*Caution!*). Keep the beakers in the fume hood, covered loosely with plastic wrap or a watchglass when not in use. Place a cotton ball in each beaker.

Using forceps, transfer the cotton balls from the beakers to the opposite ends of the glass tubing apparatus. Do this as *quickly* as possible so that one gas will not get too much of a "head start" over the other (perhaps ask a friend to help you to insert the cotton balls simultaneously). Stopper the ends of the tubing and do *not* disturb or move the glass tube.

Allow the gases to diffuse toward each other until a white ring of ammonium chloride is evident in the tube. Mark the *first* appearance of the ring of ammonium chloride with a china marker before the gases diffuse too much and blur the location of the ring.

Remove the rubber stoppers. Using forceps, remove the cotton balls, and transfer them to a beaker of tap water to dilute the reagents (dispose of the cotton balls in the wastebasket after soaking in water).

Measure the distance diffused by each of the gases to the nearest millimeter, measuring from the respective ends of the glass tube to the center of the ammonium chloride ring.

Rinse out the glass tube with distilled water, allow it to dry, and repeat the determination twice.

Calculate the mean distance diffused by HCl(g) and by NH$_3$(g) from your three sets of data.

Realizing that the *distances* diffused by two gases in the same time period should be directly related to the *rates of diffusion* of the gases, determine the percent error in your mean experimental data, compared to Graham's law as expressed in the introduction.

$$\frac{r_1}{r_2} = \frac{\sqrt{M_2}}{\sqrt{M_1}}$$

Wear gloves during the following procedure, and work in the exhaust hood.

In separate small beakers, obtain 3–4 mL each of concentrated HCl and concentrated NH$_3$ (aqueous). Keep the beakers in the fume hood, covered loosely with plastic wrap or a watch glass when not in use. Place a cotton ball in each beaker.

Using forceps, transfer the cotton balls from the beakers to the opposite ends of the glass tubing apparatus. Do this as quickly as possible so that one gas will not get too much of a head start over the other (perhaps use a friend to help you to insert the cotton balls simultaneously). Support the ends of the tubing so it won't disturb or move the glass tube.

Allow the gases to diffuse toward each other until a white ring of ammonium chloride is visible in the tube. Mark the first appearance of the ring of ammonium chloride with a china marker before the gases diffuse too much and blur the location of the ring.

Remove the rubber stopper. Using forceps, remove the cotton balls, and transfer them to a beaker of tap water to dilute the reagent. Dispose of the cotton balls in the wastebasket after soaking in water.

Measure the distance diffused by each of the gases to the nearest millimeter, measuring from the respective ends of the glass tube to the center of the ammonium chloride ring.

Rinse out the glass tube with distilled water, allow it to dry, and repeat the determination twice.

Calculate the mean distance diffused by HCl(g) and by NH$_3$(g) from your three sets of data.

$$\frac{r_1}{r_2} = \frac{\sqrt{M_2}}{\sqrt{M_1}}$$

EXPERIMENT 16

Gas Laws 2: Graham's Law

Pre-Laboratory Questions

1. Imagine that hydrogen molecules (H_2) and nitrogen molecules (N_2) decided to have a race. If the temperature was the same for both gases, which gas would win the race? Why?

2. How many times faster will helium gas effuse through a pinhole than will carbon dioxide gas under the same conditions? Show your calculations.

3. Uranium consists primarily of two isotopes ^{238}U and ^{235}U. They differ in the number of neutrons present in the nuclei. Natural uranium consists of 99.28% ^{238}U and only 0.72% ^{235}U. The ^{235}U isotope is said to be *fissile*, which means it can sustain a nuclear chain reaction, whereas ^{238}U is non-fissile and cannot sustain such a chain reaction. During World War 2, the United States embarked on a large-scale secret project – the *Manhattan Project* – to produce the world's first atomic weapons. It was necessary to process naturally occurring uranium so as to *concentrate* the amount of the fissile ^{235}U isotope to the point where a chain reaction could be sustained. Use a scientific encyclopedia to determine how gaseous diffusion was used to separate these two isotopes based on the difference in their masses.

Name: ______________________________ **Section:** __________

Lab Instructor: ______________________________ **Date:** __________

EXPERIMENT 16

Gas Laws 2: Graham's Law

Results/Observations

Observation for appearance of NH_4Cl ring:

Distance of NH_4Cl ring from ends of tube

	from NH_3 end	*from HCl end*
Trial 1	______________	______________
Trial 2	______________	______________
Trial 3	______________	______________
Mean	______________	______________

Ratio predicted from Graham's law for rates of diffusion for NH_3/HCl (show calculations)

Experimental ratio for diffusion of NH_3/HCl based on mean distances (show calculations)

Percent error

Questions

1. Methlyamine, CH_3NH_2, is a derivative of ammonia that is also able to escape as a gas from its solutions. If methylamine solution had been used in this experiment instead of ammonia, how would the position of the white ring have been affected? Would the white ring have formed closer to the HCl end of the tube than when ammonia was used, or farther away from the HCl end of the tube? Explain.

2. You were told to try to introduce the cotton balls containing HCl and NH_3 into the glass tube simultaneously. What error would have resulted if the HCl had been introduced to the glass tube substantially before the ammonia?

3. Would the position of the white ring in your experiment have been affected if there had been a change in temperature in the laboratory during the course of your several trials? Why or why not? Explain.

4. An unknown gas is found to diffuse at 0.302 times the speed at which helium diffuses. What is the approximate molar mass of the unknown gas?

EXPERIMENT 17

Calorimetry

Objective

A simple "coffee cup" calorimeter will be used to measure the quantity of heat that flows in several physical and chemical processes.

Introduction

Chemical and physical changes are always accompanied by a change in *energy*. Most commonly, this energy change is observed as a flow of heat energy either into or out of the system under study. Heat flows are measured in an instrument called a **calorimeter**. There are specific types of calorimeters for specific reactions, but all calorimeters contain the same basic components. They are *insulated* to prevent loss or gain of heat energy between the calorimeter and its surroundings. For example, the simple calorimeter you will use in this experiment is made of a heat-insulating plastic foam material. Calorimeters contain a *heat sink* that can absorb or provide the energy for the process under study. The most common material used as a heat sink for calorimeters is water because of its ready availability and large heat capacity. Calorimeters also must contain some device for the measurement of *temperature*, because it is from the temperature change of the calorimeter and its contents that the magnitude of the heat flow is calculated. Your simple calorimeter will use an ordinary thermometer for this purpose.

To determine the heat flow for a process, the calorimeter typically is filled with a weighed amount of water. The process that releases or absorbs heat is then performed within the calorimeter, and the temperature of the water in the calorimeter is monitored. From the *mass of water* in the calorimeter, and from the *temperature change* of the water, the quantity of heat transferred by the process can be determined.

When a sample of *any* substance changes in temperature, the quantity of heat, Q, involved in the temperature change is given by

$$Q = m \times C \times \Delta T$$

where m is the mass of the substance, ΔT is the temperature change, and C is a quantity called the **specific heat** of the substance. The specific heat represents the quantity of heat required to raise the temperature of one gram of the substance by one degree Celsius. (Specific heats for many substances are tabulated in handbooks of chemical data.) Although the specific heat is not constant over all temperatures, it remains constant for many substances over fairly broad ranges of temperatures (such as in this experiment). Specific heats are quoted in units of kilojoules per gram per degree, kJ/g-°C (or in molar terms, in units of kJ/mol-°C).

The calorimeter for this experiment is pictured in Figure 17-1. The calorimeter consists of two nested plastic foam coffee cups and a cover, with thermometer and stirring wire inserted through holes punched in the cover. As you know, plastic foam does not conduct heat well and will not allow heat generated by a chemical reaction in the cup to be lost to the room. (Coffee will not cool off as quickly in such a cup as in a china or paper cup.) Other sorts of calorimeters might be available in your laboratory; your instructor will demonstrate such calorimeters if necessary. The simple coffee-cup calorimeter generally gives quite acceptable results, however.

Although the plastic material from which your calorimeter is constructed does not conduct heat well, it does still absorb some heat. In addition, a small quantity of heat may be transferred to or from the metal wire used for stirring the calorimeter's contents or to the glass of the thermometer used to measure temperature changes. Some heat energy may also be lost through the openings for these devices. Rather than determining the influence of each of these separately, a function called the **calorimeter constant** can be determined for a given calorimeter. The calorimeter constant represents what portion of the heat flow from a chemical or physical process conducted in the calorimeter goes to the apparatus itself, rather than affecting the temperature of the water in the calorimeter. Once the calorimeter constant has been determined for a given apparatus, the value determined can be applied whenever that calorimeter is employed in subsequent experiments.

Although determining the calorimeter constant for a coffee cup calorimeter is not particularly difficult, the process does take rather a long time to obtain reproducible results. For this reason, we will use an average value from many experiments as the approximate value of the calorimeter constant, 40.5 J/°C. This means that every time a reaction that evolves or absorbs heat is conducted in the calorimeter, for every degree the temperature of the contents of the calorimeter changes, the calorimeter absorbs or evolves 40.5 J of the energy being transferred. Since plastic foam is a fairly good insulator, this turns out to be a relatively small quantity compared to the energy changes undergone by the contents of the calorimeter.

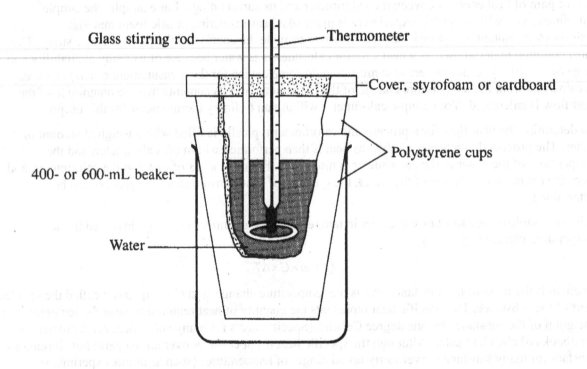

Figure 17-1. A simple calorimeter made from two nested plastic foam coffee cups.
Make certain the stirring wire can be agitated easily

In this experiment you will measure the heat flows for two acid–base neutralization reactions, and for the dissolution of two salts in water. For clarity, a discussion of the reaction/process to be studied is given with the procedure for each portion of the experiment.

Safety Precautions	• **Protective eyewear approved by your institution must be worn at all times while you are in the laboratory.** • **Although the acids and bases used in this experiment are in relatively dilute solution, the acids/bases may concentrate if they are spilled on the skin and the water evaporates from them. Wash immediately if these substances are spilled. Clean up all spills on the bench-top.** • **The salts used may be toxic. Wash after handling them, and dispose of them as directed by the instructor.**

Apparatus/Reagents Required

Calorimeter/thermometer/stirrer apparatus (see Figure 17-1), 2 *M* hydrochloric acid, 2 *M* nitric acid, 2 *M* sodium hydroxide, 2 *M* potassium hydroxide, calorimeter/thermometer/stirrer apparatus, salt samples

Procedure

1. **Strong Acid/Strong Base Heat of Neutralization**

   Many chemical reactions are performed routinely with the reactant species dissolved in water. The use of such solutions has many advantages over the use of "dry chemical" methods. The presence of a solvent matrix permits easier and more intimate mixing of the reactant species, solutions may be measured by volume rather than mass, and the presence of water may act as a moderating agent in the reaction.

   Heats of reaction between species dissolved in water are especially easy to measure because the measurement may be performed in a simple calorimeter as described above. A measured volume of solution of one of the reagents is placed in the calorimeter cup and the temperature is determined. The second reagent solution is prepared in a separate container and is allowed to come to the same temperature as the solution in the calorimeter. When the two solutions have come to the same temperature, they are combined in the calorimeter, and the temperature of the calorimeter contents is monitored as the reaction takes place. If the chemical reaction is **exergonic**, the temperature of the water in the calorimeter will *increase* as energy is transferred to it from the reagents. If the chemical reaction is **endergonic**, the temperature of the water in the calorimeter will *decrease* as thermal energy is drawn from the water into the reactant substances.

   For the purpose of tabulating heats of reaction in the chemical literature, such heats are usually converted to the basis of the number of *kilojoules of heat energy* that flows in the reaction *per mole* of reactant (or product). A typical experimental determination may only use a small fraction of a mole of reactant, and so only a few joules of heat energy will be involved in the experiment, but the results are converted to the basis of one mole. When such a determination of heat flow is conducted in a calorimeter that is equilibrated with the constant pressure of the atmosphere, the heat flow in kilojoules per mole is given the symbol ΔH, and is referred to as the **enthalpy change** for the reaction. Handbooks of chemical data list the enthalpy changes for many reactions.

You will measure the heat of reaction for two "different" reactions:

$$HCl + NaOH \rightarrow$$

$$HNO_3 + KOH \rightarrow$$

Both of these reactions represents the neutralization of an acid by a base, and although the reactions formally appear to involve different substances, the net reaction occurring in each case is the same:

$$H^+ \ \textit{(from the acid)} + OH^- \ \textit{(from the base)} \rightarrow H_2O$$

The actual reaction that occurs in each situation is the combination of a proton with a hydroxide ion, producing a water molecule. For this reason, the heat flows should be the same for all of these reactions.

The procedure that follows is written in terms of the reaction between hydrochloric acid and sodium hydroxide. Perform this determination first; then repeat the procedure for the nitric acid/potassium hydroxide reaction.

Obtain 75 mL of 2 M NaOH and place it in the calorimeter. Record the exact concentration indicated on the reagent bottle. Obtain 75 mL of 2 M HCl in a clean, dry beaker and record the exact concentration indicated on the reagent bottle. Allow the two solutions to stand until their temperatures are the same (within ± 0.5 °C). Be sure to rinse off and dry the thermometer when transferring between the solutions to prevent premature mixing of the reagents. Record the temperature(s) of the solutions to the nearest 0.2°C.

Add the HCl from the beaker *all at once* to the calorimeter, cover the calorimeter quickly, stir the mixture for 30 seconds, and record the highest temperature reached by the mixture (to the nearest 0.2°C).

From the change in temperature undergone by the mixture upon reaction, the total mass (volume) of the combined solutions, and the calorimeter constant for the apparatus (see Choice I), calculate the quantity of heat that flowed from the reactant species into the water of the solution.

Calculate the number of moles of water produced when 75 mL of 2 M HCl reacts with 75 mL of 2 M NaOH. Calculate ΔH in terms of the number of kilojoules of heat energy transferred when 1 mol of water is formed in the neutralization of aqueous HCl with aqueous NaOH.

Example:

Suppose 50.0 ml of 3.01 M HCl solution at 25.2°C is added to 50.0 mL of 3.00 M NaOH solution also at 25.2°C, whereupon it is observed that the temperature of the mixture rises to a maximum of 45.6°C. Calculate the heat flow, Q, for this experiment and ΔH for the process. Assume that the density of the solutions is 1.00 g/mL and that the specific heat capacity of the solution is 4.184 J/g °C

The temperature change undergone by the system is $\Delta T = (45.6°C - 25.2°C) = 20.4°C$

In general, $Q = m \times C \times \Delta T = (100. \text{ g})(\dfrac{4.184 \text{ J}}{\text{g} \ ^\circ\text{C}})(20.4^\circ\text{C}) = 8535 \ (8.54 \times 10^3) \text{ J} = 8.54 \text{ kJ}$

Since the reaction is of 1:1 stoichiometry, 50.0 mL of ~3 M HCl (or NaOH) reacting will result in the production of 0.150 mol of water

Therefore, $\Delta H = -\dfrac{8.54 \text{ kJ}}{0.150 \text{ mol}} = -56.9 \text{ kJ/mol}$ (heat is evolved).

Repeat the determination of ΔH for the HCl/NaOH reaction twice, and calculate a mean value for ΔH for the reaction.

Repeat the process above for the reaction between $HNO_3(aq)$ and $KOH(aq)$, making a total of three determinations of the heat flow for that reaction.

Compare the values of ΔH for the two neutralization reactions. Why is ΔH the same for both reactions?

2.   **Heat of Solution of a Salt**

When salts dissolve in water, the positive and negative ions of the salt *interact* with water molecules. Water molecules are highly polar and arrange themselves in a layer around the ions of the salt so as to maximize electrostatic attractive forces. Such a layer of water molecules surrounding an ion is called a **hydration sphere.**

For example, consider dissolving the salt potassium bromide, KBr, in water. As the ions entered solution, water molecules would orient their dipoles in a particular manner. The potassium ions become surrounded by a layer of water molecules in which the *negative* ends of the water dipoles would be oriented toward the positive potassium ions. Similarly, the bromide ions become surrounded by a layer of water molecules in which the *positive* ends of the dipoles would be oriented toward the *negative* bromide ions. The development of such hydration spheres can have a large effect on the properties of ions in solutions, which you will study in depth if you go on to take a course in physical chemistry.

Formation of a hydration sphere around an ion necessarily involves an energy change. In this choice, you will determine the energy change for two examples of such a process.

Place 75.0 mL of distilled water in the calorimeter. Determine and record the temperature of the water to the nearest 0.2°C. Monitor the temperature of the water for 3 minutes to make certain that its temperature does not change.

Obtain the first salt sample from your instructor. If the salt's identity is known, record the name and formula of the salt. If the salt is presented as an unknown, record the code number of the unknown sample. If an unknown is presented, the instructor will also tell you the molar mass of the unknown salt (record).

Using a clean, dry beaker or weighing boat, weigh out an approximately 5-g sample of the salt, recording the exact mass taken to the nearest 0.01 g.

Remove the lid from the calorimeter and quickly add the weighed salt sample. Replace the lid of the calorimeter and immediately agitate the solution with the stirring wire. Monitor the temperature of the solution while continuing to stir the solution. Record the highest—or lowest—temperature reached as the salt dissolves in the water.

From the mass of salt taken and the mass (volume) of water used, and from the temperature change and calorimeter constant, calculate the quantity of heat that flowed during the dissolving of the salt. From this quantity of heat, and from the molar mass of the salt, calculate the enthalpy change, ΔH, for the dissolving of the salt (heat of solution).

   Example: A 5.01 g sample of a salt with molar mass 63.2 causes the temperature of 75.0 mL of water to rise from 18.2°C to 27.5°C when the salt is dissolved. Calculate the heat evolved by this process and ΔH for the dissolving of the salt. Assume that the density of water and the solution is 1.00 g/mL and that the specific heat capacity of the solution is 4.184 J/g °C.

   The temperature change undergone by the system is $\Delta T = (27.5°C - 18.2°C) = 9.3°C$

In general, $Q = m \times C \times \Delta T = (80. \text{ g})(\dfrac{4.184 \text{ J}}{\text{g }^\circ\text{C}})(9.3^\circ\text{C}) = 3113 \ (3.1 \times 10^3) \text{ J} = 3.1 \text{ kJ}$

$5.01 \text{ g salt} \times \dfrac{1 \text{ mol salt}}{63.2 \text{ g salt}} = 0.0793 \text{ mol salt}$

Therefore $\Delta H = \dfrac{3.1 \text{ kJ}}{0.0793 \text{ mol}} = -39 \text{ kJ/mol (heat is evolved)}$

Dispose of the salt solution as directed by the instructor.

Clean out and dry the calorimeter, and repeat the determination twice more. Calculate a mean value for the heat of solution of your salt from your three determinations.

Obtain from your instructor a sample of the second salt to be studied, and repeat the process above to determine the enthalpy change, ΔH, for the dissolving of the second salt in water.

Name: ______________________________ **Section:** __________

Lab Instructor: ______________________________ **Date:** __________

EXPERIMENT 17

Calorimetry

Pre-Laboratory Questions

1. What does the specific heat capacity of a substance or material represent? What are the units of specific heat capacity?

2. Use a handbook or other reference to find the specific heat capacities (in J/g °C or J/mol °C) of the following materials:

   a. chromium metal ______________________

   b. boron ______________________

   c. nickel metal ______________________

   d. tungsten metal ______________________

   e. beryllium metal ______________________

3. For each of the substances in Question 2, calculate by how much the temperature of the sample will rise if 100. J of heat is applied to 5.00 g samples of each substance at 20°C.

   a. chromium metal

   b. boron

   c. nickel metal

    d.   tungsten metal

    e.   beryllium metal

4.   What does a *calorimeter constant* represent? What value for the calorimeter constant will be used in this experiment for a "coffee cup" calorimeter?

5.   If the temperature of 96.5 g of water increases from 25.2°C to 37.6°C, how much heat did the water absorb?

6.   What is the *enthalpy change* for a reaction, and how does this differ from the experimental *heat flow* measured for an experiment involving the reaction.

EXPERIMENT 17

Calorimetry

Results/Observations

1. **Strong Acid/Strong Base Heat of Neutralization**

    Reaction of HCl and NaOH

	Trial 1	*Trial 2*	*Trial 3*
Volume of 2 M NaOH used, mL	_____	_____	_____
Initial temperature of NaOH, °C	_____	_____	_____
Volume of 2 M HCl used, mL	_____	_____	_____
Initial temperature of HCl, °C	_____	_____	_____
Final temperature reached, °C	_____	_____	_____
Total mass (volume) of mixture, g	_____	_____	_____
Temperature change, ΔT	_____	_____	_____
Heat flow, J	_____	_____	_____
Moles of water produced	_____	_____	_____
ΔH (kJ/mol water)	_____	_____	_____
Mean value of ΔH	_____		

    Reaction of HNO₃ and KOH

	Trial 1	*Trial 2*	*Trial 3*
Volume of 2 M KOH used, mL	_____	_____	_____
Initial temperature of KOH, °C	_____	_____	_____
Volume of 2 M HNO₃ used, mL	_____	_____	_____
Initial temperature of HNO₃, °C	_____	_____	_____
Final temperature reached, °C	_____	_____	_____
Total mass (volume) of mixture, g	_____	_____	_____
Temperature change, ΔT	_____	_____	_____
Heat flow, J	_____	_____	_____
Moles of water produced	_____	_____	_____
ΔH (kJ/mol water)	_____	_____	_____
Mean value of ΔH	_____		

2. Heat of Solution of a Salt

first salt   Identity/Code Number of first salt _____

	Trial 1	*Trial 2*	*Trial 3*
Volume of water used, mL	_____	_____	_____
Initial temperature of water, °C	_____	_____	_____
Mass of salt taken, g	_____	_____	_____
Moles of salt taken, mol	_____	_____	_____
Highest/lowest temperature, °C	_____	_____	_____
Temperature change, ΔT	_____	_____	_____
Heat flow, J	_____	_____	_____
Heat of Solution (ΔH), kJ/mol	_____	_____	_____
Mean value of ΔH	_____		

second salt   Identity/Code Number of second salt _____

	Trial 1	*Trial 2*	*Trial 3*
Volume of water used, mL	_____	_____	_____
Initial temperature of water, °C	_____	_____	_____
Mass of salt taken, g	_____	_____	_____
Moles of salt taken, mol	_____	_____	_____
Highest/lowest temperature, °C	_____	_____	_____
Temperature change, ΔT	_____	_____	_____
Heat flow, J	_____	_____	_____
Heat of Solution (ΔH), kJ/mol	_____	_____	_____
Mean value of ΔH	_____		

Questions

1. How did your values for the mean heats of neutralization compare for the two acid–base reactions? Why would you expect the heats of neutralization for these reactions to be the same, within experimental error? What is the net ionic reaction for both acid–base reactions?

2. The heat flows measured in the first part of this experiment were actually not for the simple neutralization of a proton and hydroxide ion; rather, they include contributions based on the fact that these species are *hydrated* in aqueous solution. What does it mean to say that a proton is *hydrated*?

3. The acids and bases used in the first part of this experiment were *strong* acids and *strong* bases. What is a *weak* acid or base?

4. If the identity of the salts was provided by the instructor, use a chemical handbook to find the literature value of the heat of solution for your salts. Calculate the percent error between the literature value and your mean experimental value for ΔH of solution for each of your salts.

   Salt _____   Literature value for ΔH _____   % error _____

   Salt _____   Literature value for ΔH _____   % error _____

   Reference_____

   If your salts were unknown samples, then use a chemical handbook to find two salts whose heats of solution are endothermic, and two salts whose heats of solution are exothermic, and list those examples.

   Salt _____   Heat of solution *exo/endo*thermic (circle)

   Salt _____   Heat of solution *exo/endo*thermic (circle)

   Salt _____   Heat of solution *exo/endo*thermic (circle)

   Salt _____   Heat of solution *exo/endo*thermic (circle)

5. Use your textbook, a chemical encyclopedia, or an online reference to write a specific definition of the term *salt*.

6. What does it mean to say that an ion becomes *hydrated* when a salt is dissolved in water?

Experiment 18

Heats of Reaction and Hess' Law

Objective

To verify Hess' Law of Additivity of Reaction Enthalpies

Introduction

Reactions generally involve the transfer of energy either into or out of the reaction system. When heat is released during the course of a reaction, we say the reaction is *exothermic*; when the system absorbs energy, the reaction is *endothermic*. The energy represents the difference in bond energies between the product molecules and the reactant molecules. As a simple example, consider the combustion of methane, CH_4. The reaction is shown first as molecular formulas, then using structural formulas to show the bonds between atoms.

$$CH_4(g) + 2 O_2(g) \rightarrow CO_2(g) + 2 H_2O(g)$$

In order for the reaction to occur, the four C–H bonds in methane must be broken, as must the two double bonds joining the pairs of oxygen atoms; this requires an input of energy at least equal to the potential energy represented by those bonds. When the new bonds form, energy is released. As you might expect, each type of bond (C=O, C–H, H–O, and O=O) involves a unique quantity of energy, thus the energy released during bond formation will generally be different from the quantity absorbed in breaking the bonds of the reactants. Since the amount of energy depends directly on the number of molecules involved in the reaction, reaction energies are generally expressed in units of kilojoules per mole (kJ/mol) of some reactant or product. This quantity of energy is represented by ΔH, and the sign of ΔH reveals whether the reaction is endothermic (ΔH is positive) or exothermic (ΔH is negative). For our example reaction, the accepted value of the heat of reaction is – 890 kJ/mol, meaning that 890 kJ of energy are released for each mole (16 g) of methane burned. In other words, there were 890 kJ more energy in the bonds of CH_4 and O_2 than are stored in the bonds of the CO_2 and H_2O molecules.

Here's an important point: If the reaction between methane, CH_4, and oxygen gas, O_2, releases 890 kJ in producing CO_2 and H_2O, then the reverse reaction:

$$CO_2(g) + 2 H_2O(g) \rightarrow CH_4(g) + 2 O_2(g)$$

must absorb exactly 890 kJ. Why? Because all the bonds that were broken when methane burned are now being formed. And the bonds that were formed in the production of CO_2 and H_2O are now being broken. In other words, if you reverse the direction of a reaction, you must change the sign of the energy change that accompanies that reaction.

In 1840, G.H. Hess recognized that the energy change that accompanies a reaction depends only on what the reactants and products are, not in the way in which one is converted to the other. This came to be known as Hess' Law. It will be your goal in this experiment to demonstrate for yourself that this statement is a valid one. To do that, you will measure the energy change for two different reactions, then combine

159

them to predict the energy change for another reaction. Here are the three reactions; notice that all species that ionize in water are shown as ions, while solids and non-ionizing substances are shown in molecular form.

(1) The dissolving of sodium hydroxide in water:

$$NaOH(s) \quad \rightarrow \quad Na^+(aq) + OH^-(aq) \qquad (18\text{-}1)$$

(2) The reaction between solid sodium hydroxide and acetic acid:

$$NaOH(s) + HC_2H_3O_2(aq) \quad \rightarrow \quad H_2O(l) + Na^+(aq) + C_2H_3O_2^-(aq) \qquad (18\text{-}2)$$

(3) The reaction between aqueous solutions of sodium hydroxide and acetic acid:

$$Na^+(aq) + OH^-(aq) + HC_2H_3O_2(aq) \quad \rightarrow \quad H_2O(l) + Na^+(aq) + C_2H_3O_2^-(aq) \qquad (18\text{-}3)$$

As noted above, you will measure the heats of reactions for the first two then combine them to predict the quantity of heat that will be transferred in the third reaction. You will then measure the heat transfer of the third reaction directly to see how close your prediction came. The method of combining the first two reactions is the subject of the Prelaboratory Questions. The apparatus may be familiar to you from other experiments.

Procedure Note: This experiment can be done with a digital thermometer, or with a temperature probe interfaced to a computer or to a hand-held device. Your teacher will advise you of the setup to be used.

Safety Precautions	• **Protective eyewear approved by your institution must be worn at all times while you are in the laboratory.** • **Sodium hydroxide is extremely caustic; avoid contact with skin and clothing. If it gets on your skin, rinse the affected area with water until it no longer feels slippery. Use large amounts of water to clean up spills on surfaces.** • **While acetic acid is identified as a weak acid, it should be considered corrosive. Handle skin contact and spills as you would with any acid spill.** • **Wash your hands thoroughly with soap and water before leaving the laboratory.**

Apparatus/Reagents Required

Digital thermometer or interfaced temperature probe, expanded polystyrene cups (3), sodium hydroxide pellets, sodium hydroxide solution (~2.0 M), acetic acid solutions (1.0 M and 2.0 M).

Procedure

Note: You will have to work quickly and efficiently while carrying out the first two reactions, each involving solid sodium hydroxide. The NaOH pellets will absorb water rapidly from the air. This will make them tend to stick to the weighing container.

Assemble the calorimeter as shown at the right. Two cups are placed in the beaker for stability then the third cup is trimmed about 1 cm below the rim. This permits it to be inverted and used as a lid. It should fit snugly into the opening of the other two cups.

Make a small hole in the lid, just large enough to accept the thermometer or probe. Most digital thermometers have pointed tips, so they can make their own opening. The tighter the fit, the better the calorimeter will contain the heat that is released by the reactions.

Once the temperature probe has been inserted into the calorimeter lid, avoid moving it any more than necessary, to maintain the snug fit.

Cable to interface

Thermistor probe

Calorimeter Apparatus, with Interfaced Temperature Probe
(The 250-mL beaker is used only for support, so it is not shown here.)

If you are using an interfaced temperature probe, set up the software to take readings every second for a period of 240 seconds. If you are asked for a temperature range, select 0°C to 50°C. Do not start collecting data yet.

If you are using a digital thermometer, you will have to monitor the temperature on your own. Record the initial temperature of the liquid in the calorimeter, then watch it steadily, noting the highest temperature reached. A run of four minutes should be enough for the temperature to reach a maximum, and perhaps even start back down.

Use the large graduate to place 50.0 mL of distilled water into the calorimeter. Place the lid containing the thermometer or temperature probe on the calorimeter.

Into your small beaker or weighing boat, weigh 2.00 g of NaOH pellets.

Note: The pellets are of varying size, and you may not be able to get 2.00 grams exactly. Get as close as you can and record the actual mass of NaOH used, ± 0.01 g. Do not spend time trying to get 2.00 grams exactly.

Reaction 1: Dissolving Solid NaOH

Interfaced Temperature Probe Method – Reaction 1

Allow the unit to record the acetic acid temperature a few times, then remove the lid as you quickly and all at once add all of the sodium hydroxide pellets. Carefully replace the lid, returning the probe to the solution. Hold the apparatus in both hands and gently swirl the calorimeter. This is necessary to ensure that the pellets dissolve entirely. If some solid remains when you open the calorimeter you will have to repeat that trial.

At the conclusion of the run examine your data, looking for the initial temperature of the water and the highest temperature reached. If your interface can plot the data, it will probably give you a plot with the following shape:

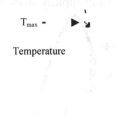

The temperature rises quickly to a maximum, levels briefly, and then starts to return toward room temperature. A best-fit straight line through the portion that starts with the maximum on the plot can be extrapolated to give a more accurate value for the actual maximum temperature reached by the system. See the dashed line in the sketch. This extrapolation technique can be used for each of the three reactions.

Thermometer Method – Reaction 1

Note and record the temperature of the distilled water in the calorimeter. Once temperature is constant, **quickly and all at once** add all of the sodium hydroxide pellets. Quickly and carefully replace the lid, returning the thermometer to the solution. Holding the apparatus with both hands, gently swirl the entire assembly to help the pellets dissolve in the acid. If some solid remains when you open the calorimeter you will have to repeat that trial.

Lift the calorimeter lid and use a wash bottle to rinse the thermometer or probe, collecting the rinsings in the calorimeter cup. The solution is essentially 1 M NaOH, so it must be neutralized. See **Disposal: Reaction (1)** for instructions.

Reaction 2: Solid NaOH Reacts with Acetic Acid.

Rinse the inside of the calorimeter with distilled water, blot it dry lightly and gently with paper towel, then repeat the experiment using 1.0 M acetic acid, $HC_2H_3O_2(aq)$, in place of the distilled water. Again, weigh out the sodium hydroxide quickly, getting as close as you can to 2.00 g, but do not waste time trying to get that mass precisely. Record the actual mass used.

Follow the appropriate procedure of step 5 to determine the quantity of heat released when solid sodium hydroxide dissolves in, and reacts with, 1.0 M acetic acid. At the conclusion of the run, lift the calorimeter lid and use a wash bottle to rinse the probe or thermometer, catching the rinsings in the calorimeter cup, then follow the clean-up instructions in **Disposal, Reactions (2) and (3)**.

Reaction 3: Solutions of Acetic Acid and Sodium Hydroxide React

Repeat the experiment, using 25.0 mL each of 2.0 M sodium hydroxide solution and 2.0 M acetic acid. The 2.0 M $HC_2H_3O_2$ acid is placed in the clean, dried calorimeter. The experiment begins when the 2.0 M NaOH is added quickly and quantitatively. This time you can expect the temperature to rise more sharply, then begin to fall after about 2 minutes.

Disposal

Reaction 1: To the contents of the calorimeter, add 50. mL of 1.0 M acetic acid to neutralize the sodium hydroxide. Flush the solution down the drain with large amounts of water.

Reactions 2 and 3: Rinse the temperature probe with distilled water, catching the rinsings in the calorimeter. The solution remaining in the calorimeter is essentially neutral and may be rinsed down the drain with water.

If you used an interfaced temperature probe, your teacher will direct you regarding what to do with the probe and interface. Be sure the thermometer or temperature probe has been cleaned and dried before putting it away.

The calorimeter lids may be discarded in the wastebasket unless your teacher directs you to leave them for another class or another use. The two calorimeter cups are to be rinsed thoroughly and blotted dry. They can be saved for another use.

Processing the Data

1. Use the equation $q = m \times C \times \Delta T$ to calculate the quantity of heat, q, released in each of the three processes. In each case, assume that the solution in the calorimeter has the same specific heat capacity as water, 4.18 J/g °C. Assume that the acetic acid and NaOH solutions have densities of 1.0 g/mL. Remember to include the mass of solid sodium hydroxide when determining m for reactions (1) and (2).

2. In the first two reactions, you used solid sodium hydroxide. Convert the actual masses of sodium hydroxide used to moles. Show your work.

3. The quantities and concentrations of the two solutions used in Reaction 3 were chosen so that 0.0500 moles of sodium hydroxide would be present. Use this information and the numbers of moles of NaOH you calculated in question 2 to convert your values of q from question 1 to ΔH values, in kJ/mol NaOH.

Experiment 18

Hess's Law

Pre-laboratory Questions

1. Rewrite equation *(18-3)* of the introduction in net ionic form.

2. Recopy equations *(18-1)* and *(18-2)* from the introduction, but reverse equation *(18-1)* so that solid sodium hydroxide appears on the products side of the arrow and the dissolved ions appear as reactants. Now add the two equations, just as you would in algebra when solving two equations in two unknowns.

3. Show that the result you obtained by combining the reverse of equation *(18-1)* with *(18-2)* is identical to the net ionic equation version of equation *(18-3)*.

4. All three reactions that you will carry out in this experiment will result in temperature increases. Are the reactions themselves endothermic or exothermic? Explain.

5. You will use approximately 2.00 g of solid sodium hydroxide in two of the three reactions. Convert 2.00 g of NaOH to moles of NaOH.

Name: ____________________ **Section:** ________

Lab Instructor: ____________________ **Date:** ________

Experiment 18

Hess's Law

Results /Observations

Data Table

	Reaction 1	Reaction 2	Reaction 3
Initial T of water/solution(s)			
Highest T of solution			
Volume of water/solution used			
Mass of NaOH used			-----

1. Calculate *q*, the quantity of heat released by each of the reactions. Show your work. Be sure your answers are clearly identified and have the proper units. Assume all solutions have the same density and specific heat capacity as water.

   Bear in mind that you have measured the temperature change of the surroundings, not the system you are investigating. Thus, a temperature increase shown by the temperature probe represents heat that is flowing out of the reaction between hydrogen ions, $H^+(aq)$, and hydroxide ions, $OH^-(aq)$. As noted in the Procedure,

$$q_{calorimeter} = -q_{reaction}.$$

Reaction 1: Solid sodium hydroxide dissolves in water.

Reaction 2: Solid sodium hydroxide reacts with 1.0 *M* acetic acid solution.

Reaction 3: 2.0 *M* sodium hydroxide reacts with 2.0 *M* acetic acid.

2. In the first two reactions, you used solid sodium hydroxide. Convert the actual masses of sodium hydroxide used to moles. Show your work.

   Reaction 1:

   Reaction 2:

3. The quantities and concentrations of the two solutions used in Reaction III were chosen so that 0.0500 moles of sodium hydroxide would be present. Use this information and the numbers of moles of NaOH you calculated in question 2 to convert your values of q from question 1 to ΔH values, in kJ/mol.

   ΔH for Reaction 1

   ΔH for Reaction 2

   ΔH for Reaction 3

4. Recall that when you combined the equations for Reactions (I) and (II) in the second Prelaboratory Question, you reversed the direction of Reaction (I). If a reaction is exothermic in one direction, reversing the direction (products $\longrightarrow$ reactants) changes the sign, but not the magnitude, of ΔH. Use your results from question 3 to determine values for ΔH_1, ΔH_2, and ΔH_3, *as identified below*. Watch signs and be sure to include appropriate units. Space has been left to show your work. Notice that ΔH_1 is for the reverse of Reaction (I).

Reaction	ΔH
$Na^+(aq) + OH^-(aq) \longrightarrow NaOH(s)$	$\Delta H_1 =$
$NaOH(s) + HC_2H_3O_2(aq) \longrightarrow H_2O(l) + Na^+(aq) + C_2H_3O_2^-(aq)$	$\Delta H_2 =$
$Na^+(aq) + OH^-(aq) + HC_2H_3O_2(aq) \longrightarrow H_2O(l) + Na^+(aq) + C_2H_3O_2^-(aq)$	$\Delta H_3 =$

5. Calculate the sum of ΔH_1 and ΔH_2. Show your work.

6. Because the sum of the first two equations of question 4 is the same as the third equation, the combined values of ΔH_1 and ΔH_2 should equal ΔH_3. Determine the percent deviation between the sum of ΔH_1 and ΔH_2, and the value you obtained for ΔH_3 by dividing the difference by the value for ΔH_3. Note the absolute value signs in the equation. The minus sign in the denominator is simply there to ensure that the percent deviation has a positive value.

$$\frac{|\{\Delta H_1 + \Delta H_2\}|}{|-\Delta H_3|} \times 100\% =$$

7. In this experiment some loss of heat energy is unavoidable. Identify some of the ways in which heat energy may not have been fully accounted for in this experiment.

EXPERIMENT 19

Spectroscopy 1:
Spectra of Atomic Hydrogen and Nitrogen

Objective

The emission and absorption of light energy of particular wavelengths by atoms and molecules are common phenomena. The emissions/absorptions are characteristic for each element's atoms and arise from transitions of electrons among the various energy levels of the each type of atom. The apparatus used to study the wavelengths of light emitted/absorbed by atoms is called a **spectroscope**. In this experiment, you will first calibrate a spectroscope, and then you will use the calibrated spectroscope to determine the wavelengths of the emission lines of hydrogen and nitrogen.

Introduction

The radiant energy emitted by the sun (or other stars) contains *all* possible wavelengths of electromagnetic radiation. The portion of this radiation to which the retina of the human eye responds is called the **visible light region** of the electromagnetic spectrum. The fact that the radiation emitted by the sun contains a *mixture* of radiation wavelengths may be demonstrated by passing sunlight through a prism. A prism *bends* light; the *degree* to which light is bent is related to the wavelength of the light. When sunlight (or other "white" light) containing all possible wavelengths is passed through a prism, each component color – wavelength – of light is bent, with higher energy violet being bent to the greatest extent, and lower energy red light being bent the least. The result is the beam of white light being spread out into a complete rainbow of colors. Such a rainbow pattern is called a **continuous spectrum**. It was discovered that the use of a narrow *slit* in the spectroscope between the prism and the source of white light sharpened and improved the quality of spectra produced by the beam of white light. See Figure 21-1.

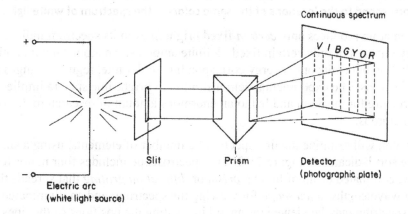

Figure 19-1. The spectrum of "white" light. When light from the sun or from a high-intensity incandescent bulb is passed through a prism, the component wavelengths are spread out into a continuous rainbow spectrum

Most substances will emit light energy if heated to a high enough temperature. For example, a fireplace poker will glow red if left in the fireplace flame for several minutes. Similarly, neon gas will glow with a bright red color when excited with a sufficiently high electrical voltage; this is made use of in neon signs.

When energy is applied to a substance, the atoms present in the substance may *absorb* some of that energy. Electrons within the atoms of the substance move from their normal positions to positions of higher potential energy, farther away from the nuclei of the atoms. Later, atoms that have been excited by the application and absorption of energy will "relax" and will *emit* the excess energy they had gained. When atoms re-emit energy, more often than not, at least a portion of this energy is visible as light.

However, atoms do *not* emit light energy *randomly*. In particular, the atoms of a given element do *not* generally emit a continuous spectrum but, rather, emit visible radiation at only certain discrete, well defined, fixed wavelengths. For example, if you have ever spilled common table salt, NaCl, in a flame, you have seen that sodium atoms emit a characteristic yellow/orange wavelength of light.

If the light being emitted by a particular element's atoms is passed through a prism and is viewed with a spectroscope, only certain sharp, bright-colored *lines* are seen in the resulting spectrum. The positions of these colored lines correspond to the locations (wavelength regions) of the same colors in the spectrum of white light. See Figure 21-2, which illustrates the bright-line spectrum of hydrogen.

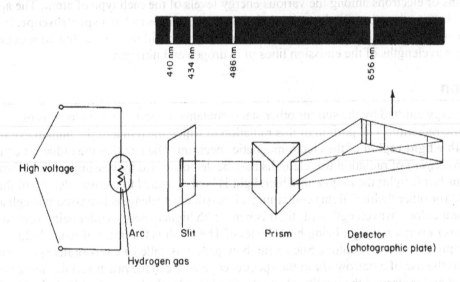

Figure 19-2. The line spectrum of hydrogen. The locations of the colored lines in the spectrum correspond to the locations of the same colors in the spectrum of white light.

The fact that a given atom produces only *certain* fixed bright *lines* in its spectrum indicates that the atom can undergo energy changes of only certain fixed, definite amounts. An atom cannot continuously or randomly emit radiation but emits only energy corresponding to definite, regular changes in the energies of its component electrons. The experimental demonstration of bright-line spectra implied a regular, fixed electronic microstructure for the atom and led to an enormous amount of research to discover exactly what that microstructure is.

In this experiment, you will examine the line spectra of a number of elements, using a simple spectroscope of the sort indicated in Figure 21-3. The spectroscope includes four major features: a *slit* for admitting a narrow, collimated beam of light; a *prism* or *diffraction grating* that spreads the incident light into its component wavelengths; a *telescope* for viewing the spectrum; and an illuminated *reference scale* against which the spectrum may be viewed (as an aid in locating the positions of the lines in the spectrum).

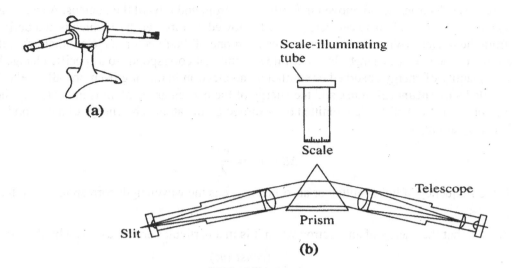

Figure 19-3. (a) A view of the type of spectroscope to be used. (b) A schematic representation of the spectroscope, showing its component parts. When viewed through the telescope, the spectrum will appear superimposed on the numerical scale.

The scale of the spectroscope is provided merely as a convenience, and the divisions on the scale are arbitrary. For this reason, the spectroscope must be *calibrated* before it is used to determine the spectrum of an unknown element. This is accomplished by viewing a known element that produces especially sharp lines in its spectrum and whose spectrum has been previously characterized (with the emission wavelengths being known with great precision). The positions on the spectroscope scale of the lines in the spectrum of the known element are recorded. Then a **calibration curve** that relates the wavelength of a spectral line of the known element to its position when viewed against the spectroscope scale is prepared. This calibration curve may then be used to calculate the wavelengths of emission lines in the spectra of unknown elements when viewed through the same spectroscope under the same conditions.

In this option, you will calibrate the spectroscope by viewing the spectrum of mercury. The spectrum of mercury has been intensively studied. It has several bright lines in the visible region (400–700-nanometer wavelength):

Spectral lines of mercury

404.7 nm	violet
435.8 nm	blue
546.1 nm	green
579.0 nm	yellow

A calibration graph will then be prepared, with the apparent position of the spectral line on the scale of the spectroscope plotted against the known wavelength of that spectral line. This calibration graph should be kept if any additional measurements are to be made with the spectroscope.

Hydrogen is the simplest of the atoms, consisting of a single proton and a single electron. The emission spectrum of hydrogen is of interest because this spectrum was the first to be completely explained by a theory of atomic structure, that of Danish scientist Niels Bohr.

As described above, we know that atoms absorb and emit radiation as light of fixed, characteristic wavelengths when excited. Bohr interpreted this absorption and emission of light as corresponding to electrons within the atom moving away from the nucleus (energy absorbed) or closer to the nucleus (energy emitted). Atoms emit and absorb energy of only certain wavelengths (bright or dark lines in the spectrum) because electrons do not move randomly away from and toward the nucleus. According to Bohr's model, they move only between certain fixed, allowed "orbits," each of which is at a definite fixed distance from the nucleus. When an electron moves from one of the fixed orbits to another orbit, the attractive force of the nucleus changes by a definite amount that corresponds to a specific change in energy. The quantity of energy absorbed or emitted by an electron in moving from one allowed orbit to another is called a **quantum** (**photon**), and the energy of the particular quantum is indicated by the wavelength (or frequency) of the light emitted or absorbed by the atom. The energy of a photon is given by the Planck equation:

$$\Delta E = h\upsilon = \frac{hc}{\lambda}$$

where υ is the frequency of light emitted or absorbed and λ is the wavelength corresponding to that frequency.

Bohr postulated that the energy of an electron when it is in a particular orbit was given by the formula

$$E = -\frac{(constant)}{n^2}$$

where n is the number of the orbit as counted out from the nucleus ($n = 1$ means the first (innermost) orbit, $n = 2$ means the second orbit, etc.). The negative sign reflects our convention that the closer the electron is to the nucleus, the lower its potential energy, and zero potential energy would be the state at which the electron has been ionized – removed from the atom entirely.

The constant, n, is called the **principal quantum number**. The proportionality constant in Bohr's theory is called the Rydberg constant (given the symbol R_H) and has the value 2.18×10^{-18} J. According to Bohr's theory, if an electron were to move from one orbit (designated as $n_{initial}$) to a different orbit (designated as n_{final}), a photon of light should be emitted, having energy given by

$$\Delta E = E_{final} - E_{initial} = -2.18 \times 10^{-18} J \times \left[\left(\frac{1}{n_{final}^2}\right) - \left(\frac{1}{n_{initial}^2}\right)\right]$$

The wavelength (λ) of this photon would be given by the Planck formula as

$$\lambda = \frac{hc}{\Delta E}$$

Bohr performed calculations of wavelengths for various values of the principal quantum number, n, and found that the predicted wavelengths from theory agreed exactly with experimental wavelengths measured with a spectroscope. Bohr even went so far as to predict emissions by hydrogen atoms in other regions of the electromagnetic spectrum (ultraviolet, infrared) that had been reported by Paschen and Lyman. Bohr's predictions were confirmed almost immediately.

Bohr's simple atomic theory of an electron moving between fixed orbits helped greatly to explain observed spectra and formed the basis for the detailed modern atomic theory for more complex atoms with more than one electron. The spectra of larger, multi-electron atoms are considerably more complicated than that of hydrogen, but generally a *characteristic* spectrum is seen. Bohr's theory for hydrogen accounted on a microscopic basis for the macroscopic phenomena of spectral emission lines.

In this experiment, you will measure the wavelengths of the lines in the emission spectrum of hydrogen with a spectroscope and then determine by calculation to which atomic transition (of the electron between the various orbits) each of these spectral lines corresponds. You will also examine the emission spectra of nitrogen, which as a multi-electron atom is considerably more complicated to interpret.

Safety Precautions	• Protective eyewear approved by your institution must be worn at all times while you are in the laboratory. • In addition to visible light, the lamps used in this experiment emit radiation at ultraviolet wavelengths. Ultraviolet radiation is *damaging* to the eyes. Wearing your safety glasses while taking readings with the spectroscope will absorb most of this radiation. Refrain from looking at the source of radiation for any extended period. • The power supply used with the lamps develops a potential of several thousand volts. *Do not touch* any portion of the power supply, wire leads, or mercury lamp unless the power supply is unplugged from the wall outlet. • Always unplug the power supply before adjusting the position of the mercury lamp or any other part of the apparatus.

Apparatus/Reagents Required

Spectroscope with illuminated scale, mercury vapor lamp (discharge tube), hydrogen lamp, nitrogen lamp, high-voltage power pack with lamp holder

Procedure

1. **Calibration of the Spectroscope**

   Record all data and observations directly in your notebook in ink.

   Check to make sure that the power supply pack is *unplugged*. Remember that this pack operates at high voltages and is dangerous.

   Turn on the illuminated scale of the spectroscope, and look through the eyepiece to make sure that the scale is visible but not so brightly lighted as to obscure the mercury spectral lines.

   Position the power supply pack containing the mercury vapor lamp so that the lamp is directly in front of the slit opening of the spectroscope.

   With the instructor's permission, plug in the power supply and turn on the power supply switch to illuminate the mercury lamp.

   Look through the eyepiece, and adjust the slit opening of the spectroscope so that the mercury vapor spectral lines are as bright and as sharp as possible. If necessary, adjust the illuminated scale of the spectroscope so that the numbered scale divisions are easily read but do not obscure the mercury spectral lines.

   Record the *color* and *location* on the numbered scale of the spectroscope for each line in the visible spectrum of mercury.

   Using the line positions you have recorded with the spectroscope and the known wavelengths of the emissions of the mercury atom given earlier, construct a calibration graph for the spectroscope on fine-scale graph paper. Plot the *observed scale reading* for each bright line versus the *wavelength* of the line. This calibration graph will be used to convert observed scale readings of emission lines into their wavelengths.

2.   **Spectra of Atomic Hydrogen and Nitrogen**

Check to make sure that the power supply pack is *unplugged*. Remember that this power supply operates at high voltages and is dangerous.

Turn on the illuminated scale of the spectroscope. Look through the eyepiece to make sure that the scale is visible but not so brightly lighted as to obscure the hydrogen spectral lines.

Position the power supply pack containing the hydrogen vapor lamp so that the lamp is directly in front of the slit opening of the spectroscope.

With the instructor's permission, plug in the power supply and turn on the power supply switch to illuminate the hydrogen lamp.

Look through the eyepiece, and adjust the slit opening of the spectroscope so that the hydrogen spectral lines are as bright and as sharp as possible. If necessary, adjust the illuminated scale of the spectroscope so that the numbered scale divisions are easily read but do not obscure the hydrogen spectral lines.

Record the color and location on the numbered scale of the spectroscope for each line in the visible spectrum of hydrogen. You should easily observe red, blue-green, and violet lines. A second, very faint violet line may also be visible if the room lighting and illuminated scale lights are not too bright.

Use the calibration plot prepared in Part 1 for mercury vapor to determine the wavelengths of the lines in the hydrogen emission spectrum. Look up the true wavelengths of these lines in your textbook or a chemical handbook, and calculate the percent error in the determination of each line's wavelength.

Use the equations provided in the Introduction to calculate the predicted wavelengths in nanometers (according to Bohr's theory) from the electronic transitions in the hydrogen atom corresponding to the following: $n = 3 \rightarrow n = 2$; $n = 4 \rightarrow n = 2$; $n = 5 \rightarrow n = 2$; $n = 6 \rightarrow n = 2$. How do these predicted wavelengths correspond to those you have measured for hydrogen?

Turn off and unplug the power pack containing the hydrogen lamp. Position the power pack containing the nitrogen lamp in front of the slit opening of the spectroscope. Adjust the spectroscope and scale, and record the location of the bright lines in the nitrogen spectrum. A portion of the emission spectrum of nitrogen appears as a band of several colors. Record the position of this band on the spectroscope scale. Using the calibration curve prepared in Part 1, determine the wavelengths of the bright lines in the nitrogen spectrum and also the approximate wavelength range covered by the band.

Name: _____ Section: _____

Lab Instructor: _____ Date: _____

EXPERIMENT 19

Spectroscopy 1:
Spectra of Atomic Hydrogen and Nitrogen

Pre-Laboratory Questions

1. Given the wavelengths and colors of the major mercury emission lines listed in the introduction to this experiment, construct a graph (on a sheet of graph paper from the end of this manual) that will approximate the spectrum to be determined. Lay out the horizontal axis of the graph in terms of the wavelengths of the lines (in nanometers), and make the vertical axis about 2 inches high to approximate the appearance of the lines to be measured. Using colored pens that correspond to those colors listed, sketch in the emission lines. Compare your sketch of the spectrum to the actual spectrum recorded. Attach your sketch to this page.

2. a. What are the approximate *wavelengths* (in nanometers) in the spectrum of visible sunlight that correspond to the following colors?

    Blue   _____

    Green   _____

    Yellow _____

    Red   _____

    b. What are the approximate *frequencies* (in Hertz, Hz) corresponding to the wavelengths you have listed?

    Blue   _____

    Green   _____

    Yellow _____

    Red   _____

3.   In addition to the visible light region studied in this experiment, excited atomic hydrogen also emits radiation in the ultraviolet region (Lyman series) and the infrared region (Paschen series) of the electromagnetic spectrum. Use a scientific encyclopedia or online reference to find the wavelengths at which hydrogen emits energy in these other regions.

Ultraviolet   _____nm _____nm _____nm _____nm _____nm

Infrared   _____nm _____nm _____nm _____nm _____nm

4.   The study of spectral lines emitted from excited atoms is important to our study of the structure of matter here on the Earth, but the study of spectral lines is also very important in the science of astronomy in the study of the sun and stars. Use a scientific encyclopedia or online reference to list several ways in which the study of spectral lines emitted by astronomical bodies contributes to our understanding and knowledge of those bodies.

Name: _____ Section: _____

Lab Instructor: _____ Date: _____

EXPERIMENT 19

Spectroscopy 1:
Spectra of Atomic Hydrogen and Nitrogen

Results/Observations

1. **Calibration of the Spectroscope**

    Description of the spectrum

    Lines observed:

Color	*Location on spectroscope scale*
_____	_____
_____	_____
_____	_____
_____	_____
_____	_____
_____	_____

    Attach your calibration graph to this report sheet.

2. **Spectra of Atomic Hydrogen and Nitrogen**

    Hydrogen Emission Spectrum

    Description of the spectrum

Lines observed:

Color	Location on spectroscope scale
_____	_____
_____	_____
_____	_____
_____	_____
_____	_____
_____	_____

Wavelengths of the lines (from calibration graph, Part 1)

Color	Wavelength
_____	_____
_____	_____
_____	_____
_____	_____
_____	_____

Calculate the percent error in the determination of the wavelengths (comparing the wavelengths you have determined with those given in your textbook).

Color	Percent error in wavelength
_____	_____
_____	_____
_____	_____
_____	_____
_____	_____

Nitrogen Emission Spectrum

Description of the spectrum

Lines observed:

Color	Location on spectroscope scale
_____	_____
_____	_____
_____	_____
_____	_____
_____	_____
_____	_____

Wavelengths of the lines (from calibration graph, Part 1)

Color	Wavelength
_____	_____
_____	_____
_____	_____
_____	_____
_____	_____

Questions

1. Why were the emission lines of mercury used to calibrate the spectroscope? Could the emission lines of another element have been used?

2. What is the purpose of the *slit* in the spectroscope?

3. Why should you *not* look at the emissions from a spectrum tube for a prolonged time period?

4. The spectral lines observed in the visible spectrum of hydrogen arise from transitions from upper states back to the $n = 2$ principal quantum level. Calculate the predicted wavelengths for the spectral transitions of the hydrogen atom from the $n = 6$ to $n = 2$, for the $n = 5$ to $n = 2$, for the $n = 4$ to $n = 2$, and for the $n = 3$ to $n = 2$ levels in atomic hydrogen.

5. The line spectrum of nitrogen differs in appearance from the line spectrum of hydrogen, in that there tends to be a region with a continuous band of color in addition to several bright lines. What do you think might account for such a continuous spectral region?

EXPERIMENT 20

Spectroscopy 2:
Emission Spectra of Metallic Elements

Objective

Metallic elements emit light at characteristic wavelengths when heated in a flame because of the unique internal electronic structure of the metal atoms. The characteristic wavelengths are observed on a macroscopic basis as characteristic colors imparted to the flame.

Introduction

A number of common metallic elements emit light strongly in the visible light region when ions of the metals are excited. The spectra can be studied using a simple spectroscope.

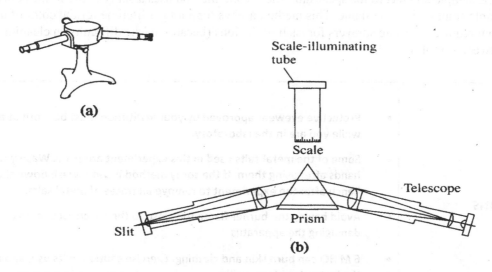

Figure 20-1. (a) A view of the type of spectroscope to be used. (b) A schematic representation of the spectroscope, showing its component parts. When viewed through the telescope, the spectrum will appear superimposed on the numerical scale.

The spectroscope includes four major features: a *slit* for admitting a narrow, collimated beam of light; a *prism* or *diffraction grating* that spreads the incident light into its component wavelengths; a *telescope* for viewing the spectrum; and an illuminated *reference scale* against which the spectrum may be viewed (as an aid in locating the positions of the lines in the spectrum). The scale of the spectroscope is provided merely as a convenience, and the divisions on the scale are arbitrary. If precise numerical data is desired, the scale of the spectroscope can be calibrated using the spectrum of a known element (see Experiment 19). In this experiment, however, we will only qualitatively determine the relative location, color, and intensity of the bright lines emitted by the metallic elements under study

A number of metallic elements from Groups IA and IIA have especially bright emission lines in the visible light region. The emissions are so strong and characteristically colored that these elements can often be recognized by the gross color they impart when aspirated into a burner flame, even without use of a spectroscope. For example, lithium ions impart a red color to a burner flame, sodium ions a yellow/orange color, potassium ions a violet color, calcium ions a brick red color, strontium ions a brighter red color, and barium ions a green color. Upon examination with the spectroscope, it is noted that the spectra of these ions contain several additional lines, but generally the *brightest* line when the spectrum is viewed through the spectroscope corresponds to the gross color imparted to a flame when the spectroscope is not used. Naturally, when several of these ions are present together in a sample (as happens with real samples very commonly), one color may mask another so that direct visual identification of the ions may not be possible. In such cases, only a calibrated spectroscope can determine which elements are present. This is done by comparing the positions of lines in the spectrum of the unknown mixture with the positions of lines in known single samples of the ions in question.

In this experiment, you will excite the ions of each of the elements listed earlier, using a Bunsen burner. You will observe both the gross color imparted to the flame and the spectral lines emitted by the elements as viewed through the spectroscope. You will then determine what elements of those tested are present in an unknown mixture.

Two alternative procedures may be available for introducing the metal ion samples into the burner flame. In the first method, a wire loop is used to pick up a drop of metal solution, and the drop is then placed into the flame for vaporization. This method is simple and requires little equipment, but it produces only a brief burst of color before the sample evaporates completely. This makes it difficult for one person to both introduce the sample and record the spectrum. The second method uses a spray bottle to introduce a fine mist of sample solution into the flame. This method allows for a longer lifetime for the color of the ion in the flame but requires separate sprayers for each of the ions (because of the difficulty in cleaning the sprayer between samples).

Safety Precautions	• **Protective eyewear approved by your institution must be worn at all times while you are in the laboratory.** • **Some of the metal salts used in this experiment are *toxic*. Wash your hands after using them. If the spray method is used, wash down the lab bench after the experiment to remove all traces of metal salts.** • **Avoid having the burner flame too close to the spectroscope, to avoid damaging the apparatus.** • **6 *M* HCl can burn skin and clothing. Exercise caution in its use, and inform the instructor of any spills.**

Apparatus/Reagents Required

Spectroscope with illuminated scale; burner; Nichrome® wire or spray bottles; 6 *M* HCl; 0.10 *M* solutions of the following metal chlorides: lithium, sodium, potassium, calcium, and strontium; unknown solution containing one of these ions; unknown mixture containing two of these ions; combination solutions (lithium chloride + sodium chloride, potassium chloride + sodium chloride, calcium chloride + potassium chloride)

Procedure

Record all data and observations directly in your notebook in ink.

Your instructor will tell you which method to use for introducing the metal ions into the flame. In either case, you will be determining the spectra of solutions of lithium chloride, sodium chloride, potassium chloride, calcium chloride, and strontium chloride.

If the wire method is to be used, obtain several 6-inch lengths of Nichrome® wire and a small amount of 6 M HCl (*Caution!*). Bend the last quarter-inch of each wire into a small loop for picking up the sample solutions. Dip the loops into 6 M HCl to remove any oxides present, rinse in distilled water, and then heat in the oxidizing portion of a burner flame until no color is imparted to the flame by the wires.

If the spray method is to be used, check in the sink to make sure that the sprayers deliver a very fine mist. If the sprayer nozzle is adjustable, try adjusting the nozzle to improve the character of the spray. If the nozzle cannot be adjusted, consult your instructor for a method of cleaning the sprayer.

Set up a laboratory burner directly in front of the slit of the spectroscope but far enough away from the spectroscope to avoid damaging the instrument. Adjust the flame of the burner so that it is as hot as possible. Adjust the illuminated scale of the spectroscope so that approximate positions of the spectral lines can be noted (exact measurements will not be made).

Using the wire method, introduce a drop of one of the metal ion solutions into the oxidizing portion of the flame, and note the gross color imparted to the flame by the solution. Then introduce a second drop of the same metal ion solution into the flame while looking at the flame through the eyepiece of the spectroscope. Note the color, intensity, and approximate scale position of the brightest few lines in the metal ion's spectrum. It may be necessary to repeat the introduction of a drop of the metal ion solution to the flame to permit recording of all the spectral lines shown by the metal ion.

Using the spray method, spray a fine mist of one of the metal ion solutions into the flame, and note the gross color imparted to the flame by the metal ion. Then, while looking through the eyepiece of the spectroscope, spray additional bursts of the same metal ion solution into the flame, noting the color, intensity, and approximate scale position of the brightest few lines in the metal ion's spectrum.

Repeat the determinations (by either method) using the other metal ion samples. If using the wire method, use a new length of wire for each successive sample. If using the spray method, allow the burner to heat for a few minutes between samples to make sure that the previous sample has been evaporated completely so that there will not be a build-up of metal cations.

Several combination solutions, containing two of the metal ions studied, may be available. Observe the spectrum of the combination solution or solutions. Record the approximate scale positions, color, and relative intensity of the lines in the spectra. Does the presence of a second ion appear to affect the wavelengths, color, or intensity of the lines in the spectrum of the first ion?

When each of the known samples of metal ions has been determined, obtain an unknown sample containing just one of the metal ions. Determine the spectrum of the unknown sample, and by matching the colors, intensities, and positions of the spectral lines, identify the unknown sample.

Procedure

Record all data and observations directly in your notebook in ink.

Name: _____ Section: _____

Lab Instructor: _____ Date: _____

EXPERIMENT 20

Spectroscopy 2:
Emission Specta of Metallic Elements

Pre-Laboratory Questions

1. What are the four basic components of the spectroscope to be used in this experiment? What is the purpose (use) of each of these components?

2. Use a chemical dictionary, encyclopedia, or online source to find the distinction between atomic *emission* and atomic *absorption* spectroscopy.

3. Atomic spectroscopy is used in police laboratories for the identification of samples collected at crime scenes—bullet fragments, for example. A bullet typically might consist of an alloy of several metals (copper, zinc, lead, etc.). How would you expect the atomic spectrum of a *mixture* of elements to compare to the individual spectra of the constituent elements in the sample? Would you expect the spectrum of a mixture to be a *superimposition* of the individual spectra, or would you expect the spectral emissions of *one* element's atoms to influence the spectrum of *another* element's atoms? Discuss.

Name: _____ Section: _____

Lab Instructor: _____ Date: _____

EXPERIMENT 20

Spectroscopy 2:
Emission Spectra of Metallic Elements

Results/Observations

Which method did you use for introducing the metal ion samples (wire or spray)?

Metal ion	Gross color	Lines observed
Li	_____	_____
Na	_____	_____
K	_____	_____
Ca	_____	_____
Sr	_____	_____
Ba	_____	_____
Combination (Li/Na)	_____	_____
Combination (K/Na)	_____	_____
Combination (Ca/K)	_____	_____
Unknown	_____	_____
ID number of Unknown	_____	Identity? _____

Questions

1. How did the *gross* colors imparted to the flame by the metal ion solutions compare with the *individual spectral lines* you observed for each element? Did the gross color result from a *combination* of spectral lines or from *one* particularly intense spectral line?

2. Of the metal cations tested, sodium usually gives the brightest and most persistent color to the flame. What problems would this introduce if a real mixture containing both sodium and other cations were to be analyzed by the technique used in this experiment? How could these problems be solved?

3. One of the *Pre-Laboratory Questions* for this experiment asked you to speculate about how the spectrum of a mixture of elements would compare to the individual spectra of the components (would the spectrum of the mixture be altogether different, or just a superposition of the spectra of the components of the mixture). Now that you have actually seen the spectra of mixtures in this experiment, how would you answer this question?

Molecular Shapes and Structures

Objective

The VSEPR theory will be used to predict the 3-dimenional shape of some simple molecules.

Introduction

The shapes exhibited by molecules are often very difficult for beginning chemistry students to visualize, especially since most students' training in geometry is limited to *plane* geometry. To understand the geometric shapes exhibited by molecules, a course in solid geometry, which covers the shapes of three-dimensional figures, would be useful. In this experiment, you will encounter some unfamiliar geometric arrangements that will help you to appreciate the complexity and importance of molecular geometry.

The basic derivation and explanation of molecular shapes arise from the valence shell electron-pair repulsion theory, usually known by its abbreviation, **VSEPR**. This theory considers the environment of the most central atom in a molecule and imagines first how the valence electron pairs of that central atom must be arranged in three-dimensional space around the atom to minimize repulsion among the electron pairs. The general principle is as follows: For a given number of pairs of electrons, the pairs will be oriented in three-dimensional space to be as *far away from each other as possible*. For example, if a central atom were to have only two pairs of valence electrons around it, the electron pairs would be expected to be 180° from each other.

The VSEPR theory then also considers which of the electron pairs around the central atom are **bonding pairs** (with atoms attached) and which are **nonbonding pairs** (lone pairs). The overall geometric shape of the molecule as a whole is determined by *how many* pairs of electrons are on the central atom and by which of those pairs are used for *bonding* to other atoms.

It is sometimes difficult for students to distinguish between the orientation of the electron pairs of the central atom of a molecule and the overall geometric shape of that molecule. A simple example that clearly makes this distinction concerns the case in which the central atom of the molecule has four valence electron pairs. Consider the Lewis structures of the following four molecules: hydrogen chloride, HCl; water, H_2O; ammonia, NH_3; and methane, CH_4.

The central atom in each of these molecules is surrounded by four pairs of valence electrons. According to the VSEPR theory, these four pairs of electrons will be oriented in three-dimensional space to be as far away from each other as possible. The four pairs of electrons point to the corners of the geometric figure known as a **tetrahedron**. The four pairs of electrons are said to be *tetrahedrally oriented* and are separated by angles of approximately 109.5°.

However, three of the molecules shown are *not* tetrahedral in *overall shape*, because some of the valence electron pairs in the HCl, H_2O, and NH_3 molecules are not *bonding* pairs. The angular position of the bonding pairs (and hence the overall shape of the molecule) is determined by the *total* number of valence electron pairs on the central atom, but the non-bonding electron pairs are *not* included in the description of the molecules' overall shape. For example, the HCl molecule could hardly be said to be tetrahedral in shape, since there are only two atoms in the molecule. HCl is linear even though the valence electron pairs of the chlorine atom are tetrahedrally oriented. Similarly, the H_2O molecule cannot be tetrahedral. Water is said to be V-shaped (bent, or nonlinear), the nonlinear shape being a result of the tetrahedral orientation of the valence electron pairs of oxygen. Ammonia's overall shape is said to be that of a trigonal (triangular) pyramid. Of the four molecules used as examples, only methane, CH_4, has both tetrahedrally oriented valence electron pairs and an overall geometric shape that can be described as tetrahedral (since all four pairs of electrons about the central atom are bonding pairs).

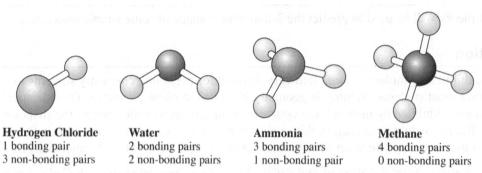

Hydrogen Chloride
1 bonding pair
3 non-bonding pairs

Water
2 bonding pairs
2 non-bonding pairs

Ammonia
3 bonding pairs
1 non-bonding pair

Methane
4 bonding pairs
0 non-bonding pairs

During this experiment, your instructor will construct a large-scale demonstration model of each of the molecular structures to be studied. At your desk, you will construct a smaller model of the structure, measure the bond angles in the structure with a protractor, sketch the structure on paper, and suggest a real molecule that would be likely to have that structure.

Safety Precautions	• Protective eyewear approved by your institution must be worn at all times while you are in the laboratory.
	• Although this experiment does not involve any chemical substances, you should exercise normal caution while in the laboratory.

Apparatus/Reagents Required

Demonstration molecular model kit, student model kit, protractor

Procedure

1. **Lewis Electron Dot Structures**

    As a preliminary exercise, draw Lewis electron dot structures for the following molecules. How many *bonding* and how many *nonbonding* electron pairs are located on the central atom in each molecule?

    SiH_4, NBr_3, PF_3, BF_3, $GeCl_4$, SF_6, ICl_5, C_2H_6, C_2H_4, C_2H_2

2. Molecular Models

Record all data and observations directly in your notebook in ink. Your instructor will build large models of each of the shapes listed in the accompanying *Table of Geometries*. Examine these large-scale models; then build a similar model with the student kit. With the protractor, measure all bond angles in your models.

Sketch a representation of the models, and indicate the measured bond angles. Your sketches do not have to be fine artwork, but the overall shape of the molecule, as well as the position of all electron pairs on the central atom (both bonding and nonbonding), must be clear. For each structure you build and sketch, use your textbook to suggest a real molecule that would be expected on the basis of its Lewis electron dot structure to have that shape. (Give the Lewis dot formula for each of the molecules you suggest.)

Build models for each of the molecules for which you drew Lewis electron dot structures in Part A and sketch the models, showing all electron pairs on the central atom and the approximate bond angles.

Table of Geometries

	No. valence pairs on central atom	Arrangement of valence pairs	No. bonding pairs on central atom	Molecular shape	Type formula
a.	2	linear	2	linear	AB_2
b.	3	trigonal planar	1	linear	—
c.	3	trigonal planar	2	bent	—
d.	3	trigonal planar	3	trigonal planar	AB_3
e.	4	tetrahedral	1	linear	AB
f.	4	tetrahedral	2	bent	AB_2
g.	4	tetrahedral	3	trigonal pyramid	AB_3
h.	4	tetrahedral	4	tetrahedral	AB_4
i.	5	trigonal bipyramid	1	linear	—
j.	5	trigonal bipyramid	2	linear	AB_2
k.	5	trigonal bipyramid	3	T-shape	AB_3
l.	5	trigonal bipyramid	4	see-saw	AB_4
m.	5	trigonal bipyramid	5	trig. bipyramid	AB_5
n.	6	octahedral	1	linear	—
o.	6	octahedral	2	linear	—
p.	6	octahedral	3	T-shape	—
q.	6	octahedral	4	square planar	AB_4
r.	6	octahedral	5	square pyramid	AB_5
s.	6	octahedral	6	octahedral	AB_6

Note. A "—" in the *Type formula* column indicates that no simple molecules with this structure are known (or are likely to be discussed in an introductory general chemistry text).

Name: ______________________________ Section: __________

Lab Instructor: ______________________________ Date: __________

EXPERIMENT 21

Molecular Shapes and Structures

Pre-Laboratory Questions

1. Give a short summary of the main points of the VSEPR theory.

2. Draw Lewis electron dot structures for the following molecules/ions, and predict the bond angles and overall geometry for each.

   a. NCl_3

   b. SCl_6

c. ICl_5

d.   HF

e. H_2Se

f. SO_3^{2-}

g. CH_4

i. CO_2

EXPERIMENT 21

Molecular Shapes and Structures

Results/Observations

1. **Lewis Electron Dot Structures**

   SiH_4 NBr_3

   PF_3 BF_3

   $GeCl_4$ SF_6

   ICl_5 C_2H_6

   C_2H_4 C_2H_2

2.   **Molecular Models**

	Sketch (showing bond angles)	*Lewis dot formula*
a.		
b.		
c.		
d.		
e.		
f.		
g.		
h.		
i.		
j.		

Sketch (showing bond angles)	*Lewis dot formula*

k.

l.

m.

n.

o.

p.

q.

a.

s.

Questions

1. The models you have built do not consider those molecules having double or triple bonds (as predicted for their Lewis dot structures). What effect might a second or third pair of electrons shared between two individual atoms have on the overall shape of a molecule?

2. The models you have built and sketched do not take into account the fact that bonding and nonbonding pairs of electrons do not repel each other to exactly the same extent; repulsion by nonbonding pairs is stronger than repulsion by bonding pairs. How will this affect the true geometric shape of the molecules you have drawn? How will the bond angles you have measured be changed by this effect? Use specific examples.

Properties of Some Representative Elements

Objective

In this experiment, you will examine the properties of some of the more common representative elements.

Introduction

Even in the earliest studies of chemistry, it became evident that certain elemental substances were very much like other substances in their physical and chemical properties. For example, the more common alkali metals (Na, K) were almost indistinguishable to early chemists. Both metals are similar in appearance and undergo reaction with the same reagents (giving formulas of the same stoichiometric ratio). As more and more elemental substances were separated, purified, and identified, more and more similarities between the new elements and previously known elements were detected. Chemists began to wonder why such similarities existed.

In the mid-1800s, Mendeleev and Meyer independently proposed that the properties of the known elements seemed to vary systematically when arranged in the order of their atomic masses. That is, when a list of the elements is made in order of increasing atomic mass, a given set of properties would repeat at regular intervals among the elements. Mendeleev generally receives most credit for the development of the periodic law, because he suggested that some elements were missing from the table. That is, based on the idea that the properties of the elements should repeat at regular intervals, Mendeleev suggested that some elements that should have certain properties had not yet been discovered. Mendeleev went so far as to predict the properties of the elements yet undiscovered, basing his predictions on the properties of the elements already known. Mendeleev's periodic law became widely accepted when the missing elements were finally discovered and had the very properties that Mendeleev had predicted for them.

Mendeleev arranged the elements in order of increasing atomic mass, but the modern periodic table is arranged in order of increasing atomic number. At the time of Mendeleev's work, the structure of the atom had not yet been determined. Generally, the arrangement of the elements by atomic mass is very similar to the arrangement by atomic number, with some notable exceptions. Some elements in Mendeleev's arrangement were out of order and did not have the properties expected. When the arrangement is made by atomic number, however, the properties of the elements do fall into a completely regular order.

Group 1A Elements

The Group 1A elements are commonly referred to as the **alkali metals**. These substances are among the most reactive of elements, are never found in nature in the uncombined state, and are relatively difficult and expensive to produce and store. In particular, these elements are very easily oxidized by oxygen in the air; usually they are covered with a layer of oxide coating unless the metal has been freshly cut or cleaned. When obtained from a chemical supply house, these elements are usually stored under a layer of kerosene to keep them from contact with the air.

The elements of Group 1A are very low in density and are soft enough to be easily cut with a knife or spatula. The only real reaction undergone by these elements involves the loss of their single valence electron to some other species:

$$M \longrightarrow M^+ + e^-$$

In compounds, the elements of Group 1A are invariably found as the unipositive (1+) ion. The least reactive of these elements is lithium (the topmost member, with the valence electron held most tightly by the nuclear charge), whereas the most reactive is cesium (the bottommost member). Cesium is so reactive that the metal may be ionized merely by shining light on it. This property of cesium has been made use of in certain types of photocells (in which the electrons produced by the ionization of cesium are channeled into a wire as electric current). The most common elements of this group, sodium and potassium, are found naturally in great abundance in the combined state on the earth. The elements of Group 1A are rather low in density (for example, they float on the surface of water while reacting) and are very soft (they can be easily cut with a knife).

Group 2A Elements

The elements of Group 2A are commonly referred to as the **alkaline earth elements**. The "earth" in this sense indicates that the elements are not as reactive as the alkali metals of Group 1A and are found (occasionally) in the free state in nature. Beryllium, the topmost element of this group, is not very common. But some of the other members of the group, such as magnesium and calcium, are found in fairly high abundance. As with the Group 1A elements, the relative reactivity of the Group 2A elements increases from top to bottom in the periodic table. For example, metallic magnesium can be kept in storage for quite a while without developing an oxide coating (by reaction with air), whereas a freshly prepared sample of pure metallic calcium will develop a layer of oxide much more quickly. In compounds, the elements of Group 2A are usually found as dipositive ions (through loss of the two valence electrons).

Group 3A Elements

Whereas all the elements of Group 1A and Group 2A are metals, the first member of Group 3A (boron) is a nonmetal. The dividing line between metallic elements and nonmetallic elements forms the "stairstep" region indicated at the right-hand side of most periodic charts. Because boron is to the right of the stairstep, boron shows many properties that are characteristic of nonmetals. For example, boron is not a very good conductor of heat and electricity, whereas aluminum (which is the element beneath boron in the group) shows the much higher conductivities associated with metallic elements. Boron is generally found covalently bonded in its compounds, which further indicates its nonmetallic nature. Aluminum is the only relatively common element from Group 3A.

In nature, aluminum is generally found as the oxide, and most metallic aluminum is produced by electrolysis of the molten oxide. This is a relatively difficult and expensive process, which is part of the reason why the aluminum in cans was commonly recycled even before the present emphasis on preserving the environment. Aluminum is a relatively reactive element; however, this reactivity is sometimes masked under ordinary conditions. Aluminum used for making cooking utensils or cans becomes very quickly coated with a thin layer of aluminum oxide. This oxide layer serves as a protective coating, preventing further oxidation of the aluminum metal underneath. (This is why aluminum pans are seldom as shiny as copper or stainless steel pans.)

Group 4A Elements

Group 4A contains some of the most common, useful, and important elements known. The first member of the group, carbon, is a nonmetal. The second and third members, silicon and germanium, show both some metallic and some nonmetallic properties (and because of this in-between nature, these elements are referred to as **metalloids** or **semimetals**). The last two members of the group, tin and lead, show mostly metallic properties.

The element carbon forms the framework for life. Virtually all biological molecules are, in fact, carbon compounds. Carbon is unique among the elements in that it is able to form long chains of many hundreds or even thousands of similar carbon atoms. Pure elemental carbon itself is usually obtained in either of two forms, graphite or diamond, each of which has a very different structure. Graphite contains flat, two-dimensional layers of covalently bonded carbon atoms, with each layer effectively being a molecule of graphite, independent of the other layers present in the sample. The layers of carbon atoms in graphite are able to *move* relative to each other. This makes graphite slippery and useful as a lubricant for machinery and locks. Diamond contains three-dimensional extended covalent networks of carbon atoms, making an entire diamond effectively a single molecule. Because all the atoms present in diamond are covalently bonded to each other in three dimensions, diamond is relatively stable and is also the hardest substance known. Natural diamonds used as gemstones are produced deep in the earth under high pressures over eons, but synthetic diamonds are produced in the laboratory by compressing graphite to several thousand atmospheres of pressure. Because of its hardness, diamond is used as an abrasive in industry.

Silicon and germanium are used in the semiconductor industry. Because these elements have some of the properties of nonmetals (brittleness, hardness), as well as some of the properties of metals (slight electrical conductivity), they have been used extensively in transistors.

The last two elements of Group 4A, tin and lead, show mostly metallic properties (shiny luster, conductivity, malleability) and frequently form ionic compounds. However, covalently bonded compounds of these metals are not rare. For example, leaded gasoline contains tetraethyllead, which is effectively an organic lead compound. Tin has been used in the past to make cooking utensils and is used extensively as a liner for steel cans ("tin" cans), because it is less likely to be oxidized than iron and is also less likely to impart a bad taste to foods. Whereas tin is nonpoisonous and is used in the food industry, lead compounds are very dangerous poisons. Formerly, pigments for white paint were made using lead oxide. Lead compounds are remarkably sweet tasting, and many children have been seriously poisoned by eating flecks of peeling lead paints.

Group 5A Elements

The first few elements of Group 5A are nonmetals, with nitrogen and phosphorus being the two most abundant elements of the group. The earth's atmosphere is nearly 80% N_2 gas by volume. This large percentage of nitrogen is believed to be a remnant of the primordial composition of the earth's atmosphere. Astronomers have determined that several of the planets of the solar system contain large quantities of ammonia, NH_3, in their atmospheres. It is known that ammonia actually comes to equilibrium with its constituent elements:

$$2NH_3 \longrightarrow N_2 + 3H_2$$

Astronomers theorize that the atmosphere of the early earth was also mostly ammonia but that the lighter hydrogen gas has escaped the earth's relatively weak gravitational field over the eons, leaving the atmosphere with a large concentration of nitrogen gas. Nitrogen is also an important constituent of many of the molecules synthesized and used by living cells (amino acids, for example).

Phosphorus, the second member of Group 5A, is a typical solid nonmetal. Three allotropic forms of elemental phosphorus exist: red, white, and black phosphorus. White phosphorus is the most interesting of the three forms and is the form that is produced when phosphorus vapor is condensed. White phosphorus is very unstable toward oxidation. Below about 35°C, white phosphorus reacts with oxygen of the air, emitting the energy of the oxidation as light rather than heat (phosphorescence); above 35°C, white phosphorus will spontaneously burst into flame. (For this reason, white phosphorus is stored under water to keep it from contact with air.) Whereas elemental nitrogen consists of diatomic molecules, white phosphorus contains tetrahedrally shaped P_4 molecules. When white phosphorus is heated to high temperatures in the absence of air, it is converted to the red allotrope. The red allotrope is much less subject to oxidation and is the form usually available in the laboratory. The lower reactivity is due to the fact that the phosphorus atoms have polymerized into one large molecule, which is far less subject to attack by oxygen atoms than the individual P_4 molecules of the white allotrope.

Phosphorus is essential to life, being a component of many biological molecules, most notably the phosphate compounds used by the cell for storing and transferring energy (AMP, ADP, and ATP). Phosphorus also forms part of the structural framework of the body, bones being complex calcium phosphate compounds.

The other members of Group 5A are far less common than are nitrogen and phosphorus. For example, arsenic (As) is known primarily for its severe toxicity and has been used in various rodenticides and weed killers (though As has been supplanted by other agents for most of these purposes).

Group 6A Elements

The most abundant elements of Group 6A (sometimes referred to as the **chalcogens**) are oxygen and sulfur, which, along with selenium, are nonmetallic. Elemental oxygen makes up about 20% of the atmosphere by volume and is vital not only to living creatures but also to many common chemical reactions. Most of the oxygen in the atmosphere is produced by green plants, especially by plankton in the oceans of the earth. On a microscopic basis, oxygen is used in living cells for the oxidation of carbohydrates and other nutrients. Oxygen is needed also for the oxidation of petroleum-based fuels for heating and lighting purposes and for myriad other important industrial processes. Oxygen has two allotropic forms: the normal elemental form of oxygen is O_2 (dioxygen), whereas a less stable allotrope, O_3(ozone), is produced in the atmosphere by high-energy electrical discharges.

Sulfur is obtained from the earth in a nearly pure state and needs little or no refining before use. In certain areas of the earth, there are vast underground deposits of pure sulfur. High-pressure steam is pumped into such deposits, liquefying the sulfur and allowing it to be pumped to the surface of the earth for collection. The main use of sulfur is in the manufacture of sulfuric acid, which is the industrial chemical produced each year in the largest amount. When sulfur is burned in air, the sulfur is converted to sulfur dioxide, notable for its choking, irritating odor (similar to that of a freshly struck match). If sulfur dioxide is further oxidized with a catalyst, it is converted to sulfur trioxide, the anhydride of sulfuric acid.

Sulfur has several allotropic forms. At room temperature, the normal allotrope is orthorhombic sulfur, which consists of S_8 cyclic molecules. At higher temperatures, a slight rearrangement of the shape of these rings occurs, producing a solid allotropic form of sulfur: monoclinic sulfur. This substance has a different crystalline structure than the orthorhombic form. (Monoclinic sulfur slowly changes into orthorhombic sulfur at room temperature.) The most interesting allotrope of sulfur, however, is produced when boiling sulfur is rapidly lowered in temperature. This technique produces a plastic (amorphous) allotrope. The properties of plastic sulfur are very different from those of orthorhombic or monoclinic sulfur. Whereas plastic sulfur is soft and pliable (being able to be stretched and pulled into strings), orthorhombic and monoclinic sulfur are both hard, brittle solids. Sulfur is insoluble in water, but the orthorhombic form is fairly soluble in the nonpolar solvent carbon disulfide, CS_2.

The most common oxygen compound is, of course, water, H_2O. The analogous sulfur compound, hydrogen sulfide, H_2S, is a noxious gas at room temperature. (The gas smells like rotten eggs; eggs contain a protein involving sulfur that releases hydrogen sulfide as the egg spoils.) Selenium, which occurs in Group 6A just beneath sulfur, is the only other member of the group that occurs in any abundance. Selenium compounds are generally very toxic. Selenium is used in certain prescription dandruff shampoos.

Group 7A Elements

The Group 7A elements are more commonly referred to as the **halogens** ("salt-formers"). The elementary substances are very reactive; these elements are, therefore, usually found in the combined state, as the negative halide ions. Elemental fluorine is the most reactive nonmetal and is seldom encountered in the laboratory because of problems of toxicity and handling of the gaseous element. Most compounds of fluorine are very toxic, although tin(II) fluoride is used in toothpastes and mouthwashes as a decay preventative.

Hydrogen fluoride is very different from the other hydrogen halides, since aqueous solutions of HF are weakly acidic, rather than strongly acidic (HCl, HBr). Hydrogen fluoride is able to attack and dissolve glass and is always stored in plastic bottles.

Gaseous elemental chlorine is somewhat less reactive than elemental fluorine and is used commercially as a disinfectant. Chlorine will oxidize biological molecules present in bacteria (thereby killing the bacteria), while it is itself reduced to chloride ion (which is nontoxic). For example, drinking water is usually chlorinated, either with gaseous elemental chlorine or with some compound of chlorine that is capable of releasing elemental chlorine when needed. Gaseous hydrogen chloride, when dissolved in water, forms the solution known as hydrochloric acid, which is one of the most commonly used acids.

Elemental bromine is a strikingly dark red liquid at room temperature. Elemental bromine is used as a common laboratory test for the presence of double bonds in carbon compounds. Br_2 reacts with the double bond, and its red color disappears, thereby serving as an indicator of reaction.

Elemental iodine is a dark gray solid at room temperature but undergoes sublimation when heated slightly, producing an intensely purple vapor. Elemental iodine is the least reactive of the halogens. Iodine is very important in human metabolism, being a component of various hormones produced by the thyroid gland. Since many people's diets do not contain sufficient iodine from natural sources, commercial table salt usually has a small amount of sodium iodide added as a supplement ("iodized" salt). Iodine is also used sometimes as a topical antiseptic, since it is a weak oxidizing agent that is capable of destroying bacteria. Iodine for this purpose (called "tincture" of iodine) is usually sold in alcohol solution.

	• Protective eyewear approved by your institution must be worn at all times while you are in the laboratory.
	• The reactions of lithium, sodium, and potassium with water are very dangerous because of the hydrogen gas and quantity of heat released by the reactions (which may be enough to ignite the hydrogen). Your instructor will *demonstrate* these reactions for you. Do *not* attempt these reactions yourself.
	• Hydrogen gas is extremely flammable and forms explosive mixtures with air. Small amounts of hydrogen will be generated in some of the student reactions. The room must be well ventilated. Avoid flames and use caution.
	• The flames produced by burning magnesium and calcium are intensely bright and can damage the eyes. Do *not* look directly at the flame while these substances are burning.
Safety Precautions	• Oxides of nitrogen are toxic and are extremely irritating to the respiratory system. Generate these gases only in the fume exhaust hood.
	• Hydrochloric acid and nitric acids will burn the skin. Wash immediately if the acids are spilled and inform the instructor. These acids will also damage clothing.
	• Vapors of methylene chloride and of the halogens are toxic and irritating. Keep these substances in the fume exhaust hood during use. Dispose of methylene chloride as directed by the instructor.
	• Hydrogen sulfide and sulfur dioxide are toxic and have noxious odors. Confine them to the fume exhaust hood at all times.
	• Salts of the Group 1A and 2A metals may be toxic. Wash after handling these compounds.

Apparatus/Reagents Required

Lithium, sodium, potassium, universal indicator solution, calcium, magnesium ribbon, samples of solid $LiCl$, $NaCl$, KCl, $CaCl_2$, $BaCl_2$, and $SrCl_2$, flame test wires, 6 M hydrochloric acid, boric acid, aluminum oxide, sodium carbonate, copper wire, concentrated nitric acid, 0.1 M NaOH, chlorine water, bromine water, iodine water, methylene chloride, 0.1 M NaCl, 0.1 M NaBr, 0.1 M NaI, ferrous sulfide, solid iodine

Procedure

Record all data and observations directly in your notebook in ink.

1.   Some Properties of the Alkali and Alkaline Earth Elements

The reactive metals of Group 1A and Group 2A react with cold water to liberate elemental hydrogen, leaving a solution of the strongly basic metal hydroxide

Group 1A: $2M(s) + 2H_2O \longrightarrow 2M^+(aq) + 2OH^-(aq) + H_2(g)$

Group 2A: $M(s) + 2H_2O \longrightarrow M^{2+}(aq) + 2OH^-(aq) + H_2(g)$

Generally, the reactivity of the metals increases going from the top of the group toward the bottom of the group. The valence electrons of the metal are less tightly held by the nucleus toward the bottom of the group.

a.   Lithium, Sodium, and Potassium (Instructor Demonstration)

The reactions of the Group 1A metals with water are very dangerous and will be performed by your instructor as *demonstrations*. Do *not* handle the Group 1A elements yourself. In the exhaust hood, behind a safety shield, the instructor will drop small pellets of lithium, sodium, and potassium into beakers containing a small amount of cold water. Compare the speed and vigor of the reactions.

When the reactions have subsided, obtain, in separate clean test tubes, small portions of the water from the beakers in which the reactions were conducted. Add 2–3 drops of universal indicator to each test tube; refer to the color chart provided with the indicator to determine the pH of the solution. Write equations for the reactions of the metals with water.

b.   Reactions of Magnesium and Calcium (Student Procedure)

Repeat the procedure that you saw in the demonstration, using pieces of the Group 2A metals magnesium and calcium in place of lithium, sodium, and potassium. Work in the exhaust hood, with the safety shield pulled down. Account for differences in reactivity between these two Group 2A metals. Test the pH of the water with which the metals were reacted.

Add a single turning of magnesium to 10 drops of 6 M HCl in a small test tube. Why does magnesium react with *acid* but not with cold water?

c.   Flame Tests

Obtain a 6–8-inch length of nichrome wire for use as a flame test wire. Form one end of the wire into a small loop that is no more than a few millimeters in diameter.

Obtain about 20 mL of 6 M HCl in a small beaker, and immerse the loop end of the wire for 2–3 minutes to clean the wire. Ignite a Bunsen burner flame, and heat the loop end of the wire in the flame until it no longer imparts a color to the flame.

Obtain a few crystals each of LiCl, NaCl, KCl, CaCl$_2$, BaCl$_2$, and SrCl$_2$. Dip the loop of the flame test wire into one of the salts and then into the oxidizing portion of the burner flame. Record the color imparted to the flame by the salt.

Clean the wire in 6 M HCl, heat in the flame until no color is imparted, and repeat the flame test with the other salts. The colors imparted to the flame by the metal ions are so intense and so characteristic that they are frequently used as a test for the presence of these elements in a sample.

2. **Oxides/Sulfides of Some Elements**

Oxygen compounds of most elements are known, and they are generally referred to as **oxides**. However, there are great differences between the oxides of metallic elements and those of nonmetals. The oxides of metallic elements are *ionic* in nature: they contain the oxide ion, O^{2-}. Ionic metallic oxides form *basic* solutions when dissolved in water

$$O^{2-} + H_2O \longrightarrow 2OH^-$$

Oxides of nonmetals are generally *covalently* bonded and form *acidic* solutions when dissolved in water.

$$SO_3 + H_2O \longrightarrow H_2SO_4$$

Sulfur compounds of both metallic and nonmetallic substances are known. Sulfur compounds of nonmetallic compounds are generally covalently bonded, whereas with metallic ions, sulfur is usually present as the sulfide ion, S^{2-}. Most sulfur compounds have characteristic unpleasant odors or decompose into compounds that have the characteristic odors.

a. **Metallic Oxides**

Place a small amount of distilled water in a beaker and add 2–3 drops of universal indicator. Refer to the color chart provided with the indicator, and record the pH of the water.

Obtain a piece of magnesium ribbon about 1 inch in length. Hold the magnesium ribbon with tongs above the beaker of water and ignite the metal with a burner flame. Let the oxide of magnesium fall into the water in the beaker. Do *not* look directly at the flame produced by burning magnesium. It is intensely bright and *damaging* to the eyes. When the flame has expired, stir the liquid in the beaker for several minutes and record the pH of the solution. Write an equation for the reactions occurring.

Place approximately 10 drops of distilled water in each of three test tubes, and add a drop of universal indicator to each. Add a *very* small quantity of sodium peroxide to one test tube, calcium oxide (lime) to a second test tube, and aluminum oxide (alumina) to the third test tube. Stir each test tube with a clean glass rod, record the pH of the solution, and write an equation to account for the pH.

b. **Nonmetallic Oxides**

Place approximately 10 drops of water into each of two test tubes, and add 1 drop of universal indicator to each. In one test tube, dissolve a small amount of boric acid. To the second test tube, add a small chunk of dry ice (solid carbon dioxide–*Caution!*). Stir the test tubes with a clean glass rod, record the pH, and write an equation to account for the pH of the solutions.

c. **Sulfur Compounds**

In the exhaust hood set up a small beaker of cold water and a burner. Ignite a *tiny* amount of powdered sulfur on the tip of a spatula in the burner flame and *cautiously* note the odor. Hold the burning sulfur just above the surface of the water in the beaker until the flame disappears, then dip the tip of the spatula into the water to cool it. Determine the pH of the water. Why is it acidic?

Obtain a *tiny* portion of iron(II) sulfide in a test tube, and add 2 drops of dilute hydrochloric acid. *Cautiously* note the odor of the hydrogen sulfide generated. Transfer the test tube to the exhaust hood to dispose of the hydrogen sulfide.

3.   **Some Properties of the Halogen Family (Group 7)**

All of the members of Group 7A are nonmetallic. The halogen elements tend to gain electrons in their reactions, and the attraction for electrons is stronger if the electrons are closer to the nucleus of the atom. The activity of the halogen elements therefore decreases from top to bottom in Group 7 of the periodic table. As a result of this, elemental chlorine is able to replace bromide and iodide ion from compounds, and elemental bromine is able to replace iodide ion from compounds:

$$Cl_2 + 2NaBr \longrightarrow Br_2 + 2NaCl$$

$$Cl_2 + 2NaI \longrightarrow I_2 + 2NaCl$$

$$Br_2 + 2NaI \longrightarrow I_2 + 2NaBr$$

a.   **Identification of the Elemental Halogens by Color**

The elemental halogens are nonpolar and are not very soluble in water. Solutions of the halogens in water are not very brightly colored. However, if an aqueous solution of an elemental halogen is shaken with a nonpolar solvent, the halogen is preferentially extracted into the nonpolar solvent and imparts a characteristic, relatively bright color to the nonpolar solvent. Because the nonpolar solvent is not miscible with water, the halogen color is evident as a separate colored layer.

In the following tests, dispose of the methylene chloride samples as directed by the instructor.

Obtain about 1 mL of chlorine water in a small test tube. Note the color. In the exhaust hood, add 10 drops of methylene chloride to the test tube. Stopper and shake.

Allow the solvent layers to separate, and note the color of the lower (methylene chloride) layer. Elemental chlorine is not very intensely colored and imparts only a pale yellow/green color.

Repeat the procedure using 1 mL of bromine water in place of the chlorine water. Bromine imparts a red color to the lower layer.

Repeat procedure using 1 mL of iodine water. Iodine imparts a purple color to the methylene chloride layer.

b.   **Relative Reactivity of the Halogens**

Add 2 mL of chlorine water to each of two test tubes. Add 2 mL of 0.1 M NaBr to one test tube and 2 mL of 0.1 M NaI to the second tube. In the exhaust hood, add 10 drops of methylene chloride to each test tube.

Stopper, shake, and allow the solvents to separate. Record the colors of the lower layers, and identify which elemental halogens have been produced.

Add 2 mL of bromine water to each of two test tubes. Add 2 mL of 0.1 M NaI to one test tube and 2 mL of 0.1 M NaCl to the other test tube.

In the exhaust hood, add 10 drops of methylene chloride to each test tube. Stopper, shake, and allow the layers to separate. Record the colors of the lower layers. Indicate where a reaction has taken place.

Add 2 mL of iodine water to each of two test tubes. Add 2 mL of 0.1 M NaBr to one test tube and 2 mL of 0.1 M NaCl to the other test tube.

In the exhaust hood, add 10 drops of methylene chloride to each test tube. Stopper, shake, and allow the layers to separate. Record the colors of the lower layers. Did any reaction take place?

c.   **Sublimation of Iodine**

One property of elemental iodine that makes it relatively unusual among pure substances is its ability to undergo *sublimation* readily. When heated, most solids are first converted to the liquid state (melting) before they vaporize (boiling). Iodine, when heated, goes directly from the solid state to the gaseous state.

Using forceps, place a few crystals of elemental iodine in a small beaker. Set the beaker on a ring stand in the exhaust hood, and cover with a watch glass containing some ice.

Heat the iodine crystals in the beaker with a small, gentle flame for 2–3 minutes, and watch the sublimation of the iodine. Extinguish the flame, and examine the iodine crystals that have formed on the lower surface of the watch glass.

EXPERIMENT 22

Properties of Some Representative Elements

Pre-Laboratory Questions

1. Mendeleev originally thought that the properties of the elements seemed to vary systematically with their *atomic masses*, whereas our modern understanding is that the properties of the elements vary regularly with *atomic numbers*. How do these two concepts differ? Why do we feel the modern concept is the correct one?

2. For each of the following elements, list the atomic number, the average atomic mass, in which group (vertical column) of the periodic table the element can be found, and in which period (horizontal row) the element is located.

   a. B _____

   b. Rn _____

   c. Ba _____

   d. K _____

   e. N _____

   f. Al _____

   g. Cs _____

   h. Xe _____

   i. Se _____

   j. Sr _____

   k. Sb _____

   l. P _____

3. Describe how the halides (Cl⁻, Br⁻, and I⁻) will be determined in this experiment, and how they can be distinguished from one another.

4. Sublimation is the transition directly from the solid phase of a substance to the vapor phase, without passing through a liquid phase. In this experiment, the sublimation of iodine will be studied. Suggest some substances from everyday life that undergo sublimation.

Name: ______________________ **Section:** __________

Lab Instructor: ______________________ **Date:** __________

EXPERIMENT 22

Properties of Some Representative Elements

Results/Observations

1. **Some Properties of the Alkali and Alkaline Earth Elements**

   a. Lithium, Sodium, and Potassium (Instructor Demonstration)

Metal	*Evidence of reaction*	*Relative vigor*	*Indicator color*	*Equation for reaction*
Li				
Na				
K				

   b. Magnesium and Calcium (Student Procedure)

Mg				
Ca				

   Reaction of Mg with acid ______________________

   c. Flame tests

Metal	*Color observed*

2. **Oxides/Sulfides of Some Elements**

Substance	*Results/equations – universal indicator test*
Magnesium oxide	
Sodium peroxide	
Calcium oxide	
Aluminum oxide	
Boric acid	

Dry ice   _____

Sulfur dioxide   _____

FeS + HCl   _____

3.   **Some Properties of the Halogen Family**

    a.   Identification of the elemental halogens

Halogen	Color of water solution	Color of CH_2Cl_2 extract
Cl_2	_____	_____
Br_2	_____	_____
I_2	_____	_____

    b.   Relative reactivity of the halogens

Elemental Halogen	Halide Ion	Color of Lower Layer	Which halogen is in the lower layer?	Did a reaction take place?
Cl_2	Br^-	_____	_____	_____
Cl_2	I^-	_____	_____	_____
Br_2	I^-	_____	_____	_____
Br_2	Cl^-	_____	_____	_____
I_2	Br^-	_____	_____	_____
I_2	Cl^-	_____	_____	_____

Equations for reactions that occurred

_____

_____

_____

_____

_____

_____

_____

Observations of sublimation of iodine

Questions

1. In comparing the reactions of lithium, sodium, and potassium with water, as demonstrated by your instructor, you undoubtedly observed that the reaction of potassium metal was much more vigorous than the reaction of lithium metal. Based on the location of these elements in the periodic table and their electronic structures, explain why this difference in vigor might be expected for these elements.

2. In your experiment, you observed that the oxides of metallic elements produce *basic* solutions when dissolved in water, whereas the oxides of nonmetallic elements produce *acidic* solutions. For the oxide of chlorine, Cl_2O_7, and the oxide of potassium, K_2O, write equations showing how these materials form acidic and basic solutions, respectively, when dissolved in water.

3. You demonstrated the *sublimation* of iodine in this experiment. However, another substance you used in the experiment also underwent sublimation. What substance was this?

4. Flame tests for some of the metal ions are used in several experiments in this manual. Why does a metal like sodium always produce the *same characteristic color* when introduced into a flame? Explain.

5. Methylene chloride was used to *extract* the halogen elements from aqueous solution into a separate layer, the color of which was then used to *identify* the halogen. Suggest a reason why the free halogen elements (Cl_2, Br_2, and I_2) were so easily transferred to the methylene chloride layer rather than remaining dissolved in the aqueous layer.

Classification of Chemical Reactions

Objective

Chemical reactions may be classified as precipitation, acid–base, complexation, or oxidation–reduction. Several examples of each will be examined in this experiment.

Introduction

There are over 100 known elements and millions of known compounds. Throughout history, chemists have sought to organize the wealth of data and observations that have been recorded for these substances. In this experiment, you will perform examples of each of the reaction types mentioned in the objective.

1. Precipitation Reactions

Certain substances are not very soluble in water. Frequently, such sparingly soluble substances may be generated *in situ* in a reaction vessel. For example, silver chloride is not soluble in water. If an aqueous solution of silver nitrate (soluble) is mixed with an aqueous solution of sodium chloride (soluble), the *combination* of silver ions from one solution and chloride ions from the other solution generates silver chloride, which then forms a precipitate that settles to the bottom of the container. The solution that remains above the precipitate of silver chloride effectively becomes a solution of sodium nitrate. Silver ions and sodium ions have switched partners, ending up in a compound with the negative ion that originally came from the opposite substance:

$$AgNO_3(aq) + NaCl(aq) \longrightarrow AgCl(s) + NaNO_3(aq)$$

The silver ion and sodium ion have replaced each other in this process; this sort of reaction is sometimes referred to as a double-displacement reaction. Precipitation reactions are often emphasized by write the *net-ionic equation* for the reaction, showing only the ions that combine to form the precipitate:

$$Ag^+(aq) + Cl^-(aq) \longrightarrow AgCl(s)$$

This net ionic equation implies that silver ion from *any* source and chloride ion from *any* source would combine to form a precipitate of silver chloride.

2. Acid–Base Reactions

There are many theories and definitions that attempt to explain what constitutes an acid or a base. An early but still useful theory, developed by Arrhenius in the late 1800s, defines an acid as a substance that produces hydrogen ions, H^+, when dissolved in water. A base, according to Arrhenius, is a species that produces hydroxide ions, OH^-, when dissolved in water. For example, hydrogen chloride is an acid in the Arrhenius theory, because hydrogen chloride ionizes when dissolved in water:

$$HCl(aq) \longrightarrow H^+(aq) + Cl^-(aq)$$

and releases hydrogen ions. In particular, hydrogen chloride is called a *strong* acid because virtually *every* HCl molecule ionizes when dissolved. Other substances, although they do in fact produce hydrogen ions when dissolved, do not *completely* ionize when dissolved and hence are called *weak* acids.

217

For example, the acidic component of vinegar (acetic acid) is a weak acid:

$$CH_3COOH(aq) \rightleftarrows H^+(aq) + CH_3COO^-(aq)$$

The equation for the ionization of acetic acid indicates that this substance reaches equilibrium when dissolved in water, at which point a certain maximum apparent concentration of hydrogen ion is present. The concentration of hydrogen ion produced by dissolving a given amount of weak acid is typically several orders of magnitude *less* than the concentration produced when the same concentration of strong acid is prepared.

In a similar manner, there are both strong and weak bases in the Arrhenius scheme. Sodium hydroxide, for example, is a strong base:

$$NaOH(s) \longrightarrow Na^+(aq) + OH^-(aq)$$

For every mole of NaOH that might be dissolved in water, an equivalent number of moles of OH^- is produced. In contrast, ammonia, NH_3, is a weak base:

$$NH_3(g) + H_2O(l) \rightleftarrows NH_4^+(aq) + OH^-(aq)$$

Ammonia in effect reacts with water and comes to equilibrium, at which point a certain maximum concentration of OH^- exists. As with weak acids, the concentration of hydroxide ion in a solution of a weak base is much smaller than if a strong base had been used.

The most important reaction of acids and bases is neutralization. The hydrogen ion from an aqueous acid combines with the hydroxide ion from an aqueous base, producing water. For example,

$$HCl(aq) + NaOH(aq) \longrightarrow NaCl(aq) + H_2O(l)$$

$$HNO_3(aq) + NaOH(aq) \longrightarrow NaNO_3(aq) + H_2O(l)$$

The net reaction is the same in each of these and is typical of the reaction between acids and bases in aqueous solution:

$$H^+(aq) + OH^-(aq) \longrightarrow H_2O(l)$$

Although the Arrhenius definitions of acids and bases have proved very useful, the theory is restricted to the situation of *aqueous* solutions. Aqueous solutions are most common, but it is reasonable to question whether HCl and NaOH will still behave as an acid or base when dissolved in some other solvent. The Brønsted–Lowry theory of acids and bases extends the Arrhenius definitions to more general situations. An acid is defined in the Brønsted–Lowry theory as any species that provides hydrogen ions, H^+, in a reaction. Since a hydrogen ion is nothing more than a simple proton, acids are often referred to as being proton donors in the Brønsted–Lowry scheme. A base in the Brønsted–Lowry system is defined as any species that receives hydrogen ions from an acid. Bases are referred to then as proton acceptors. The similarity between the definitions of an acid in the Arrhenius and Brønsted–Lowry schemes is obvious. On first glance, however, it would seem that the definition of what constitutes a base differs between the two systems. Realize, however, that hydroxide ion in aqueous solution (Arrhenius definition) will very readily accept protons from any acid that might be added. Therefore, hydroxide ion is in fact a base in the Brønsted–Lowry system as well.

3. Complexation Reactions

Metal ions, especially in solution, react with certain neutral molecules and many negatively charged ions to form species known as coordination complexes or complex ions. Such compounds are particularly common with ions of the transition metals. The neutral or anionic species that reacts with the metal ion is called a ligand. In order for a molecule or anion to be able to function as a ligand, the species must have at least one pair of nonbonding valence electrons. Most metal ions have empty, relatively low-energy atomic *d*-orbitals. When a suitable ligand is added to a solution of a metal ion, a coordinate covalent bond can form between the metal ion and the ligand molecule or ion. The nonbonding valence electron pair of the ligand provides the electron pair needed for the covalent bond.

In their complex ions, many metals characteristically bind with the same number of ligand species, regardless of the particular identity of the ligands. The number of ligands bound to a metal ion is called the coordination number of the metal. For example, copper(II) ion usually exhibits a coordination number of four, whereas nickel(II) usually is found bound with six ligand molecules.

Coordination complexes of transition metals are often very striking in color and for this reason are used as pigments in paints. These colors arise from transitions of electrons among the *d*-orbitals of the coordinated metal ion. In the presence of a set of ligands, the *d*-orbitals of the metal ion are displaced somewhat from their normal energies. These energy differences between the displaced *d*-orbitals correspond to the absorption/emission of visible light. For example, an aqueous solution of copper(II) ion is pale blue [because of the presence of water molecules coordinated to the copper(II) ion]. If ammonia is added to aqueous copper(II) ion solution, ammonia molecules displace coordinated water molecules and themselves coordinate with the copper(II) ion.

$$Cu(H_2O)_4^{2+}(aq) + 4NH_3(aq) \longrightarrow Cu(NH_3)_4^{2+}(aq) + 4H_2O(l)$$

The resulting copper/ammonia complex is a striking dark blue. Coordination complexes are very important in biological systems. Important molecules such as hemoglobin and chlorophyll contain coordinated metal ions.

The formation of coordination complexes actually represents a special case of acid–base chemistry. G. N. Lewis extended the study of acid–base processes beyond the theories of Arrhenius and Brønsted–Lowry to include species that had no protons but that demonstrated reactions very similar to or suggestive of other acid–base reactions. Lewis defined a base as a species capable of providing a valence electron pair for formation of a bond between atoms. According to this definition, ligand molecules in coordination compounds would be bases. Similarly, Lewis defined an acid as a species capable of receiving an electron pair in the formation of a bond. Lewis's definitions did not contradict earlier definitions of what constituted an acid or base. Any species that could receive a proton in the Brønsted–Lowry scheme would have to have at least one nonbonding pair of valence electrons for donating to a coordinate covalent bond. However, in the Lewis definition, a species would not have to be able to donate a proton to be an acid. Metal ions that receive pairs of electrons in the formation of a coordination compound could be looked upon as Lewis acids.

4. Oxidation–Reduction Reactions

A large and important class of chemical reactions can be classified as oxidation–reduction (or redox) processes. Oxidation–reduction processes involve the transfer of electrons from one species to another. For example, in the reaction

$$Zn(s) + CuSO_4(aq) \longrightarrow ZnSO_4(aq) + Cu(s)$$

$Zn(s)$ represents uncharged elemental zinc atoms, whereas in $ZnSO_4(aq)$, zinc exists in solution as Zn^{2+} ions. Each zinc atom has effectively lost two electrons in the process and is said to have been oxidized:

$$Zn(s) \longrightarrow Zn^{2+}(aq) + 2e^- \qquad \text{oxidation}$$

Oxidation is defined as a loss of electrons by an atom or ion, or, better, as a transfer of electrons from one species to another. Similarly, in the preceding reaction, each Cu^{2+} ion in solution has gained two electrons, becoming uncharged elemental copper atoms:

$$Cu^{2+}(aq) + 2e^- \longrightarrow Cu(s) \qquad \text{reduction}$$

Copper ions are said to have been reduced in the process. Reduction is defined as a gain of electrons by a species, or, better, as transfer of electrons to one species from another.

To clarify the transfer of electrons between the species undergoing oxidation and the species undergoing reduction, redox reactions are usually divided into two half-reactions, one for each process. This was done for the zinc/copper reaction discussed earlier. Some redox reactions and their corresponding half-reactions are shown following.

$$2HgO \longrightarrow 2Hg + O_2 \qquad \textit{overall reaction}$$

$$2Hg^{2+} + 4e^- \longrightarrow 2Hg \qquad \textit{reduction}$$

$$2O_2^- \longrightarrow O_2 + 4e^- \qquad \textit{oxidation}$$

$$Cl_2 + 2I^- \longrightarrow 2Cl^- + I_2 \qquad \textit{overall reaction}$$

$$Cl_2 + 2e^- \longrightarrow 2Cl^- \qquad \textit{reduction}$$

$$2I^- \longrightarrow I_2 + 2e^- \qquad \textit{oxidation}$$

Be careful to distinguish between oxidation–reduction reactions (in which electrons are completely transferred from one species to another) and complexation reactions (in which one species provides both electrons for a covalent bond between itself and another species).

Safety Precautions	<ul><li>Protective eyewear approved by your institution must be worn at all times while you are in the laboratory.</li><li>Chromium compounds are toxic, mutagenic, and can cause burns to the skin. Wash thoroughly after use.</li><li>Lead and barium compounds are toxic if ingested.</li><li>Silver nitrate and potassium permanganate will discolor the skin if spilled. The discoloration requires several days to wear off.</li><li>All acids and bases should be assumed to be damaging to skin, eyes, and clothing if spilled. Wash immediately, and inform the instructor if any spills occur.</li><li>Copper, nickel, and thiocyanate ion solutions are toxic. Wash after use.</li><li>Concentrated ammonia solution is a severe respiratory irritant and cardiac stimulant. Confine all use of ammonia to the fume exhaust hood.</li><li>Ethylene diamine is toxic and its vapor is harmful. Confine the use of ethylene diamine to the fume exhaust hood.</li><li>The dimethylglyoxime (DMG) solution is flammable. No flames are permitted during its use.</li><li>Hydrogen gas is extremely flammable and forms explosive mixtures with air.</li><li>The flame produced by burning magnesium is intensely bright and can damage the retina. Do not look directly at the flame of burning magnesium.</li><li>Hydrogen peroxide solutions are unstable and will burn the skin if spilled.</li><li>"Chlorine water" may burn the skin if spilled and may emit toxic chlorine gas. Wash hands after use, and confine its use to the fume exhaust hood.</li><li>Methylene chloride is toxic if inhaled or absorbed through the skin, and is a possible mutagen. Wash after use, and confine its use to the fume exhaust hood.</li></ul>

Apparatus/Reagents Required

0.1 M solutions of the following: silver nitrate, potassium chloride, lead acetate, potassium sulfate, barium chloride, sulfuric acid, hydrochloric acid, acetic acid, sodium hydroxide, ammonia, sodium thiocyanate; conductivity tester; universal indicator; 0.5 M solutions of copper(II) and nickel(II) sulfate; concentrated ammonia; 1 M ammonium nitrate; 0.5 M iron(III) nitrate in 0.1 M nitric acid; 1 M sodium chloride; 1 M hydrochloric acid; 1% dimethylglyoxime; magnesium ribbon; steel wool; oxygen gas; 3% hydrogen peroxide; 1 M lead acetate; metallic zinc strips; metallic copper strips; 10% potassium iodide; 10% potassium bromide; chlorine water; methylene chloride; 0.5 M NaCl; 0.5 M NaBr

Procedure

Record all data and observations directly in your notebook in ink.

1. **Precipitation Reactions**

   In a spot plate or small test tubes, combine approximately 5-drop portions of the 0.1 M aqueous solutions in the list that follows. Use the table of solubilities from your textbook or Appendix E of this manual to predict what precipitate forms in each double displacement reaction. Write equations for the reactions:

   Silver nitrate and sodium chloride $\longrightarrow$

   Lead acetate and potassium sulfate $\longrightarrow$

   Barium chloride and dilute sulfuric acid $\longrightarrow$

2. **Acid–Base Reactions**

   Obtain 10-drop samples of 0.1 M HCl and 0.1 M acetic acid in separate clean 75-mm test tubes. Using pH paper, determine the pH of each solution. How do you account for the fact that the pH of the acetic acid solution is *higher* than that of the HCl solution, although both have the same 0.1 M concentration?

   Obtain 10-drop samples of 0.1 M NaOH and 0.1 M ammonia in separate clean test tubes. Using pH paper, determine the pH of each solution. How do you account for the fact that the pH of the ammonia solution is *lower* than that of the NaOH solution, although both have the same 0.1 M concentration?

   One easy method for demonstrating the difference in the degree of ionization between solutions of strong acids and weak acids (or of other electrolytes) is to measure the relative electrical conductivity of the solutions. Solutions of strong acids (or other strongly ionized electrolytes) conduct electrical currents well, whereas solutions of weak acids (or other weak electrolytes) conduct electricity poorly. Your instructor has set up a light bulb conductivity tester and will demonstrate the electrical conductivity of the 0.1 M acid and base solutions. The light bulb will glow brightly when the electrodes of the conductivity tester are immersed in strong acid or base solution but will glow only dimly when the electrodes are immersed in solutions of weak acid or base. Record which of the acids/bases are strong electrolytes, and which are weak.

   Place 2.0 ± 0.1 mL of 0.1 M HCl in a clean test tube, and add 1 drop of universal indicator. Record the color of the indicator and the pH of the solution. This indicator changes color gradually with pH and can be used to monitor the pH of a solution during a neutralization reaction. Keep handy the color chart provided with the indicator. It shows the color of the indicator under different pH conditions.

Obtain a sample of 0.1 M NaOH in a small beaker, and begin adding NaOH to the sample of HCl and indicator, 1 drop at a time, with a medicine dropper or plastic micropipet. Record the color and pH after 5 drops of NaOH have been added.

Continue adding NaOH dropwise, recording the color and pH at 5-drop intervals until the pH has risen to above 10. Approximately how many drops of NaOH were required to reach pH 7 (neutral)?

Given that 1 mL of liquid is approximately equivalent to 20 average drops, approximately how many milliliters of 0.1 M NaOH were required to neutralize the 2-mL sample of 0.1 M HCl?

Repeat the process, using 2 mL of 0.1 M acetic acid in place of HCl. How does the change in pH during the addition of NaOH differ when a weak acid is neutralized?

3. **Complexation Reactions**

   a. **Ammonia Complexes**

      Place approximately 10 drops of 0.5 M copper(II) sulfate and 0.5 M nickel(II) sulfate solutions in separate clean semi-micro test tubes. Record the colors of the solutions.

      In the exhaust hood, add concentrated ammonia solution dropwise to each of the metal ion solutions. Initially, a precipitate of the metal hydroxide may form, but on further addition of ammonia, brightly colored complex ions will form. Record your observations. Write equations for the complexation reactions that occurred.

      Repeat the tests on fresh samples of nickel ion and copper ion, using a solution of 1 M ammonium sulfate in place of the concentrated ammonia solution. Why is there no change in color when the ammonium ion is added?

   b. **Iron Complexes**

      Place 10-drop samples of 0.5 M Fe(NO$_3$)$_3$ (in 0.1 M nitric acid) in each of three clean semi-micro test tubes. Record the color of the solution.

      Add 1 M NaCl solution dropwise to one iron(III) solution until the color of the resulting complex is evident.

      Add 1 M HCl to a second sample of iron solution until the color of the complex is evident.

      Which ion (Na$^+$, H$^+$, or Cl$^-$) in the preceding two tests is the ligand that reacts with iron(III) ion?

      Add a few drops of 0.1 M NaSCN to the third iron(III) sample. Record the color of the resulting iron/thiocyanate complex. The intense color of the iron/thiocyanate complex is used commonly as a test to detect the presence of iron in a sample.

   c. **An Insoluble Nickel Complex**

      Place 10 drops of 0.5 M nickel(II) sulfate solution in a clean semi-micro test tube. In the exhaust hood, add 1 drop of concentrated ammonia solution, followed by 5 drops of 1% dimethylglyoxime reagent. Record the appearance of the nickel/dimethylglyoxime complex. Dimethylglyoxime is an organic ligand used as a standard qualitative and quantitative test for the presence of nickel ion.

   d. **Copper Complexes**

      Place approximately 10 drops of 0.5 M CuSO$_4$ solution to each of three semi-micro test tubes.

      To one of the test tubes, add 0.5 M NaCl solution dropwise until the color of the complex is evident.

      To the second copper(II) sample, add 0.5 M NaBr dropwise until the complex has formed.

In the exhaust hood, add 10% ethylenediamine solution dropwise to the third sample of copper(II) ion until the color of the complex is evident. How does this complex compare to the copper/ammonia complex prepared in Part 1?

4.   Oxidation–Reduction Reactions

Oxidation–reduction reactions are very common and take many forms. Reactions in which elemental substances combine to form a compound or in which compounds are decomposed to elemental substances (or simpler compounds, in many cases) are redox reactions. The reactions of metallic substances with acids, or with corrosive gases, and the reaction of many species in solution are all examples of oxidation–reduction processes.

a.   Oxidation of Magnesium

Obtain a small strip of magnesium ribbon approximately 1 cm long. When magnesium is ignited, the flame produced is dangerous to the eyes. *Caution!* Do not look directly at burning magnesium. Hold the magnesium ribbon with tongs, and have ready a beaker of distilled water to catch the product of the reaction as it is formed. Ignite the magnesium ribbon in a burner flame, and hold the burning ribbon over the beaker of water. The unbalanced reaction is

$$Mg + O_2 \longrightarrow MgO$$

Allow the beaker of water containing the magnesium oxide produced by the reaction to stand for 5–10 minutes; then test the water with pH paper. Why is the solution basic? Magnesium is oxidized in the reaction, whereas oxygen is reduced.

b.   Oxidation of Iron

Your instructor will demonstrate (in the exhaust hood) the burning of iron (steel wool) in pure oxygen. Iron oxidizes only slowly in air at room temperature (forming rust), but the reaction is considerably faster in pure oxygen at elevated temperatures. Iron has two common oxidation states. The unbalanced reactions are

$$Fe(s) + O_2(g) \longrightarrow FeO(s)$$

$$Fe(s) + O_2(g) \longrightarrow Fe_2O_3(s)$$

c.   Oxidation–Reduction of Hydrogen Peroxide

Place about 20 mL of 3% hydrogen peroxide in a large test tube. Obtain a single crystal of potassium permanganate. Ignite a wooden splint; then blow out the flame so that the splint is still glowing. Add the crystal of permanganate to the hydrogen peroxide sample to catalyze the decomposition of the hydrogen peroxide. Insert the glowing wood splint partially into the test tube, being careful not to wet the wood splint. The glowing wood splint should immediately burst into flame as it comes in contact with the pure oxygen being generated by the decomposition reaction

$$H_2O_2(aq) \longrightarrow H_2O(l) + O_2(g)$$

d.   Oxidation–Reduction of Copper and Zinc

Obtain about 2 mL of 0.5 *M* copper(II) sulfate solution in a small test tube. Add a small strip or turning of metallic zinc and allow the solution to stand for 15–20 minutes. Then remove the zinc and examine the coating that has formed on the strip. The reaction is

$$Zn(s) + CuSO_4(aq) \longrightarrow ZnSO_4(aq) + Cu(s)$$

e. **Oxidation–Reduction of Lead and Zinc**

Obtain about 2 mL of 1 M lead acetate solution in a small test tube, and place the test tube in a place where it will not be disturbed. Add a small strip of metallic zinc to the solution. Allow the solution to stand for 10–15 minutes. After this period, examine the "tree" of metallic lead that has formed. The reaction is

$$Zn(s) + Pb(CH_3COO)_2(aq) \longrightarrow Pb(s) + Zn(CH_3COO)_2(aq)$$

f. **Oxidation–Reduction of the Halogens**

Obtain about 10 drops each of 10% potassium iodide solution and 10% potassium bromide solution in separate semimicro test tubes. Add 10 drops of "chlorine water" to each test tube; stopper and shake briefly. Record any color changes. Add 5 drops of methylene chloride to each test tube, stopper, and shake.

Elemental iodine and elemental bromine have been produced by replacement; these substances are more soluble in methylene chloride than in water and are preferentially extracted into this solvent. The colors of the methylene chloride layer are characteristic of these halogens. The unbalanced reactions are

$$Cl_2 + KI \longrightarrow KCl + I_2$$

$$Cl_2 + KBr \longrightarrow KCl + Br_2$$

g. **Oxidation–Reduction of Copper and Silver**

Scrub a 2-inch length of copper wire with steel wool and rinse with water. Place 2 mL of 0.1 M silver nitrate solution in a test tube, and add the copper wire. Allow the test tube to stand undisturbed for 15–20 minutes, and then examine the copper wire. The unbalanced reaction is

$$Cu(s) + AgNO_3(aq) \longrightarrow Cu(NO_3)_2(aq) + Ag(s)$$

EXPERIMENT 23

Classification of Chemical Reactions

Pre-Laboratory Questions

1. In the Arrhenius acid–base theory, what constitutes an acid? How does the Arrhenius definition of an acid compare with the Brønsted–Lowry definition? Write equations showing the ionization of HCl, HNO_3, and $HC_2H_3O_2$ in water that demonstrate both of the acid–base definitions.

2. What is a *precipitation* reaction? Using aqueous solutions of the substances listed below, write five overall balanced equations and balanced net ionic equations demonstrating precipitation reactions:

   $BaCl_2(aq)$, $AgNO_3(aq)$, $Na_2CO_3(aq)$, $NaOH(aq)$, $NiCl_2(aq)$, $Na_2S(aq)$, $CuSO_4(aq)$, $CaCl_2(aq)$

3. What is a *coordination complex*? Give two examples from your textbook of important coordination complexes.

4. What is meant by a *redox* reaction? Give two examples, including balanced chemical equations, of redox reactions. What is meant by a *half-reaction* for a redox reaction? For the two examples of redox reactions that you have given, also write the half-reactions, indicating which is the oxidation and which is the reduction.

Name: ______________________ **Section:** __________

Lab Instructor: ______________________ **Date:** __________

EXPERIMENT 23

Classification of Chemical Reactions

Results/Observations

1. **Precipitation Reactions**

Silver nitrate + sodium chloride

Observations ______________________

Equation ______________________

Lead acetate + potassium sulfate

Observations ______________________

Equation ______________________

Barium chloride + sulfuric acid

Observations ______________________

Equation ______________________

2. **Acid–Base Reactions**

pH of 0.1 *M* HCl ____________ pH of 0.1 *M* acetic acid ____________

Explanation

____

____

____

pH of 0.1 *M* NaOH ____________ pH of 0.1 *M* ammonia ____________

Explanation

____

____

____

Conductivity tests:

0.1 *M* HCl ______________________

0.1 *M* acetic acid ______________________

0.1 *M* NaOH ______________________

0.1 *M* ammonia ______________________

Why do HCl and NaOH allow the light bulb to glow brightly, but ammonia and acetic acid only allow the bulb to glow dimly (if at all)?

_____

_____

_____

_____

Universal indicator tests:

HCl + NaOH

_____

_____

Acetic acid + NaOH

_____

_____

What difference did you notice in the response of the Universal indicator between the two reactions? Explain.

_____

_____

_____

_____

_____

_____

3.   **Complexation Reactions**

   a.   Ammonia complexes

      Copper sulfate + ammonia

_____

_____

      Nickel sulfate + ammonia

_____

_____

      Copper sulfate + ammonium sulfate

_____

_____

Nickel sulfate + ammonium sulfate

_____

_____

b.   Iron complexes

Iron nitrate + NaCl

_____

_____

Iron nitrate + HCl

_____

_____

Iron nitrate + NaSCN

_____

_____

c.   Insoluble nickel complex

Nickel sulfate + DMG

_____

_____

d.   Copper complexes

Copper(II) sulfate + NaCl

_____

_____

Copper(II) sulfate + NaBr

_____

_____

Copper(II) sulfate + ethylene diamine

_____

_____

Comparison with the preceding $CuSO_4/NH_3$ reaction

_____

_____

4. **Oxidation–Reduction Reactions**

    a.   Magnesium + oxygen

_____

_____

    b.   Iron + oxygen

_____

_____

    c.   Hydrogen peroxide decomposition

_____

_____

    d.   Zinc strip + copper sulfate solution

_____

_____

    e.   Zinc strip + lead acetate solution

_____

_____

    f.   Potassium iodide + chlorine water

_____

_____

        Potassium bromide + chlorine water

_____

_____

    g.   Copper strip + silver nitrate solution

_____

_____

Questions

1. Balance all the chemical equations given in the Introduction and Procedure sections of this experiment.

    _____

    _____

    _____

    _____

    _____

    _____

    _____

    _____

    _____

    _____

    _____

    _____

    _____

    _____

    _____

    _____

    _____

    _____

2. A crystal of potassium permanganate was added to the hydrogen peroxide as a *catalyst* to speed up the decomposition of the hydrogen peroxide. Use your textbook to find a definition of *catalyst* and two examples in which a catalyst is used to speed up a reaction.

3.   Identify each of the following reactions as a *precipitation reaction*, an *acid–base* reaction, a *complexation* reaction, or an *oxidation–reduction* reaction. Explain your classifications.

   a. $Zn(s) + 2AgNO_3(aq) \rightarrow Zn(NO_3)_2(aq) + 2Ag(s)$

   b. $HCl(aq) + AgNO_3(aq) \rightarrow AgCl(s) + HNO_3(aq)$

   c. $H_2SO_4(aq) + 2CH_3NH_2(aq) \rightarrow (CH_3NH_3)_2SO_4(aq)$

   d. $CuSO_4(aq) + 4NH_3(aq) \rightarrow Cu(NH_3)_4SO_4(aq)$

   e. $4Fe(s) + 3O_2(g) \rightarrow 2Fe_2O_3(s)$

EXPERIMENT 24

Colligative Properties 1: Freezing Point Depression and the Determination of Molar Mass

Objective

The depression of the freezing point of a solution of an unknown substance in naphthalene will be used to calculate the molar mass of the unknown substance.

Introduction

When a solute is dissolved in a solvent, the properties of the solvent are *changed* by the presence of the solute. The magnitude of the change generally is proportional to the amount of solute added. Some properties of the solvent are changed only by the *number* of solute particles present, without regard to the particular *nature* of the solute. Such properties are called **colligative properties** of the solution. Colligative properties include changes in vapor pressure, boiling point, freezing point, and osmotic pressure.

For example, if a nonvolatile, nonionizing solute is added to a volatile solvent (such as water), the amount of solvent that can escape from the surface of the liquid at a given temperature is lowered, relative to the case where only the pure solvent is present. The vapor pressure above such a solution will be lower than the vapor pressure above a sample of the pure solvent under the same conditions. Molecules of nonvolatile solute physically *block* the surface of the solvent, thereby preventing as many molecules from evaporating at a given temperature (see Figure 24-1). As shown in the figure, if the vapor pressure of the solution is lowered, there is an increase in the boiling point of the solution as well as a decrease in the freezing point.

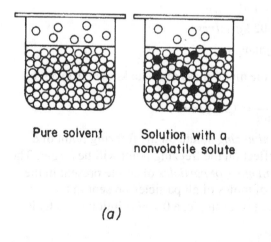

(a)

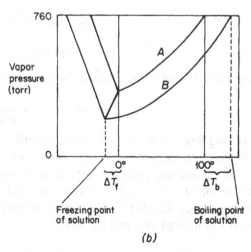

(b)

Figure 24-1. (a) The presence of a nonvolatile solute lowers the vapor pressure of the solvent by blocking the surface. (b) Plot of vapor pressure *vs.* temperature for a pure solvent and a solution. Note that the solute affects the properties of the solution.

In this experiment, you will determine the freezing points of a pure solvent (naphthalene), a solution of a *known* solute (1,4-dichlorobenzene, $C_6H_4Cl_2$) dissolved in naphthalene, and of a solution of an *unknown* solute dissolved in naphthalene. You will use this information to calculate the molar mass of the unknown solute.

The decrease in freezing point (ΔT_f) when a nonvolatile, nonionizing solute is dissolved in a solvent is proportional to the *molal* concentration (m) of solute present in the solvent:

$$\Delta T_f = K_f \times m \qquad\qquad (24\text{-}1)$$

K_f is a constant for a given solvent (called the *molal freezing point depression constant*) and represents by how many degrees the freezing point will change when 1.00 mol of solute is dissolved per kilogram of solvent. For example, K_f for water is 1.86°C kg/mol, whereas K_f for the solvent benzene is 5.12°C kg/mol.

The molal concentration of a solution represents how many *moles of solute* are dissolved per *kilogram* of the solvent. For example, if 0.50 mol of the sugar sucrose were dissolved in 100 g of water, this would be equivalent to 5.0 mol of sugar per kilogram of solvent, and the solution would be 5.0 molal. The molality of a solution is defined as

$$\text{molality}, m = \frac{moles\ of\ solute\ particles}{kilograms\ of\ solvent} \qquad\qquad (24\text{-}2)$$

The measurement of freezing point lowering is routinely used for the determination of the molar masses of unknown solutes. If the two equations above are combined, it can be derived that the *molar mass* of a solute is related to the *freezing point lowering* experienced by the solution and to the *composition* of the solution. Consider the following example:

> **Example:** The molal freezing point depression constant K_f for the solvent benzene is 5.12°C kg/mol. A solution of 1.08 g of an unknown in 10.02 g of benzene freezes at a temperature 4.60°C lower than pure benzene. Calculate the molar mass of the unknown.
>
> The molality of the solution can be obtained using Equation *24-1*:
>
> $$m = \frac{\Delta T_f}{K_f} = \frac{4.60\ ^\circ C}{5.12\ ^\circ C \cdot kg/mol} = 0.898\ mol/kg$$
>
> From the definition of molality in Equation *24-2*, the number of moles of unknown present can be calculated
>
> $$m = 0.898\frac{mol}{kg} = \frac{mol\ unknown}{0.01002\ kg\ solvent}$$
>
> $$mol\ unknown = 0.00899\ mol$$
>
> Thus, 0.00899 mol of the unknown weighs 1.08 g, and the molar mass of the unknown is given by
>
> $$molar\ mass = \frac{1.08\ g}{0.00899\ mol} = 120.\ g/mol$$

In the preceding discussion, we considered the effect of a *nonionizing* solute on the freezing point of a solution. If the solute does indeed ionize in the solvent, the effect on the freezing point will be *larger*. The depression of the freezing point of a solvent is related to the *number of particles* of solute present in the solvent. If the solute ionizes as it dissolves, the total number of moles of all particles present in the solvent will be *larger* than the formal concentration indicates. For example, a 0.1 m solution of NaCl is effectively 0.1 m in *both* Na^+ and Cl^- ions.

	• Protective eyewear approved by your institution must be worn at all times while you are in the laboratory.
Safety Precautions	• Naphthalene and 1,4-dichlorobenzene are toxic, and their vapors are harmful. Do not handle these substances. Work in an exhaust hood. Wear disposable gloves. These substances are also combustible and should be kept away from open flames.
	• Use glycerine when inserting the thermometer through the rubber stopper. Protect your hands with a towel. Wash off all glycerine.
	• Acetone is highly flammable. Use acetone only in the exhaust hood and keep away from all flames.
	• Dispose of the chemicals as directed by the instructor.

Apparatus/Reagents Required

8-inch test tube with two-hole stopper to fit, thermometer, stirring wire, beaker tongs, naphthalene, 1,4-dichlorobenzene, acetone, disposable gloves, unknown sample

Procedure

Record all data and observations directly in your notebook in ink.

1. **Determination of the Freezing Point of Naphthalene**

    Set up a 600-mL beaker about three-fourths full of water in an exhaust hood, and heat the water to boiling.

    Using glycerine as a lubricant and protecting your hands with a towel, insert your thermometer through a two-hole stopper so that the temperature can still be read from 60°C–90°C. Equip the second hole of the stopper with a length of copper wire, coiled at the bottom into a ring that will fit around the thermometer. The copper wire will be used to *stir* the solution. (See Figure 28-2.)

    Weigh a clean, dry 8-inch test tube to the nearest 0.01 g.

    Add approximately 10 g of naphthalene to the test tube, and reweigh precisely, to the nearest 0.01 g.

    Use a clamp to set the test tube vertically in the boiling-water bath. Allow the naphthalene to melt completely in the boiling water bath.

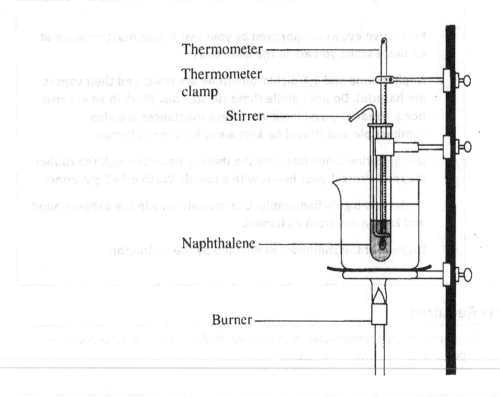

Figure 24-2. Apparatus for freezing-point determination Be certain that the stirring wire can be moved freely and that the thermometer can be read from 60°C–90°C.

When the naphthalene has melted completely, insert the rubber stopper containing the thermometer and copper wire into the test tube. (See Figure 24-2.) Make certain that the bulb of the thermometer is immersed in the molten naphthalene and that the copper stirring wire can be agitated freely.

Continue to heat the test tube to re-melt any naphthalene that might have crystallized on the thermometer or stirring wire. Move the stirring wire up and down to make certain the temperature is uniform throughout the naphthalene sample.

When the naphthalene has re-melted completely (generally the temperature will be above 95°C), remove the beaker of hot water from under the test tube (*Caution!*) using tongs or a towel to protect your hands from the heat.

Ensure that the thermometer is immersed in the center of the molten naphthalene and does *not* touch the walls of the test tube. When the temperature of the molten naphthalene has dropped to 90°C, begin recording the temperature of the naphthalene every 30 seconds. Record the temperatures to the nearest 0.1°C.

Continue taking the temperature of the naphthalene until the temperature has dropped to around 60–65°C. The naphthalene will freeze in this temperature range.

On graph paper from the back of this manual, plot a **cooling curve** for naphthalene: graph the *temperature* of the naphthalene as it cools versus the *time* in minutes. You will notice that the naphthalene *remains* at the same temperature for several minutes; this results in a horizontal, flat region in the cooling curve. This temperature represents the freezing point of pure naphthalene. (See Figure 24-3.)

Use the boiling-water bath to re-melt the naphthalene in the test tube, and remove the thermometer and wire stirrer. In the exhaust hood, rinse any adhering naphthalene from the thermometer/stirrer with acetone (*Caution: Flammable!*). Allow the thermometer and stirrer to remain in the hood until all acetone has evaporated.

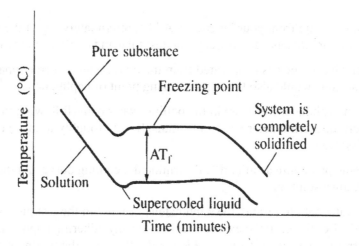

**Figure 24-3. Cooling curve for a pure solvent and for a solution involving the solvent
The flat portions of the curves represent freezing.**

2. Determination of K_f for Naphthalene

Reweigh the test tube containing the naphthalene to the nearest 0.01 g. Some of the naphthalene may have been lost on the thermometer/stirrer apparatus or may have vaporized.

Add approximately 1 g of 1,4-dichlorobenzene crystals to the test tube containing the naphthalene, and reweigh the test tube to the nearest 0.01 g.

Melt the naphthalene/dichlorobenzene mixture in boiling water.

Make certain that all acetone has evaporated from the thermometer/stirrer apparatus (any remaining acetone would *also* lower the freezing point of naphthalene).

Insert the thermometer/stirrer apparatus into the test tube containing the molten naphthalene/dichloro-benzene mixture. Stir the mixture for several minutes while it is in the boiling water bath to make certain that the solution is completely homogeneous.

Following the same procedure as in Part 1, determine the cooling curve and the freezing point of the naphthalene/1,4-dichlorobenzene solution.

From the freezing point depression of the naphthalene/dichlorobenzene solution and the composition of the solution, calculate the molal freezing point depression constant, K_f, for naphthalene. The molecular formula of 1,4-dichlorobenzene is $C_6H_4Cl_2$.

Use the boiling-water bath to re-melt the naphthalene mixture in the test tube, and remove the thermometer and wire stirrer. In the exhaust hood, rinse any adhering naphthalene from the thermometer/stirrer with acetone (*Caution: Flammable!*). Allow the thermometer/stirrer to remain in the hood until all acetone has evaporated. Dispose of the freezing point mixture as directed by the instructor.

3.   **Determination of the Freezing Point of an Unknown Solute**

Clean and dry an 8-inch test tube, and weigh it to the nearest 0.0l g.

Add approximately 10 g of naphthalene to the test tube, and reweigh the test tube and contents to the nearest 0.01 g.

Obtain an unknown solute from your instructor. Add approximately 1 g of the unknown to the test tube containing the naphthalene, and reweigh the test tube and contents to the nearest 0.01 g.

Make certain that all acetone has evaporated from the thermometer/stirrer apparatus used earlier (any remaining acetone would also lower the freezing point of naphthalene).

Melt the unknown/naphthalene mixture in the boiling-water bath (following the procedure described earlier), and insert the thermometer/stirrer apparatus. Stir thoroughly to make certain that the sulfur is completely dissolved.

Following the same procedure as in Part 1, determine the cooling curve and the freezing point of the naphthalene/unknown solution.

Use the boiling-water bath to re-melt the naphthalene mixture in the test tube, and remove the thermometer and wire stirrer. In the exhaust hood, rinse any adhering naphthalene from the thermometer/stirrer with acetone (*Caution: Flammable!*). Allow the thermometer/stirrer to remain in the hood until all acetone has evaporated. Dispose of the naphthalene/dichlorobenzene mixture as directed by the instructor.

4.   **Calculations**

Use the *depression* in the freezing point (Parts 1 and 3), the *composition* of the unknown/naphthalene solution (Part 3), and the *molal freezing point depression constant* determined for naphthalene (Part 2) to calculate the apparent molar mass of the unknown solute.

Attach your three graphs to the laboratory report.

Name: _____ Section: _____

Lab Instructor: _____ Date: _____

EXPERIMENT 24

Colligative Properties 1:
Freezing Point Depresssion
and the Determination of Molar Mass

Pre-Laboratory Questions

1. List three colligative properties of solutions, and describe how each colligative property is related to the *concentration* of solute in the solution.

2. If 1.35 g of a non-ionizing solute of molar mass 42.1 g is dissolved in 11.6 g of a particular solvent, the resulting solution has a volume of 13.2 mL. Calculate the *molality* and the *molarity* of the solution. Which concentration—molality or molarity—would be expected to vary with temperature? Why?

3.   Suppose the freezing point of a solution of 1.18 g of an unknown molecular substance in 8.31 g of the
     solvent benzene is measured. The solution is found to freeze at a temperature 6.33°C lower than that at
     which pure benzene itself freezes. Calculate the molar mass of the unknown substance.

4.   Ionic solutes have a larger effect on the colligative properties of a solution than a comparable formal
     concentration of a nonionic solute would have. Use a scientific encyclopedia or online reference to
     describe what is meant by the *Van't Hoff factor* for a solute in a solution.

Name: _____ Section: _____

Lab Instructor: _____ Date: _____

EXPERIMENT 24

Colligative Properties 1:
Freezing Point Depression
and the Determination of Molar Mass

Results/Observations

1. **Determination of the Freezing Point of Naphthalene**

   Mass of empty test tube _____

   Mass of test tube plus naphthalene _____

   Mass of naphthalene taken _____

   Freezing point of naphthalene (from graph) _____

2. **Determination of K_f for Naphthalene**

   Mass of test tube plus naphthalene _____

   Mass of naphthalene in test tube _____

   Mass of test tube/naphthalene plus dichlorobenzene _____

   Mass of 1,4-dichlorobenzene taken _____

   Freezing point of naphthalene/dichlorobenzene solution _____

   Molality of dichlorobenzene solution _____

   Change in freezing point (compared to pure naphthalene) _____

   K_f for naphthalene _____

3. **Determination of the Freezing Point of an Unknown Solute**

   Mass of empty test tube _____

   Mass of test tube with naphthalene _____

   Mass of naphthalene taken _____

   Mass of test tube after adding unknown _____

   Mass of unknown added _____

   Freezing point of naphthalene/unknown solution _____

   Change in freezing point (compared to pure naphthalene) _____

   molality of unknown in the solution _____

   molar mass of the unknown _____

243

Questions

1. A phenomenon called **supercooling** is frequently encountered in this experiment. In supercooling, a solution momentarily drops *below* its freezing point, and then warms up again, before solidification begins. What event is likely to give rise to supercooling?

2. Look up the freezing point constant, K_f, for naphthalene in a handbook. Calculate the percent error between your determination of K_f and the handbook value. What might have led to your obtaining a different value?

3. The molal freezing point depression constant for the solvent benzene is 5.12 °C/molal, whereas K_f for camphor is 40.0 °C/molal. Suppose separate solutions are made containing 1.05 g of toluene ($C_6H_5CH_3$) in 10.0 g of these solvents. Calculate by how many degrees the freezing point of these solutions will differ from the freezing points of the respective pure solvents.

Colligative Properties 2: Osmosis and Dialysis

Introduction

Plant and animal cell membranes must provide some mechanism for the transfer of materials across the membrane without the wholesale loss of the cell's fluid contents or microstructures. Although there are several mechanisms by which molecules are passed through cell walls, perhaps the simplest to understand and demonstrate are the processes of *osmosis* and *dialysis*.

Cell membranes are said to be **semipermeable**: such membranes contain pores that are of a sufficient size that they permit the passage of ions and small molecules, while *not* allowing the passage of most macromolecules or cell organelles. Osmosis represents the passage of *solvent* through a semipermeable membrane, whereas dialysis refers to the passage of *solute* molecules as well. Both osmosis and dialysis occur whenever a concentration gradient (difference) exists between the solutions on either side of a membrane. Osmosis and dialysis occur in order to *equalize* the concentrations of the solutions on either side of the membrane.

For example, suppose you had just eaten a sugary snack after a few hours without food. After digestion of the snack took place, your bloodstream would have a relatively high concentration of carbohydrates, whereas in the interior of your cells, the concentration of carbohydrates would presumably be lower; whatever sugars had been present would have been metabolized during the period without food. Dialysis of sugar molecules from the region of high concentration in the bloodstream to the region of low concentration inside the cells would take place until the concentrations on either side of the cell membrane were similar. Osmosis of water from the cell interior into the bloodstream would also take place.

Osmosis and dialysis are special examples of the process of diffusion: molecules of solute and solvent diffuse across a membrane separating two solutions in an effort to equalize the concentrations of the solutions, which is very nearly the same as would happen between the two solutions if the membrane were not present.

The *force* with which osmosis/dialysis takes place—that is, how strongly solvent or solute flows through the membrane—depends on the degree of difference in concentration across the membrane. If the two solutions separated by the membrane are of very nearly the same concentration, the force of osmosis/dialysis that takes place will be very gentle. If a *large* difference in concentration exists across the membrane, however, the osmosis/dialysis will be very strong (perhaps so forceful as to rupture the membrane). For this reason, solutions to be injected into the bloodstream, such as the dextrose solutions used for intravenous feeding in hospitals, have their concentrations adjusted to be very nearly that of normal bodily fluids: such solutions are said to be *isotonic* with the normal fluids of the body.

The force with which osmosis takes place is measured in terms of the **osmotic pressure** that exists between a solution and a sample of the pure solvent. The osmotic pressure represents the pressure that would have to be applied to the solution to just *prevent* osmosis of the solvent across a membrane.

Osmotic pressure (Π) is a colligative property of solutions and is dependent on the molarity of solute present in the solution:

$$\Pi = MRT$$

Accurate osmotic pressures are not conveniently measurable in the general chemistry laboratory, and so this experiment will investigate the processes of osmosis and dialysis only on a qualitative basis. Osmosis into both a concentrated and a dilute solution of the same solute will be examined to determine into which solution osmosis occurs with the greater force. Dialysis of sodium chloride (an ionic solute), dextrose (a molecular carbohydrate solute), and starch (a colloidal macromolecular solute) will be investigated, and the relative *time* required for dialysis determined.

Safety Precautions	• Protective eyewear approved by your institution must be worn at all times while you are in the laboratory.
	• Silver nitrate will stain the skin if spilled. The stain will require several days to wear off. Nitrates are strong oxidizing agents. Wash your hands after handling.
	• Benedict's and Fehling's solutions are toxic. Wash your hands after handling them.
	• The I_2/KI solution may stain skin if spilled. The stay is not harmful but may require several days to wear off. The solution will also stain clothing.

Apparatus/Reagents Required

Cucumber slice, NaCl(*s*), cellophane dialysis tubing, paper punch, 5% NaCl, 25% NaCl, 5% glucose, 1% starch solution, 0.1 *M* silver nitrate, Benedict's or Fehling's reagents, 0.1 *M* I_2/KI reagent, 600-mL beakers, graduated cylinder

Procedure

1.   **Qualitative Demonstration of Osmosis**

Obtain a slice of fresh cucumber (or other freshly cut fruit) and place it on a watch glass. Obtain approximately a teaspoonful of table salt (NaCl) and sprinkle it on the cucumber slice. Observe the cucumber slice periodically during the remainder of the laboratory period. Describe and explain your observations.

2.   **Demonstration of Osmotic Pressure**

Cut two strips of cellophane dialysis tubing, each about 10–12 inches long. With a paper punch, make a small hole through each piece of tubing about 3/4 inch from each end. See Figure 29-1.

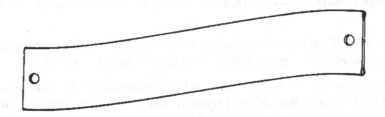

Figure 25-1. Preparation of dialysis tubing; Note the location of the punch holes.

Soak the strips of tubing in water until they become soft and the tubing walls can be spread apart.

Set up two 600-mL beakers in a location where they will not be disturbed. Place 450–500 mL of distilled water in each beaker.

Obtain 30–40 mL of 5% NaCl solution in a clean beaker, and then transfer exactly 25 mL of this solution to a graduated cylinder.

Holding one of the pieces of dialysis tubing by both ends to prevent spillage, transfer the 25 mL of 5% NaCl solution to the tubing. See Figure 25-2(a).

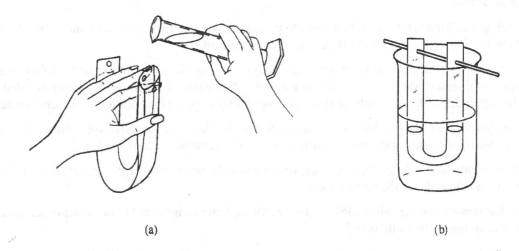

(a)   (b)

Figure 25-2. (a) Make a "sling" with a length of dialysis tubing containing the NaCl solution. (b) Suspend the solution in the tubing so that it lies beneath the water's surface in the beaker.

Insert a stirring rod through the holes in the end of the dialysis tubing, and suspend the tubing containing 5% NaCl solution in one of the beakers of distilled water. If the solution in the tubing is not covered by the water in the beaker, add additional water to the beaker. See Figure 25-2(b).

Repeat the process using the second length of dialysis tubing and the other beaker, using 25% NaCl in place of the 5% solution.

Allow the two beakers to stand for approximately one hour.

At the end of this time, carefully determine the volume of the solutions in the dialysis tubing lengths by pouring the solutions separately into a graduated cylinder.

Determine the volume of water that has entered each length of tubing by osmosis. Did more water diffuse into the concentrated or the dilute NaCl solution?

3.   **Dialysis**

In separate, clean beakers, obtain 15–18 mL each of 5% NaCl solution, 5% dextrose solution, and 1% starch.

Transfer about 1 mL of each solution to separate, clean test tubes. These test tube samples will be used for a *preliminary* test for each of the three solutes to be used for dialysis.

To the test tube containing 5% NaCl solution, add 5–6 drops of 0.1 M silver nitrate (*Caution!*). The appearance of a white precipitate of AgCl upon treatment with silver nitrate during the dialysis experiment will confirm the presence of chloride ion.

To the test tube containing starch, add 5–6 drops of 0.1 M I_2/KI reagent. The appearance of a blue-black color upon treatment with iodine reagent during the dialysis experiment will confirm the presence of starch.

To the test tube containing dextrose, add approximately 5 mL of Benedict's or Fehling's reagent, and heat briefly in a boiling-water bath. The appearance of a green or orange precipitate upon similar treatment will confirm the presence of dextrose during the dialysis.

Obtain a 10–12-inch length of dialysis tubing, and punch a hole about ¾ inch from either end. Soak the tubing in water until the walls can be spread apart.

Set up a 600-mL beaker containing 450–500 mL of distilled water in a place where the beaker will not be disturbed.

Holding the dialysis tubing by both ends to prevent spillage, transfer approximately 10 mL of the 5% NaCl, 5% dextrose, and 1% starch solutions to the tubing.

Insert a stirring rod through the holes in the dialysis tubing, rinse the outside of the tubing with a stream of distilled water from a wash bottle, and suspend the tubing in the beaker of distilled water [see Figure 25-2(b)]. Add additional water if necessary to cover the solution in the dialysis tubing.

Immediately after placing the dialysis tubing in the distilled water, remove approximately 3–4 mL of water from the beaker and perform the tests for NaCl, dextrose, and starch described earlier.

At *10-minute intervals* for the next hour, remove small portions of distilled water from the beaker and repeat the tests for the three solutes.

In what *order* were the solutes able to dialyze through the membrane? How is this order related to the size of the particles involved?

Name: _____ Section: _____

Lab Instructor: _____ Date: _____

EXPERIMENT 25

Colligative Properties 2: Osmosis and Dialysis

Pre-Laboratory Questions

1.   What is meant by a *semi-permeable* membrane?

2.   Osmosis and dialysis are often confused. Define each term and explain how these processes differ.

3.   A solution delivered into the body by intravenous infusion must be *isotonic* with the blood into which the solution is being dispensed. If the solution being injected is *hyper*tonic (has an osmotic pressure higher than that of blood), *crenation* of cells in the blood may take place. If the solution is *hypo*tonic compared to the blood, *hemolysis* of blood cells may take place. Use a scientific encyclopedia or online source to define *hemolysis* and *crenation*, and describe how these processes a result of a difference in osmotic pressure between the solution being injected and the blood.

4.   Ionic solutes produce more particles when dissolved in water than is indicated by their formal concentration. For example, when one mole of NaCl is dissolved in a large amount of water, effectively *two* moles of solute particles are introduced into the solvent. Saline solution for injection into the human body is approximately 0.9% NaCl, whereas the dextrose (glucose) solutions used for intravenous feeding are approximately 5% dextrose. Are these two solutions isotonic with each other?

Name: ______________________________ **Section:** __________

Lab Instructor: ______________________________ **Date:** __________

EXPERIMENT 25

Colligative Properties 2: Osmosis and Dialysis

Results/Observations

1. **Qualitative Demonstration of Osmosis: Observations**

2. **Demonstration of Osmotic Pressure**

   Volume of 5% sodium chloride solution taken ______________

   Volume measured after osmosis ______________

   Volume of water that passed through membrane ______________

   Volume of 25% sodium chloride solution taken ______________

   Volume measured after osmosis ______________

   Volume of water that passed through membrane ______________

   Into which NaCl solution was the osmosis stronger? ______________

   Why?

   ____

   ____

   ____

   ____

   ____

3. **Dialysis**

   Observation of $AgNO_3$ test for Cl^- ______________

   Observation of I_2 test for starch ______________

   Observation of Benedict's test ______________

After approximately what time intervals were the solutes detectable by dialysis from the tube into the beaker of water?

NaCl   _____

Dextrose   _____

Starch   _____

Questions

1.   Which solute dialyzed through the membrane faster, NaCl or dextrose? Why?

2.   If the experiment was performed correctly, you should *not* have been able to detect starch in the beaker even after a full hour. What is different about a starch "solution" that causes this?

3.   Benedict's Solution was used to detect the presence of dextrose. Use a scientific encyclopedia or online reference to discover the composition of Benedict's solution. What ion is responsible for the bright blue color of Benedict's solution? What is the identity of the red precipitate that forms if Benedict's reagent is added to a solution of a reducing sugar?

EXPERIMENT 26

Rates of Chemical Reactions

Objective

The study of the *speed* with which a chemical process takes place is crucial. Chemists and chemical engineers want reactions to take place quickly enough that they will be useful, but not so quickly that the reaction cannot be studied or controlled. Biologists use the study of reaction rates as an indication of the mechanism by which a biochemical process takes place. This experiment examines how varying the concentration of the reactants in a process affects the measured rate of the reaction. The effect of temperature in speeding up or slowing down reactions will also be examined briefly.

Introduction

The rate at which a chemical reaction occurs depends on several factors: the *nature* of the reaction, the *concentrations* of the reactants, the *temperature*, and the presence of possible *catalysts*. Each of these factors can influence markedly the observed speed of the reaction.

Some reactions at room temperature are very slow. For example, although wood is quickly oxidized in a fireplace at elevated temperatures, the oxidation of wood at room temperature is negligible. Many other reactions are essentially instantaneous. The precipitation of silver chloride when solutions containing silver ions and chloride ions are mixed is an extremely rapid reaction, for example.

For a given reaction, the rate typically *increases* with an increase in the concentrations of the reactants. The relation between rate and concentration is a remarkably simple one in many cases. For example, for the reaction

$$a\text{A} + b\text{B} \longrightarrow \text{products}$$

the rate can usually be expressed by the relationship

$$\text{Rate} = k\,[\text{A}]^m[\text{B}]^n$$

in which m and n are often small whole numbers. In this expression, called a **rate law**, [A] and [B] represent, respectively, the *concentration* of substances A and B; k is called the **specific rate constant** for the reaction (which provides the correct numerical proportionality). The exponents m and n are called the **orders** of the reaction with respect to the concentrations of substances A and B, respectively (and may bear no obvious relationship to the stoichiometric coefficients). For example, if $m = 1$, the reaction is said to be *first-order* with respect to the concentration of A. If $n = 2$, the reaction is *second-order* with respect to the concentration of B. The so-called *overall order* of the reaction is represented by the *sum* of the individual orders of reaction. For the examples just mentioned, the reaction would have overall order

$$1 + 2 = 3 \text{ (third order)}$$

The rate of a reaction is also significantly dependent on the *temperature* at which the reaction occurs. An increase in temperature increases the rate. A rule of thumb (which does have a theoretical basis) states that an increase in temperature of 10 Celsius degrees will double the rate of reaction. Although this rule is only approximate, it is clear that a rise in temperature of 100°C would affect the rate of reaction appreciably. As with concentration, there is a quantitative relationship between reaction rate and temperature, but here the relationship is less straightforward. The relationship is based on the idea that in order to react, the reactant species must possess a certain minimum amount of energy at the time the

253

appropriate reactant molecules or ions actually *collide* with each other during the rate-determining step of the reaction. This minimum amount of energy is called the **activation energy** for the reaction and generally reflects the kinetic energies of the molecules at the temperature of the experiment. The relationship between the specific rate constant (k) for the reaction, the absolute temperature (T), and the activation energy (E_a) is

$$\log_{10} k = -E_a/2.3RT + \text{constant}$$

In this relationship, R is the ideal gas constant, which has value $R = 8.31$ J/mol K. The equation therefore gives the activation energy, E_a, in units of *joules*. By experimentally determining k at various temperatures, the activation energy can be calculated from the *slope* of a plot of $\log_{10} k$ versus $1/T$. The slope of such a plot would be ($-E_a/2.3R$).

In this experiment, you will study a reaction called the "iodine clock." In this reaction, potassium iodate (KIO_3) and sodium hydrogen sulfite ($NaHSO_3$) react with each other, producing elemental iodine:

$$5HSO_3^- + 2IO_3^- \rightarrow I_2 + 5SO_4^{2-} + H_2O + 3H^+$$

This is an oxidation–reduction process, in which iodine(V) is reduced to iodine(0), and sulfur(IV) is oxidized to sulfur(VI). Because elemental iodine is *colored* (whereas the other species are colorless), the rate of reaction can be monitored simply by determining the *time required* for the appearance of the *color* of the iodine. As usual with reactions in which elemental iodine is produced, a small quantity of starch is added to heighten the color of the iodine. Starch forms an intensely colored blue-black complex with iodine. It would be difficult to detect the first appearance of iodine itself (since the solution would be colored only a very pale yellow), but if starch is present, the first few molecules of iodine produced will react with the starch present to give a much sharper color change.

The rate law for this reaction would be expected to have the general form

$$\text{Rate} = k\,[\,HSO_3^-\,]^m[\,IO_3^-\,]^n$$

in which m is the order of the reaction with respect to the concentration of bisulfite ion, and n is the order of the reaction with respect to the concentration of iodate ion. Notice that even though the stoichiometric coefficients of the reaction are known, these are *not* the exponents in the rate law. The order of the reaction must be determined *experimentally* and may bear no relationship to the stoichiometric coefficients of the balanced chemical equation. The rate law for a reaction reflects what happens in the slowest, or *rate-determining,* step of the reaction mechanism. A chemical reaction generally occurs as a *series* of discrete microscopic *steps*, called the **mechanism** of the reaction, in which only one or two molecules are involved at a time. For example, in the bisulfite/iodate reaction, it would be statistically almost impossible for five bisulfite ions and two iodate ions to all come together in the same place at the same time with the right orientations and energies for reaction. It is much more likely that one or two of these molecules will first interact with each other, forming some sort of *intermediate* perhaps, and that this intermediate will react with the rest of the ions at some later time. By careful experimental determination of the rate law for a process, information is obtained about exactly what molecules react during the slowest step in the reaction, and frequently this information can be extended to suggest what happens in all the various steps of the reaction's mechanism.

In this experiment, you will determine the order of the reaction with respect to the concentration of potassium iodate. You will perform several runs of the reaction, each time using the *same* concentration of all other reagents, but *varying* the concentration of potassium iodate in a systematic manner. By measuring the time required for reaction to occur with different concentrations of potassium iodate, and realizing that the time required for reaction is inversely proportional to the rate of the reaction, you will determine the exponent of iodate ion in the rate law.

Safety Precautions	• Protective eyewear approved by your institution must be worn at all times while you are in the laboratory.
	• Sodium hydrogen sulfite (sodium bisulfite) is harmful to the skin and releases noxious sulfur dioxide (SO_2) gas if acidified. Use with adequate ventilation in the room. Keep the solution covered with a watch glass when not in use.
	• Potassium iodate is a strong oxidizing agent and can damage skin. Wash after handling. Do not expose KIO_3 to any organic chemical substance or an uncontrolled oxidation may result.
	• Elemental iodine may stain the skin. The stains are generally not harmful at the concentrations used in this experiment, but will require several days to wear off. Iodine will stain clothing.

Apparatus/Reagents Required

Laboratory timer (or watch with second hand), Solution 1 (containing potassium iodate at 0.024 M concentration), Solution 2 (containing sodium hydrogen sulfite at 0.016 M concentration and starch)

Procedure

Record all data and observations directly in your notebook in ink.

1.   **Solutions to Be Studied**

Two solutions have been prepared for your use in this experiment. It is essential that the two solutions *not be mixed* in any way before the actual kinetic run is made. Be certain that graduated cylinders used in obtaining and transferring the solutions are *rinsed* with distilled water between solutions. Also rinse thermometers and stirring rods before transferring between solutions.

Solution 1 is 0.024 M potassium iodate. Solution 2 is a mixture containing two different solutes, sodium hydrogen sulfite and starch. Solution 2 has been prepared so that the solution contains hydrogen sulfite ion at 0.016 M concentration. The presence of starch in Solution 2 may make the mixture appear somewhat cloudy.

Obtain about 400 mL of Solution 1 in a clean, dry 600-mL beaker. Keep covered with a watch glass to minimize evaporation. Obtain about 150 mL of Solution 2 in a clean, dry 250-mL beaker. Cover with a watch glass.

2.   **Kinetic Runs**

Clean out several graduated cylinders and beakers for the reactions. It is important that Solutions 1 and 2 do not mix until the reaction time is to be measured. Use separate graduated cylinders for the measurement of each solution.

The general procedure for the kinetic runs is as follows (specific amounts of reagents to be used in the actual runs are given in a table further on in the procedure):

Measure out the appropriate amount of Solution 2 in a graduated cylinder. Take the temperature of Solution 2 while it is in the graduated cylinder, being sure to rinse and dry the thermometer to avoid mixing the solutions prematurely. Transfer the measured quantity of Solution 2 from the graduated cylinder to a clean, dry 250-mL beaker.

Measure out the appropriate amount of Solution 1 using a clean graduated cylinder, and transfer to a clean, dry small beaker. Again using a clean graduated cylinder, add the appropriate amount of distilled water for the run, and stir to mix. Measure the temperature of Solution 1, being sure to rinse and dry the thermometer to avoid mixing the solutions prematurely.

If the temperatures of Solutions 1 and 2 differ by more than one degree, wait until the two solutions come to the same temperature.

When the two solutions have come to the same temperature, prepare to mix them. Have ready a clean stirring rod for use after mixing the solutions.

Noting the time (to the nearest second), pour Solution 1 into the beaker containing Solution 2 and stir for 15–30 seconds. Watch the mixture carefully, and record the time the blue-black color of the starch/iodine mixture appears.

Repeat the same run, using the same amounts of Solutions 1 and 2, before going on to the next run.

The table that follows indicates the amounts of Solution 1, distilled water, and Solution 2 to be mixed for each run. Distilled water is added in varying amounts to keep the total volume the same for all the runs.

Table of Kinetic Runs

Run	mL Solution 1	mL distilled water	mL Solution 2
A	10.0	80.0	10.0
B	20.0	70.0	10.0
C	30.0	60.0	10.0
D	40.0	50.0	10.0
E	50.0	40.0	10.0

3. **Temperature Dependence**

A rule of thumb indicates that the rate of reaction is *doubled* for each 10°C temperature increase. Confirm this by performing two determinations of run A in the table at a temperature approximately 10°C *higher* than the temperature used previously, and then again at a temperature approximately 10°C *lower* than that used in the original determinations.

To perform the reactions at the higher temperature, set up a metal trough half filled with water on a hotplate. Adjust the controls of the hotplate so as to warm the water in the trough to approximately 10°C above room temperature. Once the temperature of the water in the trough has stabilized and no longer rises or decreases, add the two beakers of the reagents for run A, and allow them to stand in the warm water bath for 5 minutes so that they come to the temperature of the water bath. Then combine the reagents, keeping them still in the warm-water bath during the reaction. Record the time required for the iodine color to appear. Repeat the determination.

To perform the reactions at the lower temperature, set up a trough or basin half filled with water and add ice to the water in the trough/basin until the temperature is approximately 10°C below room temperature. Add ice as necessary to maintain the temperature as constant as possible at the lower temperature.

Once the temperature of the water in the trough has stabilized and no longer rises or decreases, add the two beakers of the reagents for run A, and allow the beakers to stand in the cold-water bath for 5 minutes. Add ice as needed to maintain the temperature of the cold-water bath.

Combine the reagents, keeping them still in the cold-water bath during the reaction, adding ice as needed to maintain the temperature. Record the time required for the iodine color to appear. Repeat the determination.

Does the "rule of thumb" seem to hold true?

4. **Graphs**

Data for studies of reaction rates are often treated graphically. Depending on what is being calculated, different sorts of graphs might be used. Generally the graph is set up so that the slope of the graph represents the quantity being determined. In this experiment, we have studied the rate law

$$\text{Rate} = k\,[\,HSO_3^-\,]^m[\,IO_3^-\,]^n$$

In each of the reaction runs, we kept the concentration of bisulfite ion the same each time, so that we could isolate the effect of the iodate ion concentration on the rate of reaction. We can write a modified rate law to describe that situation

$$\text{Rate} = k\,[\text{constant}]^m[\,IO_3^-\,]^n = k'[\,IO_3^-\,]^n$$

Remember that the rate represents the change in concentration with time, that is

$$Rate = -\frac{\Delta[IO_3^-]}{\Delta t} \quad \text{(or in terms of calculus)}, \quad Rate = -\frac{d[IO_3^-]}{dt}$$

Combining expressions, we can say

$$\frac{\Delta[IO_3^-]}{\Delta t} = k'[\,IO_3^-\,]^n$$

Using calculus, this equation can be expressed in a different form called an "integrated" rate law ("ln" indicates the *natural* logarithm, base e).

$$\ln[IO_3^-] = -k't + \ln[IO_3^-]_0 = -kt + \text{constant}$$

a. Construct a plot of *time required for reaction* (vertical axis) versus *concentration of iodate ion* (horizontal axis).

b. Construct a second plot, in which you plot the *reciprocal* of the time required for reaction on the vertical axis versus the concentration of iodate ion. Why is the reciprocal graph a straight line?

c. Construct a third plot, in which you plot the natural logarithm of the concentration of iodate ion, $\ln[IO_3^-]$ versus the time required for the reaction. Why is the logarithm graph a straight line? What does the slope of this line represent?

EXPERIMENT 26

Rates of Chemical Reactions

Pre-Laboratory Questions

1. Given the following data, determine the *orders* with respect to the concentrations of substances A and B in the reaction. Explain your reasoning

   A + B ⟶ products

$[A]_{initial}$	$[B]_{initial}$	Time for reaction
0.30 *M*	0.10 *M*	87 s
0.20 *M*	0.20 *M*	66 s
0.10 *M*	0.20 *M*	131 s
0.20 *M*	0.20 *M*	66 s
0.10 *M*	0.20 *M*	131 s

2. What must be the *units* of the specific rate constant in Pre-Laboratory Question 1, given that the rate of reaction would be measured in *M*/s? Explain how you arrive at your answer.

3.   What is the *mechanism* of a chemical reaction? How can a study of reaction rates be used to elucidate the mechanism of a chemical reaction?

4.   In order for two molecules to react with one another, the molecules must (1) collide with each other, (2) collide with enough kinetic energy to react with each other, and (3) collide with the proper spatial orientation for reaction. Generally, raising the temperature of a reaction increases the rate of reaction. Which factor listed above is most likely to be affected by raising the temperature at which the reaction is performed? Explain.

EXPERIMENT 26

Rates of Chemical Reactions

Results/Observations

Kinetic Runs

Time required for I_2 color to appear	*First trial*	*Second trial*
Run A (10 mL Solution 1)	_____	_____
Run B (20 mL Solution 1)	_____	_____
Run C (30 mL Solution 1)	_____	_____
Run D (40 mL Solution 1)	_____	_____
Run E (50 mL Solution 1)	_____	_____

Judging on the basis of your results, what is the *order* of the reaction with respect to potassium iodate concentration? Explain your reasoning.

Temperature Dependence

What higher temperature did you use? _____

What times were required for reaction? First trial _____ Second Trial _____

What lower temperature did you use? _____

What times were required for reaction? First trial _____ Second Trial _____

Do these times confirm the rule of thumb concerning the effect of temperature? Explain.

Questions

1.   Why was it necessary in all the kinetic runs to keep constant the total volume of the reagents after mixing, (that is, why was it necessary to add distilled water in inverse proportion to the quantity of Solution 1 that was required?)

2.   In this experiment you determined the dependence of the reaction rate on the concentration of potassium iodate. Devise an experiment for determining the dependence of the rate on the concentration of sodium sulfite.

3.   In this experiment you prepared three graphs using your experimental data. Some of these graphs should have produced straight lines. Explain in terms of the rate law for the reaction why these graphs produced straight lines.

Chemical Equilibrium 1:
Titrimetric Determination of an Equilibrium Constant

Objective

Many chemical reactions, especially those of organic substances, do *not* go to *completion*. Rather, they come to a point of **chemical equilibrium** before the reactants are fully converted to products. At the point of equilibrium, the concentrations of all reactants remain *constant* with time. The *position* of this equilibrium is described by a function called the **equilibrium constant**, K_{eq}, which is a *ratio* of the amount of product present to the amount of reactant remaining once the point of equilibrium has been reached. In this experiment, you will determine the equilibrium constant for an esterification reaction.

Introduction

Early in the study of chemical reactions, it was noted that many chemical reactions do not produce as much product as might be expected, based on the amounts of reactants taken originally. These reactions appeared to have *stopped* before the reaction was complete. Closer examination of these systems (after the reaction had seemed to stop) indicated that there were still significant amounts of all the original *reactants* present. Quite naturally, chemists wondered why the reaction had seemed to stop, when all the necessary ingredients for *further* reaction were still present.

Some reactions appear to stop because the products produced by the original reaction *themselves begin to react*, in the reverse direction to the original process. As the concentration of products begins to build up, product molecules will react more and more frequently. Eventually, as the speed of the forward reaction decreases while the speed of the reverse reaction increases, the forward and reverse processes will be going on at exactly the *same* rate. Once the forward and reverse rates of reaction are identical, there can be no further net change in the concentrations of any of the species in the system. At this point, a dynamic state of *equilibrium* has been reached. The original reaction is still taking place but is opposed by the reverse of the original reaction also taking place.

The point of chemical equilibrium for a reaction is most commonly described numerically by the **equilibrium constant**, K_{eq}. The equilibrium constant represents a *ratio* of the concentrations of all product substances present at the point of equilibrium to the concentrations of all original reactant substances (all concentrations are raised to the power indicated by the coefficient of each substance in the balanced chemical equation for the reaction).

For example, for the general reaction

$$aA + bB \rightleftarrows cC + dD \qquad \text{(Equation 27-1)}$$

the equilibrium constant would have the form

$$K_{eq} = \frac{[C]^c[D]^d}{[A]^a[B]^b}$$

A ratio is used to describe the point of equilibrium for a particular chemical reaction scheme because such a ratio is independent of the specific amounts of substance that were used initially in a particular experiment. The equilibrium constant K_{eq} is a constant for a given reaction at a given temperature.

In this experiment, you will determine the equilibrium constant for the esterification reaction between *n*-propyl alcohol and acetic acid.

$$CH_3COOH \quad + \quad CH_3CH_2CH_2OH \rightleftarrows CH_3COOCH_2CH_2CH_3 \quad + \quad H_2O$$

Acetic acid	*n*-propanol	*n*-propyl acetate	water
HOAc	PrOH	PrOAC	H_2O

Using the abbreviations indicated for each of the substances in this system, and noting that the values if *a*, *b*, *c*, and *d* in *Equation 27-1* are all 1, the form of the equilibrium constant would be

$$K_{eq} = \frac{[PrOAc][H_2O]}{[HOAc][PrOH]}$$

You will set up the reaction in such a way that the *initial* concentrations of HOAc and PrOH are *known*. The reaction will then be allowed to stand for one week to come to equilibrium. As HOAc reacts with PrOH, the acidity of the mixture will *decrease*, reaching a minimum once the system reaches equilibrium. The quantity of acid present in the system will be determined by titration with standard sodium hydroxide solution. Esterification reactions are typically catalyzed by the addition of a strong mineral acid. In this experiment a small amount of sulfuric acid will be added as a catalyst. The amount of catalyst added will have to be determined, because this amount of catalyst will also be present in the equilibrium mixture and will contribute to the total acid content of the equilibrium mixture.

By analyzing the amounts of each reagent used this week in setting up the reaction, and by determining the amount of acetic acid (HOAc) that will be present next week once the system has reached equilibrium, you will be able to calculate the concentration of all species present in the equilibrium mixture. From this, the value of the equilibrium constant can be determined.

The concentration of acetic acid in the mixture is determined by the technique of *titration*. Acetic acid reacts with sodium hydroxide (NaOH) on a 1:1 stoichiometric basis:

$$HOAc + NaOH \longrightarrow NaOAc + H_2O$$

A precise volume of the reaction mixture is removed with a pipet, and a standard NaOH solution is added slowly from a buret until the acetic acid has been completely neutralized (this is signaled by an indicator, which changes color). From the volume and concentration of the NaOH used and the volume of reaction mixture taken, the concentration of acetic acid in the reaction mixture may be calculated.

Safety Precautions	• Protective eyewear approved by your institution must be worn at all times while you are in the laboratory.
	• Glacial (pure) acetic acid and sulfuric acid burn skin badly if spilled. Wash immediately if either acid is spilled, and inform the instructor at once.
	• Acetic acid, n-propyl alcohol, and n-propyl acetate are all highly flammable, and their vapors may be toxic or irritating if inhaled. Absolutely no flames are permitted in the laboratory. The laboratory must be well ventilated.
	• The NaOH solution used in this experiment is very dilute, but will concentrate by evaporation on the skin if spilled. Was after use.
	• When pipetting solutions, use a rubber safety bulb. Do *not* pipet by mouth.

Apparatus/Reagents Required

50-mL buret, plastic wrap, standard 0.10 M NaOH solution, phenolphthalein indicator solution, 1-mL pipet and rubber safety bulb, n-propyl alcohol (1-propanol), glacial acetic acid, 6 M sulfuric acid

Procedure

Record all data and observations directly in your notebook in ink.

1.   **First Week: Set-up of the Initial Reaction Mixture**

Clean a buret with soap and water until water does not bead up on the inside of the buret. Clean a 1-mL volumetric transfer pipet. Rinse the buret and pipet with tap water several times to remove all soap. Follow by rinsing with small portions of distilled water.

Obtain approximately 100 mL of standard 0.10 M NaOH solution in a clean dry beaker. Rinse the buret several times with small portions of NaOH solution (discard the rinsings); then fill the buret with NaOH solution. Keep the remainder of the NaOH solution in the beaker covered until it is needed.

Clean two 250-mL Erlenmeyer flasks for use as titration vessels. Label the flasks as 1 and 2. Rinse the flasks with tap water; follow with small portions of distilled water. Place approximately 25 mL of distilled water in each flask, and set aside until needed.

Since glacial acetic acid and n-propyl alcohol are both liquids, it is more convenient to measure them out by *volume* than by mass.

Clean and dry a 125-mL Erlenmeyer flask. Label the flask as *reaction mixture*. Cover with plastic wrap a rubber stopper that securely fits the flask (this prevents the stopper from being attacked by the vapors of the reaction mixture).

Clean and dry a small graduated cylinder. Using the graduate, obtain 14 ± 0.2 mL of glacial acetic acid (0.25 mol) and transfer the acetic acid to the clean, dry reaction mixture Erlenmeyer flask.

Rinse the graduated cylinder with water and re-dry. Obtain 19 ± 0.2 mL of n-propyl alcohol (0.25 mol) and add it to the acetic acid in the reaction mixture Erlenmeyer flask. Stopper the flask and swirl the flask for several minutes to mix the reagents.

Using the 1-mL volumetric pipet and safety bulb, transfer exactly 1.00 mL of the reaction mixture to each of the two 250-mL Erlenmeyer flasks (1 and 2). Re-stopper the flask containing the n-propyl alcohol/acetic acid reaction mixture to prevent evaporation,

Add 3–4 drops of phenolphthalein indicator to each of the two samples to be titrated.

Record the initial level of the NaOH solution in the buret. Place Erlenmeyer flask 1 under the tip of the buret, and slowly begin adding NaOH solution to the sample. Swirl the flask during the addition of NaOH. As NaOH is added, red streaks will begin to appear in the sample because of the phenolphthalein, but the red streaks will disappear as the flask is swirled. The endpoint of the titration is when a single additional drop of NaOH causes a faint, permanent pink color to appear. Record the level of NaOH in the buret.

Repeat the titration using Erlenmeyer flask 2, recording initial and final volumes of NaOH.

Discard the samples in flasks 1 and 2.

From the average volume of NaOH used to titrate 1.00 mL of the reaction mixture and the concentration of the NaOH, calculate the concentration (in mol/L) of acetic acid in the reaction mixture:

moles of NaOH used = (concentration of NaOH, M) × (volume used to titrate, L)

moles of HAc = moles of NaOH at the color change of the indicator

$$molarity \ of \ HOAc = \frac{moles \ of \ HOAc}{liters \ of \ reaction \ mixture \ taken \ with \ pipet}$$

Because the reaction was begun using *equal molar amounts* of acetic acid and *n*-propyl alcohol (i.e., 0.25 mol of each), the concentration of acetic acid calculated also represents the initial concentration of *n*-propyl alcohol in the original mixture.

2.   **First Week: Determination of Sulfuric Acid Catalyst**

The reaction between *n*-propyl alcohol and acetic acid is slow unless the reaction is catalyzed. Mineral acids speed up the reaction considerably, but the presence of the mineral acid catalyst must be considered in determining the remaining concentration of acetic acid in the system once equilibrium has been reached. Next week, you will again titrate 1.00-mL samples of the reaction mixture with NaOH to determine what concentration of acetic acid remains in the mixture at equilibrium. However, because NaOH reacts with both the acetic acid of the reaction and with the mineral acid catalyst, some method must be found to determine the concentration of the mineral acid in the reaction mixture. Sulfuric acid will be used as the catalyst.

Refill the buret (if needed) with standard NaOH and record the initial level. Clean out Erlenmeyer flasks 1 and 2, rinse, and fill with approximately 25 mL of distilled water. Clean out and have handy the 1-mL pipet and rubber safety bulb.

Add, with swirling, 10 drops of 6 *M* sulfuric acid catalyst to the acetic acid/propyl alcohol reaction mixture. *Immediately* pipet a 1.00-mL sample of the catalyzed reaction mixture into both flasks. Do not delay pipetting, or the concentration of acetic acid will begin to change as the reaction occurs.

Recording initial and final NaOH levels in the buret, titrate the catalyzed reaction mixture in flasks 1 and 2, using 3–4 drops of phenolphthalein indicator to signal the endpoint.

Since the samples of catalyzed reaction mixture contain the same quantity of acetic acid as the samples of uncatalyzed mixture, the increase in volume of NaOH required to titrate the second set of 1-mL samples represents a measure of the amount of sulfuric acid present.

By subtracting the average volume of standard NaOH used in Part A (acetic acid only) from the average volume of NaOH used in Part B (acetic acid + sulfuric acid), calculate how many milliliters of the standard NaOH solution are required to titrate the sulfuric acid catalyst present in 1 mL of the reaction mixture. This volume represents a correction that can be applied to the volume of NaOH that will be required to titrate the samples next week, after equilibrium has been reached.

Stopper the 125-mL flask containing the acetic acid/*n*-propyl alcohol mixture. Place the reaction mixture in your locker in a safe place until next week.

3.   **Second Week: Determination of the Equilibrium Mixture**

After standing for a week, the reaction system of *n*-propyl alcohol and acetic acid will have come to equilibrium.

Clean a buret and 1-mL pipet. Rinse the buret and fill it with the standard 0.10 *M* NaOH solution.

Clean and rinse two 250-mL Erlenmeyer flasks (samples 1 and 2). Place approximately 25 mL of distilled water in each of the Erlenmeyer flasks.

Uncover the acetic acid/*n*-propyl alcohol reaction mixture. Using the rubber safety bulb, pipet 1.00-mL samples into each of the two Erlenmeyer flasks. You may notice that the *odor* of the reaction mixture has changed markedly from the sharp vinegar odor of acetic acid that was present last week.

Add 3–4 drops of phenolphthalein to each sample, and titrate the samples to the pale pink endpoint with the standard NaOH solution; record the initial and final levels of NaOH in the buret.

Calculate the mean volume of standard NaOH required to titrate 1.00 mL of the equilibrium mixture. Using the volume of NaOH required to titrate the sulfuric acid catalyst present in the mixture (from Part B), calculate the volume of NaOH that was used in titrating the acetic acid component remaining in the equilibrium mixture.

From the volume and concentration of NaOH used to titrate the acetic acid in 1.00 mL of the equilibrium mixture, calculate the concentration of acetic acid in the equilibrium mixture in moles per liter. Since the reaction was begun with equal molar amounts of acetic acid and *n*-propyl alcohol (0.25 mol of each), the concentration of *n*-propyl alcohol in the equilibrium mixture is the same as that calculated for acetic acid.

Calculate the *change* in the concentration of acetic acid between the initial and the equilibrium mixtures. From the change in the concentration of acetic acid, calculate the concentrations of the two products of the reaction (*n*-propyl acetate, water) present in the equilibrium mixture.

From the concentrations of each of the four components of the system at equilibrium, calculate the equilibrium constant for the reaction.

$$K_{eq} = \frac{[PrOAc][H_2O]}{[HOAc][PrOH]}$$

Name: _____ Section: _____

Lab Instructor: _____ Date: _____

EXPERIMENT 27

Chemical Equilibrium 1:
Titrimetric Determination of an Equlibrium Constant

Pre-Laboratory Questions

1. What is meant by *chemical equilibrium*? What characterizes a state of chemical equilibrium?

2. Write the expression for the equilibrium constants for the following reactions:

    a. $2NO_2(g) \rightleftarrows N_2O_4(g)$ _____

    b. $H_2O(g) + CO(g) \rightleftarrows H_2(g) + CO_2(g)$ _____

    c. $CO(g) + 3H_2(g) \rightleftarrows CH_4(g) + H_2O(g)$ _____

    d. $COCl_2(g) \rightleftarrows CO(g) + Cl_2(g)$ _____

3. A 1.00-mL pipet sample of dilute acetic acid required 24.2 mL of 0.10 M NaOH to titrate the sample to a phenolphthalein endpoint. Calculate the concentration of acetic acid in the sample. How many significant figures are justified in your answer? Why?

4. During the first part of this experiment, you are instructed to titrate samples both of the n-propyl alcohol/acetic acid mixture, and also of the n-propyl alcohol/acetic acid mixture after the sulfuric acid catalyst had been added. Explain why this second titration is necessary.

Name: _____ Section: _____

Lab Instructor: _____ Date: _____

EXPERIMENT 27

Chemical Equilibrium 1:
Titrimetric Determination of an Equilibrium Constant

Results/Observations

Concentration of standard NaOH solution

	Sample 1	*Sample 2*
Volume of NaOH to titrate 1 mL initial uncatalyzed mixture (*first week*)	_____	_____
Mean volume	_____	
Concentration of acetic acid, *M*, in original mixture	_____	_____
Volume of NaOH to titrate 1 mL catalyzed reaction mixture (*first week*)	_____	_____
Mean volume	_____	
Volume correction for sulfuric acid (*to be applied next week*)		_____
Volume of NaOH to titrate 1 mL of equilibrium mixture (*second week*)	_____	_____
Mean volume	_____	
Volume (*corrected*) of NaOH to titrate acetic acid in equilibrium mixture	_____	_____
Concentration of acetic acid, *M*, in equilibrium mixture	_____	_____
Change in concentration of acetic acid in reaching equilibrium	_____	_____

Complete the following table, indicating the concentration of each substance in mol/L:

	Initial mixture	*Equilibrium mixture*
Acetic acid	_____	_____
n-Propyl alcohol	_____	_____
n-Propyl acetate	_____	_____
Water	_____	_____

Calculate the equilibrium constant for the reaction.

Questions

1.   Sulfuric acid was used as a catalyst in this reaction. Would the presence of a catalyst affect the position of the equilibrium (i.e., the relative amounts of substances present once equilibrium was reached)? Why? Could some other acid have been used as the catalyst? Why?

2.   Write equilibrium constant expressions for each of the following equilibrium chemical reactions:

   a. $O_2(g) + ClNO(g) \rightleftarrows ClNO_2(g) + NO(g)$ _____

   b. $2NaHCO_3(s) \rightleftarrows Na_2CO_3(s) + CO_2(g) + H_2O(g)$ _____

   c. $CaO(s) + CO_2(g) \rightleftarrows CaCO_3(s)$ _____

   d. $SnO_2(s) + 2H_2(g) \rightleftarrows Sn(s) + 2H_2O(g)$ _____

Chemical Equilibrium 2: Spectrophotometric Determination of an Equilibrium Constant

Objective

Many chemical reactions, especially those of organic substances, do *not* go to *completion*. Rather, they come to a point of **chemical equilibrium** before the reactants are fully converted to products. The *position* of this equilibrium is described by a function called the **equilibrium constant**, K_{eq}. In this experiment you will determine the equilibrium constant for the complexation reaction of iron(III) with thiocyanate ion using a spectrophotometric method.

Introduction

Early in the study of chemical reactions, it was noted that many chemical reactions do not produce as much product as might be expected, based on the amounts of reactants taken originally. These reactions appeared to have *stopped* before the reaction was complete. Closer examination of these systems (after the reaction had seemed to stop) indicated that there were still significant amounts of all the original *reactants* present. Quite naturally, chemists wondered why the reaction had seemed to stop, when all the necessary ingredients for *further* reaction were still present.

Some reactions appear to stop because the products produced by the original reaction *themselves begin to react*, in the reverse direction to the original process. As the concentration of products begins to build up, product molecules will react more and more frequently. Eventually, as the speed of the forward reaction decreases while the speed of the reverse reaction increases, the forward and reverse processes will be going on at exactly the *same* rate. Once the forward and reverse rates of reaction are identical, there can be no further net change in the concentrations of any of the species in the system. At this point, a dynamic state of *equilibrium* has been reached. The original reaction is still taking place but is opposed by the reverse of the original reaction also taking place.

The point of chemical equilibrium for a reaction is most commonly described numerically by the **equilibrium constant**, K_{eq}. The equilibrium constant represents a *ratio* of the concentrations of all product substances present at the point of equilibrium to the concentrations of all original reactant substances (all concentrations are raised to the power indicated by the coefficient of each substance in the balanced chemical equation for the reaction).

For example, for the general reaction

$$aA + bB \rightleftarrows cC + dD \qquad \text{(\textit{Equation 28-1})}$$

the equilibrium constant would have the form

$$K_{eq} = \frac{[C]^c[D]^d}{[A]^a[B]^b}$$

A ratio is used to describe the point of equilibrium for a particular chemical reaction scheme because such a ratio is independent of the specific amounts of substance that were used initially in a particular experiment. The equilibrium constant K_{eq} is a constant for a given reaction at a given temperature.

Iron(III) and the thiocyanate ion (SCN^-) react with each other to produce an intensely red-colored complex product. The reaction is an equilibrium process:

$$Fe^{3+}(aq) + SCN^-(aq) \rightleftharpoons [FeSCN^{2+}(aq)]$$

Since the complex product is intensely colored, its concentration is conveniently measured with a *spectrophotometer*. Noting that the values if a, b, c, and d in *Equation 28-1* are all 1, the form of the equilibrium constant would be he equilibrium constant expression for this reaction is

$$K_{eq} = \frac{[FeSCN^{2+}]}{[Fe^{3+}][SCN^-]}$$

You will first prepare a calibration curve relating the absorbance of a sample to the concentration of the colored [$FeSCN^{2+}$] species. To do this, five solutions will be prepared in which there is a very large *excess* of iron(III) compared to thiocyanate. According to Le Châtelier's principle, the relatively small amount of thiocyanate should effectively be converted completely to the colored product.

Most spectrophotometers measure either absorbance (A) or percent transmittance ($\%T$). Since you will be applying the Beer-Lambert law, which states that absorbance, is directly proportional to molar concentration, you will measure the absorbance of each solution, and the calibration curve will then be plotted as "absorbance" versus concentration of the complex, [$FeSCN^{2+}$]. This calibration curve can then be used to determine the concentration of [$FeSCN^{2+}$] in any solution.

You will then prepare several equilibrium mixtures in which the initial concentrations of iron(III) and thiocyanate are *comparable*, but slightly *different* in each mixture. The spectrophotometer will be used to measure the absorbance of each mixture, and will use your calibration curve to determine the concentration of the colored product in each mixture. From the *initial* concentrations of the *reactants* Fe^{3+} and SCN^- taken in each mixture and the concentration of the *product* present *at equilibrium* in each mixture, you will be able to calculate the concentration of Fe^{3+} and SCN^- present in each mixture at equilibrium and the equilibrium constant for the reaction.

Safety Precautions	• **Protective eyewear approved by your institution must be worn at all times while you are in the laboratory.**
	• **The iron(III) and thiocyanate solutions also contain nitric acid as a preservative, which may make them corrosive to eyes, skin, and clothing. Wash immediately if any spills occur, and clean up all spills on the benchtop.**
	• **When pipetting solutions, use a rubber safety bulb. Do not pipet by mouth.**

Apparatus/Reagents Required

2.00×10^{-3} M iron(III) nitrate solution, 0.200 M iron(III) nitrate solution, 2.00×10^{-3} M potassium thiocyanate solution, 18×150-mm test tubes, spectrophotometer and sample cell (cuvet), 10-mL Mohr pipet and rubber safety bulb, china marker or marking pen, buret

Procedure

1. Preparation of the Calibration Curve

The table below provides the volumes of reactants needed to prepare the standard solutions. Notice that the concentration of the iron solution is much greater than that of the KSCN solution. This is to ensure that all of the KSCN is used up in the reaction. The concentration of the product will be determined from the volume and concentration of the KSCN used in each trial.

Set up 5 small beakers on the bench top and label them as 1 through 5.

A 10-mL Mohr pipet will be used to dispense the required amounts of iron(III) and thiocyanate solution. Clean out the pipet with soap and water until water runs in sheets down the inside of the pipet. a 50-mL buret will be used to dispense water. Clean the buret with soap and water until water runs in sheets down the inside of the buret barrel.

Rinse the 10-mL Mohr pipet with small portions of the Fe^{3+} solution (0.200 M iron(III) nitrate solution). Refer to the following table, and pipet the indicated amount of Fe^{3+} solution into each beaker. Use a rubber safety bulb at all times to provide the suction force for the pipet.

Rinse the pipet with water, and then with small portions of the SCN^- solution (2.00×10^{-3} M potassium thiocyanate solution). Refer to the following table, and pipet the indicated amount of SCN^- solution into each beaker.

Use a buret to dispense into each beaker the required amount of water. Stir the contents of each beaker thoroughly.

Calibration Solutions

Solution	0.00200 M KSCN	0.200 M Fe(NO$_3$)$_3$	water
1	5.0 mL	5.0 mL	15.0 mL
2	4.0 mL	5.0 mL	16.0 mL
3	3.0 mL	5.0 mL	17.0 mL
4	2.0 mL	5.0 mL	18.0 mL
5	1.0 mL	5.0 mL	19.0 mL

Your instructor will demonstrate the use of the spectrophotometer. Turn on the spectrophotometer and allow it to warm up. Set the wavelength to 450 nm. Set the zero of the machine with an empty chamber (no cuvet, no solution). Use a cuvet filled with water as the blank to set the 100% of the machine. Follow your instructor's directions to measure the % Transmittance values for each of the solutions prepared above. Record all the values in your data table.

Although spectrophotometers measure directly the fraction of light transmitted by a colored solution (% T), the *absorbance* of a colored solution is more commonly used in calculations because the *absorbance* of a colored solution is directly proportional to the concentration of the colored species in the solution (Beer's Law). As noted earlier, most modern spectrophotometers can be read directly in terms of absorbance. If your instrument does not offer that option, the absorbance of a colored solution can be calculated as the negative of the logarithm of the transmittance of the solution:

$$A = -\log T = -\log (\%T/100) = -\log \%T - \log (1/100) = -\log \%T - (-2) = 2 - \log \%T$$

Construct a calibration curve by plotting **absorbance** on the y-axis and **concentration of** $[Fe(SCN)^{2+}]$ on the x-axis. Draw a best-fit line though the data points. This graph will be used to determine the amount of product formed in the equilibrium mixtures.

2.   Preparation of the Equilibrium Mixtures

Clean and dry five 18×150-mm test tubes and number them (with a china marker or permanent-ink marking pen) as 1 through 5.

Rinse the 10-mL Mohr pipet with small portions of the second and more dilute Fe^{3+} solution (2.00×10^{-3} M iron(III) nitrate solution). Refer to the following table, and pipet the indicated volume of Fe^{3+} solution into each test tube. Use a rubber safety bulb at all times to provide the suction force for the pipet.

Rinse the pipet several times with distilled water, and then with small portions of the SCN^- solution (2.00×10^{-3} M potassium thiocyanate solution). Refer to the following table, and pipet the indicated amount of SCN^- solution into each test tube.

Rinse the pipet with distilled water. Refer to the table below, and pipet the indicated amount of distilled water into each test tube. Stir each test tube's contents with a clean stirring rod.

Equilibrium Mixtures

Test tube	mL 0.00200 M Fe(NO₃)₃	mL 0.00200 M KSCN	mL water
1	5.00	1.50	3.50
2	5.00	2.00	3.00
3	5.00	2.50	2.50
4	5.00	3.00	2.00
5	5.00	3.50	1.50

Determine the absorbance for each of your five reaction mixtures at 450 nm, making your determination to the nearest 0.1 unit. Record on the data page.

Calculate the absorbance of each of your solutions and record

3.   Calculations

Use your calibration curve to determine the concentration of the colored iron(III)/ thiocyanate complex in each of your five equilibrium reaction mixtures from Part 2. To do this, find the *absorbance* of a given solution on the vertical axis of your graph, go across to the line you have drawn for the data points in the table, and then go down to the concentration scale to find the *concentration* of colored product in the solution.

From the volumes and concentrations of iron(III), thiocyanate ion, and water used to prepare the five equilibrium mixtures (Part 2, above), calculates the *initial* concentrations of Fe^{3+} and SCN^- in each mixture.

> **Example:** Consider equilibrium mixture 1. The mixture was prepared using 5.00 mL of 0.000200 M iron(III) nitrate solution, to which was added a total of 5.00 mL of other reagents/solvent, making the final volume of the mixture 10.00 mL.

$$Initial\ concentration\ of\ iron(III) = \frac{5.00\ mL \times 0.00200\ M}{10.00\ mL} = 0.00100\ M$$

From these initial concentrations, and from the concentration of product present in each mixture *at equilibrium* (as determined from the calibration curve), calculate the concentrations of Fe^{3+} and SCN^- present in each mixture *at equilibrium*. Using the calculated equilibrium concentrations of Fe^{3+}, SCN^- and $FeSCN^{2+}$, calculate the value for the *equilibrium constant* for each mixture in Part 2. Calculate the average value and average deviation for the equilibrium constant.

Name: _____ Section: _____

Lab Instructor: _____ Date: _____

EXPERIMENT 28

Chemical Equilibrium 2: Spectrophotometric Determination of an Equilibrium Constant

Pre-Laboratory Questions

1. A colored solution gives a reading of 39.2% transmittance when determined in a calibrated spectrophotometer. Calculate the *absorbance* of the solution.

2. What is meant by a *calibration curve* for a colored substance being studied with a spectrophotometer? How is a calibration curve used to determine the concentration of a solution of unknown concentration of the colored substance?

3. Write expressions for the equilibrium constant for each of the following reactions:

    a. $H_2O(g) + CO(g) \rightleftarrows H_2(g) + CO_2(g)$ _____

    b. $3H_2(g) + N_2(g) \rightleftarrows 2NH_3(g)$ _____

    c. $2NO_2(g) \rightleftarrows N_2O_4(g)$ _____

    d. $CO(g) + 3H_2(g) \rightleftarrows CH_4(g) + H_2O(g)$ _____

4. What is the new concentration of Fe^{3+} if 3.00 mL of 0.00200 M iron(III) nitrate is diluted to a total volume of 10.00 mL?

5. Use a scientific encyclopedia or online reference to provide definitions of the following terms used in discussing spectrophotometers.

    White light source

    Wavelength

    Percent Transmittance

    Absorbance

Name: _____ Section: _____

Lab Instructor: _____ Date: _____

EXPERIMENT 28

Chemical Equilibrium 2: Spectrophotometric Determination of an Equilibrium Constant

Results/Observations

1. **Preparation of the Calibration Curve**

Test tube	%transmittance	absorbance	concentration of product
1	_____	_____	_____
2	_____	_____	_____
3	_____	_____	_____
4	_____	_____	_____
5	_____	_____	_____

2. **Preparation of the Equilibrium Mixtures**

Test tube	initial $[Fe^{3+}]$	initial $[SCN^-]$	equilib. $[Fe^{3+}]$	equilib. $[SCN^-]$
1	_____	_____	_____	_____
2	_____	_____	_____	_____
3	_____	_____	_____	_____
4	_____	_____	_____	_____
5	_____	_____	_____	_____

Calculated values of K_{eq}:

1 _____

2 _____

3 _____

4 _____

5 _____

Attach your calibration curve to your lab report.

Questions

1. Using the five values that you determined for the equilibrium constant, calculate the average value of the equilibrium constant and the average deviation. Given that the value of the iron/thiocyanate equilibrium constant is on the order of 9×10^2, calculate the percent error in your determination.

2. In spite of your five solutions having been prepared with different initial amounts of Fe^{3+} and SCN^-, the five values of the equilibrium constant calculated should have agreed with one another to at least the power of ten. What errors might account for any differences in the values of the constant?

3. For your data for test tube 3, show in detail how you calculated the initial and equilibrium concentrations of iron(III).

Chemical Equilibrium 3:
Stresses Applied to Equilibrium Systems

Objective

Le Châtelier's principle states that if we *disturb* a system that is already in equilibrium, then the system will *react* so as to minimize the effect of the disturbance. In this experiment, we will look at how changes made to systems at equilibrium will influence the position of the equilibrium.

Introduction

Many reactions come to equilibrium. The reactions in the previous two experiments seemed to have stopped before the full amount of product expected had been formed. When equilibrium had been reached in these systems, there were significant amounts of both products and reactants still present. In this experiment, you will study changes made in a system *already in equilibrium* affect the position of the equilibrium.

Le Châtelier's principle states that if we *disturb* a system that is already in equilibrium, then the system will *react* so as to minimize the effect of the disturbance. This is most easily demonstrated in cases where additional reagent is added to a system in equilibrium, or when one of the reagents is removed from the system in equilibrium.

Solubility Equilibria: Suppose we have a solution that has been saturated with a solute. This means that the solution has already dissolved as much solute as possible. If we try to dissolve additional solute, no more will dissolve, because the saturated solution is in equilibrium with the solute:

$$\text{Solute} + \text{Solvent} \rightleftarrows \text{Solution}$$

Le Châtelier's principle is most easily seen when an ionic solute is used. Suppose we have a saturated solution of sodium chloride, NaCl. Then

$$\text{NaCl}(s) \rightleftarrows \text{Na}^+(aq) + \text{Cl}^-(aq)$$

would describe the equilibrium that exists. Suppose we then tried adding an *additional amount* of one of the ions involved in the equilibrium. For example, suppose we added several drops of HCl solution (a source of chloride ion). According to Le Châtelier's principle, the equilibrium would shift so as to consume some of the added chloride ion. This would result in a net decrease in the amount of NaCl that could dissolve. If we watched the saturated NaCl solution as the HCl was added, we should see some of the NaCl precipitate as a solid.

Complex Ion Equilibria: Frequently, dissolved metal ions will react with certain substances to produce brightly colored species know as complex ions. For example, iron(III) reacts with the thiocyanate ion (SCN⁻) to produce a bright red complex ion:

$$\text{Fe}^{3+}(aq) + \text{SCN}^-(aq) \rightleftarrows [\text{FeNCS}^{2+}](aq)$$

This is an equilibrium process that is easy to study, because we can monitor the bright red color of $[\text{FeNCS}^{2+}]$ as an indication of the position of the equilibrium: If the solution is very red, there is a lot of $[\text{FeNCS}^{2+}]$ present; if the solution is not very red, then there must be very little $[\text{FeNCS}^{2+}]$ present.

Using this equilibrium, we can try adding additional Fe^{3+} or additional SCN^- to see what effect this has on the red color according to Le Châtelier's principle. We will also add a reagent (silver ion) that removes SCN^- from the system to see what effect this has on the red color.

As another example, you will study the solubility equilibrium of silver chloride as influenced when a complexing agent is added. Silver chloride is only slightly soluble in water:

$$AgCl(s) \rightleftarrows Ag^+(aq) + Cl^-(aq)$$

You will generate the silver chloride solubility equilibrium system by adding hydrochloric acid to a silver nitrate solution. A precipitate of silver chloride will form, which will be in equilibrium with a saturated solution of silver chloride. You will then add concentrated ammonia to the silver chloride system. Ammonia forms a coordination complex with the silver ion in the saturated solution and effectively removes silver ion from the system as a silver/ammonia complex ion:

$$Ag^+(aq) + 2NH_3(aq) \rightleftarrows [Ag(NH_3)_2^+](aq)$$

According to Le Châtelier's principle, removing silver ion from the silver chloride solubility equilibrium should cause a shift in that equilibrium. You will then try an acid to the system: acid reacts with ammonia and converts it to the ammonium ion:

$$H^+(aq) + NH_3(aq) \rightleftarrows NH_4^+(aq)$$

Acid/Base Equilibria: Many acids and bases exist in solution in equilibrium sorts of conditions: This is particularly true for the weak acids and bases. For example, the weak base ammonia is involved in an equilibrium in aqueous solution:

$$NH_3(aq) + H_2O(l) \rightleftarrows NH_4^+(aq) + OH^-(aq)$$

Once again, we can use Le Châtelier's principle to play around with this equilibrium: we will try adding more ammonium ion or hydrogen ion to see what happens. Because none of the components of this system is itself colored, we will be adding an acid–base indicator that changes color with pH, to have an index of the position of the ammonia equilibrium. The indicator we will use is phenolphthalein, which is pink in basic solution and colorless in acidic solution.

Safety Precautions	• Protective eyewear approved by your institution must be worn at all times while you are in the laboratory.
	• Concentrated ammonia is a strong respiratory irritant and cardiac stimulant; use concentrated ammonia *only in the fume exhaust hood*.
	• Concentrated hydrochloric acid is highly corrosive to skin, eyes, and clothing. Its vapor is irritating and toxic. Use concentrated HCl *only in the fume exhaust hood*. Wear disposable gloves while handling the acid to protect your hands. If HCl is spilled on the skin, wash immediately and inform the instructor.
	• Iron(III) chloride and potassium thiocyanate may be toxic if ingested. Wash hands after use.

Apparatus/Reagents Required

Saturated sodium chloride solution, 12 M (concentrated) HCl, 1 M HCl, 0.1 M FeCl$_3$, 0.1 M KSCN, 1 M NaOH, concentrated ammonia solution, ammonium chloride, 0.1 M AgNO$_3$, phenolphthalein indicator, 6 M HNO$_3$

Procedure

Record all data and observations directly in your notebook in ink.

1. **Solubility Equilibria**

   a. Saturated sodium chloride solution

   Obtain 5 mL of saturated sodium chloride solution in each of two test tubes. The system is in equilibrium

   $$NaCl(s) \rightleftarrows Na^+(aq) + Cl^-(aq)$$

   Add 10 drops of 12 M HCl (*Caution!*) to one test tube of saturated NaCl solution. A small amount of solid NaCl should form and precipitate out of the solution. Why? The crystals may form slowly and may be very small.

   Add 10 drops of 1 M HCl to the other test tube of saturated NaCl solution. Why does no precipitate form?

   On the lab report sheet, describe what happens in terms of Le Châtelier's principle.

   b. Saturated iron(III) hydroxide solution

   Most transition metal hydroxides are only very slightly soluble in water. Prepare a solution that is saturated with iron(III) hydroxide by combining 2 mL of 0.1 M iron(III) chloride solution with 2 mL of 1 M NaOH solution. The equilibrium is

   $$Fe(OH)_3(s) \rightleftarrows Fe^{3+}(aq) + 3OH^-(aq)$$

   Add 1 M HCl to the system, 1 drop at a time, until a change is evident. Although H$^+$ ions and Cl$^-$ ions are not written as part of the Fe(OH)$_3$ equilibrium reaction, how does it affect the equilibrium?

2. **Complex Ion Equilibria**

   a. Iron/Thiocyanate Equilibrium

   Prepare a stock sample of the bright red complex ion [FeNCS^{2+}] by mixing 2 mL of 0.1 M iron(III) chloride solution and 2 mL of 0.1 M KSCN solution. The color of this mixture is too intense to use as it is, so dilute this mixture with 100 mL of water. The equilibrium reaction is

   $$Fe^{3+} + SCN^- \rightleftarrows [FeNCS^{2+}]$$

   Pour about 5 mL of the diluted stock solution into each of four test tubes. Label the test tubes as 1, 2, 3, and 4.

   Test tube 1 will have *no change* made in it, so that you can use it to compare color with what will be happening in the other test tubes.

   To test tube 2, add dropwise with a medicine dropper about 1 mL of 0.1 M FeCl$_3$ solution. Watch very carefully as the FeCl$_3$ solution is added to the red solution. In particular, watch where the droplet of FeCl$_3$ being added first enters and mixes with the red solution.

To test tube 3, add dropwise with a medicine dropper about 1 mL of 0.1 M KSCN solution. Watch very carefully as the KSCN solution is poured into the red solution. In particular, watch where the droplet of KSCN being added first enters and mixes with the red solution.

To test tube 4, add 0.1 M AgNO$_3$ solution dropwise until a change becomes evident. Ag$^+$ ion removes SCN$^-$ ion from solution as a solid (silver thiocyanate, AgSCN). What happens to the color of the red solution as AgNO$_3$ is added?

Describe the intensification or fading of the red color in each test tube in terms of Le Châtelier's principle.

b.   Silver chloride equilibrium

Place about 10 drops of 0.1 M silver nitrate solution in a small test tube.

Under the fume exhaust hood, add 2 drops of concentrated HCl (*Caution!*) to the test tube containing the silver nitrate. Stir with a glass stirring rod to mix the reagents. This generates the silver chloride solubility equilibrium system. Allow the solution to settle for 5 minutes.

Under the fume exhaust hood, add concentrated ammonia solution (*Caution!*) dropwise with stirring to the test tube containing the silver chloride equilibrium system until a change is evident. How does adding ammonia affect the silver chloride system?

Add 6 M nitric acid, HNO$_3$, dropwise with stirring to the test tube until a change is evident? How does adding the nitric acid affect the silver chloride system?

3.   **Acid/Base Equilibria**

In the exhaust hood, prepare a dilute ammonia solution by adding 4 drops of concentrated ammonia (*Caution!*) to 100 mL of water. Stir the solution with a stirring rod to mix. The equilibrium reaction is

$$NH_3(aq) + H_2O(l) \rightleftarrows NH_4^+(aq) + OH^-(aq)$$

Add 3 drops of phenolphthalein to the dilute ammonia solution, which will turn pink (ammonia is a base, and phenolphthalein is pink in basic solution). The pink color is used as an index of the amount of hydroxide ion present in the solution: the more pink the solution, the more hydroxide ion it contains.

Place about 5 mL of the pink dilute ammonia solution into each of three test tubes.

To one of the test tubes, add several small crystals of ammonium chloride (which contains the ammonium ion, NH$_4^+$) and mix with a stirring rod to dissolve the ammonium chloride. What happens to the color of the solution when the ammonium chloride is added?

To a second test tube, add a few drops of 12 M HCl (*Caution!*). What happens to the color of the solution as HCl is added?

To the third test tube, add 1 drop of concentrated ammonia (*Caution!*). What happens to the color of the solution when the additional drop of concentrated ammonia is added?

Describe what happens to the pink color in terms of how Le Châtelier's principle is affecting the ammonia equilibrium.

Name: _____ Section: _____

Lab Instructor: _____ Date: _____

EXPERIMENT 29

Chemical Equilibrium 3: Stresses Applied to Equilibrium Systems

Pre-Laboratory Questions

1. What do we mean by a "stress" applied to an equilibrium system? Give some examples of such "stresses". How does an equilibrium system behave when a stress is applied to the system?

2. Given the gas-phase reaction

$$2SO_2(g) + O_2(g) \rightleftarrows 2SO_3(g).$$

   Suppose that the reaction has already taken place, and the system has come to equilibrium. If the following changes are made to the system in equilibrium, tell what effect these changes will have on the system (whether the equilibrium is shifted to the left, is shifted to the right, or is not shifted).

   a. Additional O_2 is added to the system.

   _____

   b. Sulfur trioxide is removed from the system as soon as it forms.

   _____

   c. The pressure of the system is decreased (all components are gases).

   _____

   d. A very efficient catalyst is used for the reaction.

   _____

3.   Explain why the solubility of a sparingly soluble salt represents an equilibrium system.

4.   Phenolphthalein is used in this experiment to study the equilibrium of the weak base ammonia in water. Phenolphthalein itself is a weak acid, which exists in equilibrium in solution. The protonated (acid) form of phenolphthalein is colorless, whereas the conjugate base form of phenolphthalein is colored pink (fuchsia). Use a scientific encyclopedia or online reference to find the formula of phenolphthalein, its chemical name, and at what pH phenolphthalein changes from the colorless form to the pink form.

Name: _____ Section: _____

Lab Instructor: _____ Date: _____

EXPERIMENT 29

Chemical Equilibrium 3: Stresses Applied to Equilibrium Systems

Results/Observations

1. **Solubility Equilibria**

   a. Observation of effect of adding HCl to saturated NaCl solution

   _____

   Explanation

   _____

   _____

   _____

   b. Observation of effect of adding HCl to saturated $Fe(OH)_3$ solution

   _____

   Explanation

   _____

   _____

   _____

2. **Complex Ion Equilibria**

   a. Observation of effect of adding Fe^{3+} to $[FeNCS^{2+}]$

   _____

   Explanation

   _____

   _____

   _____

   Observation of effect of adding SCN^- to $[FeNCS^{2+}]$

   _____

   Explanation

   _____

   _____

Observation of effect of adding Ag^+ to $[FeNCS^{2+}]$

_____

Explanation

_____

_____

_____

b.   Observation on adding concentrated HCl to silver nitrate solution

_____

Explanation

_____

_____

Observation on adding concentrated NH_3 to the equilibrium system

_____

Explanation

_____

_____

Observation on adding 6 M HNO_3 to the equilibrium system

_____

Explanation

_____

_____

_____

3.   **Acid–Base Equilibria**

Observation of effect of adding NH_4^+

_____

Explanation

_____

_____

_____

Observation of effect of adding HCl

_____

Explanation

_____

_____

_____

Questions

1. Explain how a saturated solution represents an equilibrium between the solution and any undissolved solute present.

2. Addition of 12 M (concentrated) HCl to saturated NaCl results in precipitation of NaCl. Yet when 1 M HCl is added to the same NaCl solution, *no* precipitate forms. Explain.

3. How do you explain why the addition of silver nitrate to the Fe^{3+}/SCN^- equilibrium had an effect on the equilibrium, even though neither silver ion nor nitrate ion is written as part of the equilibrium reaction?

4. When you added concentrated ammonia to the silver chloride equilibrium system, the silver chloride solid dissolved as the soluble silver/ammonia complex formed. When you subsequently added nitric acid, the silver chloride re-precipitated. Explain.

5. In describing the complex ion formed from iron(III) ion, $Fe^{3+}(aq)$, and thiocyanate ion, $SCN^-(aq)$, the formula for the complex is written as $[FeNCS^{2+}]$, rather than $[FeSCN^{2+}]$, as one might expect. Write the Lewis structure for the SCN^- ion and use formal charge to show why the iron(III) ion is more likely to link to the nitrogen end of thiocyanate.

Experiment 30

The Solubility Product Constant of Calcium Iodate

Objective

The extent to which a sparingly soluble salt dissolves in water is frequently indicated in terms of the salt's solubility product equilibrium constant, K_{sp}. In this experiment, you are to determine the value of K_{sp} for the salt, calcium iodate.

Introduction

Many ionic compounds, as you know, are only slightly soluble in water; we say such compounds are "sparingly soluble." The electrostatic attraction of polar water molecules for the ions in an ionic crystal means that no ionic compound is truly insoluble in water, although we commonly describe many as insoluble, using the solubility rules with which you are familiar. A chemical reference will generally list solubilities in terms of the mass of solute that will dissolve in 100 grams of water, but for many solutes such a measurement would be milligram quantities or less. What is needed in such cases is a more precise measure of solubility. When any sparingly soluble ionic compound is added to water, there will be an equilibrium between the undissolved crystalline solid and dissolved ions. Consider, for example, silver chloride in water. The equilibrium system would be

$$AgCl(s) \rightleftarrows Ag^+(aq) + Cl^-(aq)$$

As you know, equilibrium expressions exclude solids and pure liquids, since their concentrations are fixed, and are part of the calculated value. In the case of solubility equilibria, the equilibrium constant is referred to as the *solubility product* constant, or simply the solubility product, and is represented by K_{sp}, where the subscript stands for "solubility product." Thus, for the silver chloride equilibrium, the equilibrium expression would be

$$K_{sp} = [Ag^+(aq)][Cl^-(aq)]$$

For In this experiment the K_{sp} of different salt, calcium iodate will be determined using gravimetric titration. The equilibrium and the equilibrium constant expression for this system are

$$Ca(IO_3)_2(s) \rightleftarrows Ca^{2+}(aq) + 2\,IO_3^-(aq)$$

$$K_{sp} = [Ca^{2+}][IO_3^-]^2$$

The concentration of iodate ion in a saturated solution can be determined by first converting the iodate ion to triiodide ion, $I_3^-(aq)$, in acidic solution and then titrating the triiodide ion to a colorless starch end point with sodium thiosulfate. The net-ionic, *unbalanced* equations for the reactions are shown below.

$$IO_3^-(aq) + I^-(aq) \longrightarrow I_3^-(aq)$$

and

$$S_2O_3^{2-}(aq) + I_3^-(aq) \longrightarrow S_4O_6^{2-}(aq) + I^-(aq)$$

A solution of sodium thiosulfate is prepared; the masses of the solute and of the solution are known, so that the concentration of thiosulfate ion, in mol $S_2O_3{}^{2-}(aq)$/g of solution, can be determined to 3-significant figure precision. This will be the titrant.

Separately, a sample of saturated calcium iodate will be diluted and acidified, then an excess of iodide ion will be added, converting the iodate quantitatively to triiodide ion, I_3^-. This solution will then be titrated with the thiosulfate solution, just prepared. To avoid over-shooting the endpoint of this titration, a small sample of the triiodide solution will be reserved. The reserved amount is called a *titration thief*. Once the bulk of the triiodide solution has been titrated, the thief will be returned to the titration vessel and the resulting mixture will be titrated to completion.

You are to carry out three separate trials, perform the calculations for each trial, then calculate the mean value and average deviation. Use the mean value in answering #7 of **Analysis and Conclusions**.

Safety Precautions	• **Protective eyewear approved by your institution must be worn at all times while you are in the laboratory.** • **Iodate compounds are strong oxidizers. Handle with caution.** • **Sodium thiosulfate is a strong reducing agent and is mildly toxic. Wash hands after use.** • **Dispose of all reagents as directed by the instructor.**

Apparatus/Reagents Required

30-mL dropper bottle (or similar), thin-stem transfer pipets, volumetric (2.50 mL) or Mohr pipet (5.0 mL or 10.0 mL) and rubber safety bulb, sodium thiosulfate pentahydrate crystals, saturated calcium iodate solution, hydrochloric acid (2 M), solid potassium iodide, KI(s), starch solution indicator in thin-stem pipet.

Procedure

Record all data and observations directly in your notebook in ink.

1. **Preparation of sodium thiosulfate weight buret.**

   Weigh out between 0.18 and 0.20 grams (± 1 mg) of sodium thiosulfate pentahydrate in a tared 50-mL beaker. Add 10.0 mL of distilled water. Record the mass of the solution. Stir until fully dissolved. Transfer the solution to a 30 mL dropper bottle. (This is called a *weight buret*.) Record the mass of the weight buret and solution.

   Calculate the concentration of the solution in moles of sodium thiosulfate pentahydrate per gram of solution, mol $S_2O_3^{2-}$/g sol'n. Do this calculation as part of your observations, before you proceed.

2. **Preparation of samples for titration – three samples**

   Add about 20 mL of water to a 50 mL beaker. Pipette 2.50 mL of saturated calcium iodate solution into the beaker. Add 10 drops of 2 M HCl, followed by about 0.2 grams of potassium iodide and swirl until all of the solid has dissolved. Titrate each sample as described below. Use a clean thin-stem pipet for each titration thief, to avoid contamination.

3. **Titration of samples. Carry out the following steps with each of the three calcium iodate samples.**

Remove about 1 mL of the solution in a thin-stem pipette (the bulb of the pipette holds 3.5 mL). (This is the *titration thief*, mentioned in the Introduction.) Set the pipette aside for later use. Your solution should be a sort of yellow-brown color, due to presence of triiodide ion.

Weigh the weight buret and record its mass.

Slowly, and swirling constantly, add liquid from your weight buret until a light yellow color is reached. Add 10 drops of starch solution. Titrate to colorless with thiosulfate solution from your weight buret.

Add the solution from the titration thief and titrate very carefully to colorless once again. Swirl the titration mixture after each addition from the weight buret, to ensure that you don't overshoot the endpoint.

Weigh your weight buret again, to determine the mass of thiosulfate solution used. Note and record the temperature of your titration mixture; presumably the reaction was carried out isothermally at this temperature.

EXPERIMENT 30

The Solubility Product of Silver Acetate

Pre-Laboratory Questions

1. Write the balanced equation for the dissolving of calcium iodate, $Ca(IO_3)_2$, in water. Show this process as an equilibrium.

2. How will the concentrations of calcium ion and iodate ion in solution compare?

3. The titration reactions are carried out in the presence of acid. Write the equation for the reaction you would expect calcium ion to undergo in the presence of excess hydroxide ion. This is the reason for acidifying the titration mixtures.

4. Use the half-reaction method to balance the two reactions given in the Introduction, iodate ion reacting with iodide ion, and thiosulfate ion reacting with triiodide ion. You will need these in carrying out your post-lab calculations. The reactions take place in acidic solution.

4. Using a handbook of chemical data, or online reference, look up the values of K_{sp} for the following substances at 25°C:

    a. Copper(I) chloride _____

    b. Calcium phosphate _____

    c. Lead(II) carbonate _____

    d. Magnesium fluoride _____

    e. Silver chromate _____

    f. zinc hydroxide _____

    g. iron(II) sulfide _____

    h. cobalt(II) phosphate _____

    i. barium fluoride _____

EXPERIMENT 30

The Solubility Product of Silver Acetate

Results/Observations

1. **Preparation of Weight Buret**

Mass of $Na_2S_2O_3$ used (± 0.001g)	
Mass of beaker + sodium thiosulfate solution	
Moles of $Na_2S_2O_3$ in solution	
Concentration of $Na_2S_2O_3(aq)$ in mol$Na_2S_2O_3$/gram	

(Show calculations here)

2. **Titration Data and Calculations**

	Sample 1	Sample 2	Sample 3
Mass of weight buret before titration			
Mass of weight buret after titration			
Mass of $Na_2S_2O_3$ solution used			
Moles of $Na_2S_2O_3$ used			

a. Use the number of moles of $Na_2S_2O_3$ used (from the table) and the balanced equation for the reaction between thiosulfate and triiodide ion to determine the number of moles of triiodide ion in each of your three titrations. Show calculations here and enter your results in the Summary Table found below.

b. Use your result from the previous calculation and the balanced equation for the reaction between iodate and triiodide ions to determine the number of moles of iodate ion initially present in each sample. Enter your results in the Summary Table below.

c. Determine the number of moles of calcium ion that must have been in each sample. Enter your results in the Summary Table below.

d. Calculate the molar concentrations of Ca^{2+} and IO_3^- in each 2.50-mL sample of saturated $Ca(IO_3)_2$ solution. Add these results to the Summary Table below.

e. Calculate the value of K_{sp} for $Ca(IO_3)_2$ for each trial. Enter the results in the Summary Table.

3. Calculate the mean value of K_{sp} for your three trials

Summary Table

	Trial 1	Trial 2	Trial 3
Moles of $I_3^-(aq)$			
Moles of $IO_3-(aq)$			
Moles of $Ca^{2+}(aq)$			
Molarity of $Ca^{2+}(aq)$			
Molarity of $IO_3^-(aq)$			
Calculated K_{sp}			

Questions

1. In theory, the solubility product equilibrium constant should be, in fact, *constant* throughout your determinations.

    a. Calculate the relative deviation (% deviation) among your three values of K_{sp} for calcium iodate.

    b. What factors may not have been considered that might have led to different measured values for K_{sp} under the various conditions used in the experiment?

2. Compare your mean K_{sp} value with the literature value of 2.5×10^{-8}. Calculate the percent error.

3. How would you expect the value of Ksp for calcium iodate to be affected by increasing or decreasing the temperature? Explain.

4. The solubility product for cadmium sulfide, CdS, is 1×10^{-27} at 25°C. Calculate the number of grams of cadmium sulfide that will dissolve in 1.00 L of water at 25°C.

EXPERIMENT 31

Acids, Bases, and Buffered Systems

Objective

In this experiment, you will study the properties of acidic and basic substances, using indicators and a pH meter to determine pH. A buffered system will be prepared by half-neutralization, and the properties of the buffered system will be compared to those of an unbuffered medium. The relative conductivities of strong and weak acid–base systems will be demonstrated.

Introduction

One of the most important properties of aqueous solutions is the concentration of hydrogen (hydronium) ion present. The H^+ or H_3O^+ ($= H_2O \cdot H^+$) ion greatly affects the solubility of many organic and inorganic substances, the nature of complex metal cations found in solution; and the rate of many chemical reactions, and it also may profoundly affect the reactions of biochemical species.

The pH Scale

Typically, the concentration of H^+ ion in aqueous solutions may be a very small number, requiring the use of scientific notation to describe the concentration numerically. For example, an acidic solution may have $[H^+] = 2.3 \times 10^{-5}$ M, whereas a basic solution may have $[H^+] = 4.1 \times 10^{-10}$ M. Because scientific notation may be difficult to deal with, especially when you make *comparisons* of numbers, a mathematical simplification, called pH, has been defined to describe the concentration of hydrogen ion in aqueous solutions:

$$pH = -\log[H^+]$$

By use of a base-10 logarithm, the power of ten of the scientific notation is converted to a "regular" number, and use of the minus sign in the definition of pH produces a positive value for the pH. For example, if $[H^+] = 1.0 \times 10^{-4}$ M, then

$$pH = -\log(1.0 \times 10^{-4}) = -(-4.00) = 4.00$$

and similarly, if $[H^+] = 5.0 \times 10^{-2}$ M, then

$$pH = -\log(5.0 \times 10^{-2}) = -(-1.30) = 1.30$$

Although basic solutions are usually considered to be solutions of *hydroxide* ion, OH^-, basic solutions also must contain a certain concentration of *hydrogen* ion, H^+, because of the equilibrium that exists in aqueous solutions as a consequence of the autoionization of water. The concentration of hydrogen ion in a basic solution can be determined by reference to the equilibrium constant for the autoionization of water, K_w:

$$K_w = [H^+][OH^-] = 1.0 \times 10^{-14} \quad \text{(at 25°C)}$$

For example, if a basic solution contains $[OH^-]$ at the level of 2.0×10^{-3} M, then the solution also contains hydrogen ion at the level of

$$[H^+] = \frac{1.0 \times 10^{-14}}{[OH^-]} = \frac{1.0 \times 10^{-14}}{2.0 \times 10^{-3}} = 5.0 \times 10^{-12} \ M$$

In pure water, there must be *equal* concentrations of hydrogen ion and hydroxide ion (one of each ion is produced when a water molecule ionizes). Hence, in pure water,

$$[H^+] = [OH^-] = \sqrt{1.0 \times 10^{-14}} = 1.0 \times 10^{-7}\ M$$

The pH of pure water is thus 7.00; this value serves as the *dividing line* for the pH values of aqueous solutions. A solution in which there is *more* hydrogen ion than hydroxide ion ($[H^+] > 1.0 \times 10^{-7}\ M$) will be *acidic* and will have pH less than 7.00. Conversely, a solution with *more* hydroxide ion than hydrogen ion ($[OH^-] > 1.0 \times 10^{-7}\ M$) will be *basic* and will have pH greater than 7.00.

If the pH of a solution is known (or has been measured experimentally), then the hydrogen ion concentration of the solution may be calculated:

$$[H^+] = 10^{-pH}$$

For example, the hydrogen ion concentration of a solution with pH = 4.20 would be given by

$$[H^+] = 10^{-pH} = 10^{-4.20} = 6.3 \times 10^{-5}\ M$$

Experimental Determination of pH

The experimental determination of the pH of a solution is commonly performed by either of two methods. The first of these methods involves the use of chemical dyes called **indicators**. These substances are generally weak acids or bases and can exist in either of two colored forms, depending on whether the molecule is protonated or has been deprotonated. For example, let HIn represent the protonated form of the indicator. In aqueous solution, an equilibrium exists:

$$\underset{\textit{first color}}{HIn} \rightleftarrows \underset{\textit{second color}}{H^+ + In^-}$$

Depending on what other acidic/basic substances are present in the solution, the equilibrium of the indicator will shift, and one or the other colored form of the indicator will predominate and impart its color to the entire solution.

Indicators commonly change color over a relatively short pH range (about 2 pH units) and, when properly chosen, can be used to estimate the pH of a solution. Among the most common indicators are litmus (which is usually used in the form of test paper that has been impregnated with the indicator) and phenolphthalein (which is most commonly used as a solution added to samples for titration analysis). Litmus changes color from red to blue as the pH of a solution increases from pH 6 to pH 8. Phenolphthalein goes from a colorless form to red as the pH of a solution increases from pH 8 to pH 10. A given indicator is useful for determining pH only in the region in which it changes color. Indicators are available for measurement of pH in all the important ranges of acidity and basicity. By matching the color of a suitable indicator in a solution of known pH to the color of the indicator in an unknown solution, we can estimate the pH of the unknown solution.

The second method for the determination of pH involves an instrument called a **pH meter**. The pH meter contains two electrodes, one of which is sensitive to the concentration of hydrogen ion. Typically, the pH-sensitive electrode is a *glass membrane* electrode, whereas the second electrode may consist of a silver/silver chloride system or a calomel electrode. Frequently, a glass membrane pH electrode is combined in the same physical chamber as a silver/silver chloride electrode and is referred to as a "combination" pH electrode. In the solution being measured, the electrical potential between the two electrodes is a function of the hydrogen ion concentration of the solution. The pH meter has been designed so that the pH of the solution can be directly read from the scale provided on the face of the meter. It is essential always to calibrate a pH meter before use: a solution of known pH is measured with the electrodes, and the display of the meter is adjusted to read the correct pH for the known pH solution. A properly calibrated pH meter provides for much more precise determinations of pH than does the indicator method, so it is ordinarily used when a very accurate determination of pH is needed.

Strong and Weak Acids/Bases

Some acids and bases undergo substantial ionization when dissolved in water and are called **strong acids** or **strong bases**. Strong acids and bases are completely ionized in dilute aqueous solutions. Other acids and bases, because they ionize *in*completely in water (often to the extent of only a few percent), are called **weak acids** or **weak bases**. Hydrochloric acid (HCl) and sodium hydroxide (NaOH) are a typical strong acid and a typical strong base, respectively. A 0.1 M solution of HCl contains hydrogen ion at effectively 0.1 M concentration because of the complete ionization of HCl molecules.

$$HCl(aq) \longrightarrow H^+(aq) + Cl^-(aq)$$
$$0.1\ M \qquad 0.1\ M \qquad 0.1\ M$$

Similarly, a 1×10^{-2} M solution of NaOH contains hydroxide ion at the level of 1×10^{-2} M concentration because of the complete ionization of NaOH.

$$NaOH(aq) \longrightarrow Na^+(aq) + OH^-(aq)$$
$$1\times10^{-2}\ M \qquad 1\times10^{-2}\ M \quad 1\times10^{-2\ M}$$

Acetic acid ($HC_2H_3O_2$ or CH_3COOH) and ammonia (NH_3) are typical examples of a weak acid and a weak base, respectively. Weak acids and weak bases must be treated by the techniques of chemical equilibrium in determining the concentrations of hydrogen ion or hydroxide ion in their solutions. For example, for the general weak acid HA, the equilibrium reaction would be

$$HA(aq) \rightleftarrows H^+(aq) + A^-(aq)$$

and the equilibrium constant expression would be given by

$$K_a = \frac{[H^+][A^-]}{[HA]}$$

For acetic acid, as an example, the ionization equilibrium reaction and K_a are represented by the following:

$$CH_3COOH(aq) \rightleftarrows H^+(aq) + CH_3COO^-(aq)$$

$$K_a = \frac{[H^+][CH_3COO^-]}{[CH_3COOH]}$$

K_a is a constant and is *characteristic* of the acid HA. For a given weak acid, the product of the concentrations in the expression will remain constant at equilibrium, regardless of the manner in which the solution of the acid was prepared. For a general weak base B, the equilibrium reaction would be

$$B + H_2O \rightleftarrows BH^+(aq) + OH^-(aq)$$

and the equilibrium constant expression would be given by

$$K_b = \frac{[BH^+][OH^-]}{[B]}$$

As an example, the ionization and K_b for the weak base ammonia are represented by

$$NH_3 + H_2O \rightleftarrows NH_4^+(aq) + OH^-(aq)$$

$$K_b = \frac{[NH_4^+][OH^-]}{[NH_3]}$$

The value of the equilibrium constant K_a for a weak acid (or K_b for a weak base) can be determined experimentally by several methods. One simple procedure involves little calculation, is accurate, and does not require knowledge of the actual concentration of the weak acid (or base) under study. For example, a sample of weak acid (HA) is dissolved in water and is then divided into two equal-volume portions. One portion is then titrated with a sodium hydroxide solution to a phenolphthalein endpoint, thereby converting all HA molecules present into A^- ions:

$$OH^- + HA \longrightarrow H_2O + A^-$$

The number of moles of A^- ion produced by the titration is equal, of course, to the number of moles of HA in the original half-portion titrated and is also equal to the number of moles of HA in the remaining unused portion of weak acid. The two portions are then mixed, and the pH of the combined solution is measured. Because [HA] equals [A^-] in the combined solution, these terms (represented below by [n]) cancel each other in the equilibrium constant expression for K_a for the acid,

$$K_a = \frac{[H^+][A^-]}{[HA]} = \frac{[H^+][n]}{[n]} = [H^+]$$

and therefore the ionization equilibrium constant, K_a, for the weak acid will be given directly from a measurement of the hydrogen ion concentration,

$$K_a = [H^+] = 10^{-pH}$$

for the combined solution. In other words, the ionization equilibrium constant for a weak acid can be determined simply by measuring the pH of a half-neutralized sample of the acid.

pH of Salt Solutions

Salts that can be considered to have been formed from the complete neutralization of strong acids with strong bases—such as NaCl (which can be considered to have been produced by the neutralization of HCl with NaOH)—ionize completely in solution. Such strong acid/strong base salts do not react with water molecules to any appreciable degree (do not *undergo hydrolysis*) when they are dissolved in water. Solutions of such salts are neutral and have pH = 7. Other examples of such salts are KBr (from HBr and KOH) and NaNO₃ (from HNO₃ and NaOH).

However, when salts formed by the neutralization of a *weak* acid or base are dissolved in water, these salts furnish ions that tend to react to some extent with water, producing molecules of the weak acid or base and releasing some H^+ or OH^- ion to the solution. Solutions of such salts will *not* be neutral but, rather, will be acidic or basic themselves.

Consider the weak acid HA. If the *sodium salt* of this acid, Na^+A^-, is dissolved in water, the A^- ions released to the solution will react with water molecules to some extent

$$A^- + H_2O \rightleftarrows HA + OH^-$$

The solution of the salt will be *basic* because hydroxide ion has been released by the reaction of the A^- ions with water. For example, a solution of sodium acetate (a salt of the weak acid acetic acid) is basic because of reaction of the acetate ions with water molecules, releasing hydroxide ions:

$$CH_3COO^- + H_2O \rightleftarrows CH_3COOH + OH^-$$

Conversely, solutions of salts of weak bases (such as $NH_4^+Cl^-$, derived from the weak base NH₃) will be *acidic*, because of reaction of the ammonium ions of the salt with water. For example,

$$NH_4^+ + H_2O \rightleftarrows NH_3 + H_3O^+$$

Many salts of transition metal ions are acidic. A solution of CuSO₄ or FeCl₃ will typically be of pH 5 or lower. The salts are completely ionized in solution. The acidity comes from the fact that the metal cation is hydrated. For example, $Cu(H_2O)_4^{2+}$ better represents the state of a copper(II) ion in aqueous solution. The large positive charge on the metal cation attracts electrons from the O–H bonds in the water molecules, thereby weakening the bonds and producing some H+ ions in the solution:

$$Cu(H_2O)_4^{2+} \rightleftarrows Cu(H_2O)_3(OH)^+ + H^+$$

Buffered Solutions

Some solutions, called buffered solutions, are remarkably resistant to pH changes caused by the addition of an acid or base from an outside source. A **buffered solution**, in general, is a *mixture* of a weak acid and a weak base. The most common sort of buffered solution consists of a mixture of a conjugate acid–base pair. For example, a mixture of the salt of a weak acid or base and the weak acid or base itself would constitute a buffered solution. The half-neutralized solution of a weak acid described earlier as a means of determining K_a would therefore represent a buffered system. That solution contained equal amounts of weak acid HA and of the anion A^- present in its salt. If a small amount of strong acid were added to the buffered system from an external source, the H^+ ion introduced by the strong acid would tend to react with the A^- ion of the salt, thereby preventing a change in pH. Similarly, if a strong base were added to the HA/A^- system from an external source, the OH^- introduced by the strong base would tend to be neutralized by the HA present in the buffered system. If similar small amounts of strong acid or strong base were added to an unbuffered system (such as plain distilled water), the pH would be changed drastically by the addition.

Safety Precautions	• Protective eyewear approved by your institution must be worn at all times while you are in the laboratory. • The acids, bases, and salts to be used are in fairly dilute solution, but they may be irritating to the skin. Wash if they are spilled, and inform the instructor. • The pH meter is an electrical device and depending on the model, it may be connected to the wall outlet. Do not spill water or any solution near the meter; electrical shock may result.

Apparatus/Reagents Required

pH meter and electrode(s), buret and clamp, unknown sample for indicator determination, known pH 0–6 samples, indicators, unknown acid sample (for half-neutralization and buffered solution study), $0.2\ M$ NaOH, $0.1\ M$ HCl, $0.1\ M$ NaOH, salt solutions for pH determination [NaCl, KBr, $NaNO_3$, K_2SO_4, $NaC_2H_3O_2$, NH_4Cl, K_2CO_3, $(NH_4)_2SO_4$, $CuSO_4$, and $FeCl_3$], universal indicator solution, "light bulb" conductivity apparatus, $0.1\ M$ acetic acid, $0.1\ M$ ammonia

Procedure

Record all data and observations directly in your notebook in ink.

1.   pH Using Indicators

Obtain from your instructor a solution of unknown pH for indicator determination. Place about half an inch of this solution in a small test tube, and add 1–2 drops of one of the indicators in the table of indicators that follows. Record the color of the solution after addition of the indicator.

Indicator	Color Change	pH of Color Change
methyl violet	yellow to violet	–1 to 1.7
malachite green	yellow to green	0.2 to 1.8
cresol red	red to yellow	1 to 2
thymol blue	red to yellow	1.2 to 2.8
bromphenol blue	yellow to blue	3 to 4.7
methyl orange	red to yellow	3.2 to 4.4
bromcresol green	yellow to blue	3.8 to 5.4
methyl red	red to yellow	4.4 to 6.0
methyl purple	purple to green	4.5 to 5.1
bromthymol blue	yellow to blue	6.0 to 7.6
litmus	red to blue	4.7 to 6.8

Repeat the test with each of the indicators listed, using a small amount of the unknown solution and 1–2 drops of indicator.

From the information given in the table, estimate the pH of your unknown solution. Note that the color of an indicator is most indicative of the pH in the region in which the indicator changes color.

Having established the pH of your unknown to within approximately one pH unit, obtain a known solution with a pH about equal to that of your unknown. Stock solutions of standard reference buffered solutions with integral pH values from 0 to 6 have been prepared for you.

Test 1-mL samples of the known pH sample with the indicators most useful in determining the approximate pH of the unknown sample. Compare the colors of the known and unknown solutions. On the basis of the comparisons, decide whether the pH of the unknown is slightly higher or slightly lower than the pH of the known sample.

Select a second known solution, with the aim of bracketing the pH of your unknown solution between two known solutions. Treat the second known solution with the useful indicator(s) and compare the color(s) with the unknown sample(s). Continue testing known pH solutions until you find two that differ by one pH unit, one of which has a higher pH than the unknown, and one of which has a lower pH than the unknown.

For a final estimation of the pH of your unknown, use a 5-mL sample of the unknown and a 5-mL sample of each of the two known solutions that bracket its pH. Add 2 drops of the indicator to each solution. View the solutions against a well-lighted sheet of white paper, and estimate the pH of your unknown (as compared to the two known solutions) to ± 0.3 pH units. (See Figure 35-1.)

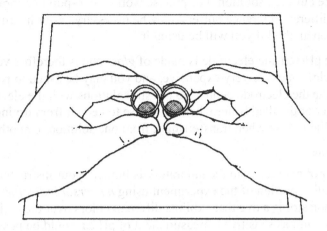

Figure 31-1. Comparison of known and unknown solution colors with indicator. The background against which the samples are viewed should be well lighted. A "light box" is ideal if available.

2. Calibration of the pH Meter

Typically, several models of pH meters are available in the laboratory (see Figure 35-2). Your instructor will provide specific instructions for the operation of the meter you will use.

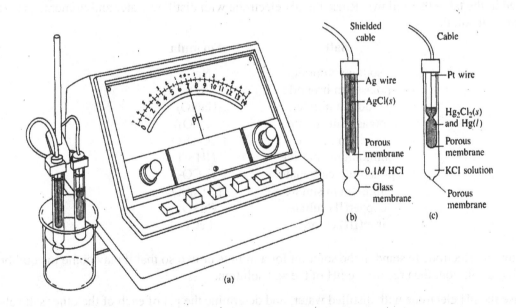

Figure 31-2. (a) A typical pH meter/electrode set-up. The two electrodes may be combined in one chamber as a "combination" pH electrode. (b) pH sensing electrode. The glass membrane measures the difference in [H^+] between the interior of the electrode and the solution being tested. (c) Calomel reference electrode. This completes the electrical circuit in the solution and is very stable and reproducible.

In all cases, the meters must be *calibrated* before being used for pH measurements. Samples of pH standard reference buffered solutions are available for calibrating the meters. Generally, the combination electrode is dipped into one of the reference buffered solutions and allowed to stand for several minutes; this permits the electrode to come to equilibrium with the buffered solution. The set or calibrate knob on the face of the meter is then adjusted until the meter display reads the correct

307

pH for the reference buffered solution. For precise work, a two-point calibration, in which the meter is checked in two different reference buffers, may be necessary. Your instructor will explain the two-point calibration method if you will be using it.

Remember that the pH-sensing electrode is made of *glass* and is therefore very *fragile*. (Combination pH electrodes are very expensive, and you may be asked to pay for the electrode if you break it.) Handle the electrode gently, do not stir solutions with the electrode, and keep the electrode in a beaker of distilled water when not in use to keep it from drying out. Rinse the electrode with distilled water when transferring it from one solution to another.

3.   pH of Salt Solutions

Note: If the number of pH meters in the laboratory is limited, your instructor may ask you to perform this qualitative portion of the experiment using *universal indicator*, rather than the pH meter. Universal indicator is a dye that exhibits different colors over the entire pH range. Universal indicator does not permit as sensitive a measurement of pH as would be possible with a pH meter or with the method using several indicators as discussed in Part A of this experiment. For a quick "ballpark" estimate of pH, however, universal indicator can be very useful.

In this case, obtain 10 drops of each salt solution, and test for pH with one drop of universal indicator. Use the color chart provided with the indicator to estimate the pH of each salt solution. Write an equation accounting for the pH observed.

If sufficient pH meters are available in the lab, first calibrate the meter using a pH reference buffered solution, and then obtain 20–25 mL (in a small beaker) of one of the 0.1 M salt solutions listed in the table that follows. Rinse the pH electrode with distilled water and immerse the electrode in the salt solution.

Salt	Formula
sodium chloride	NaCl
potassium bromide	KBr
sodium nitrate	$NaNO_3$
potassium sulfate	K_2SO_4
sodium acetate	$NaC_2H_3O_2$
ammonium chloride	NH_4Cl
potassium carbonate	K_2CO_3
ammonium sulfate	$(NH_4)_2SO_4$
copper(II) sulfate	$CuSO_4$
iron(III) chloride	$FeCl_3$

Allow the electrode to stand in the solution for a minute or two so that it may come to equilibrium with the solution; then record the pH of the salt solution.

Rinse the pH electrode with distilled water, and determine the pH of each of the other salt solutions listed in the table. On the basis of the pH measured for each salt solution, write a chemical equation that will explain the observed pH.

4.   Determination of K_a for a Weak Acid

Obtain a sample of an unknown acid for K_a determination.

If the acid is provided as a solution, use your graduated cylinder to measure out two 50-mL portions into 250-mL Erlenmeyer flasks. If the acid is provided as a solid, place the solid acid sample in 100 mL of distilled water, stir thoroughly to dissolve the acid, use your graduated cylinder to divide the acid solution into two equal volume portions and place these into two 250-mL Erlenmeyer flasks.

Clean and rinse a buret, and then fill it with 0.2 M NaOH solution.

Take one portion of your unknown acid solution, add 3–4 drops of phenolphthalein indicator, and titrate the acid solution until a faint pink color appears and persists. At this point, you have neutralized half the original acid provided and have a solution that contains the salt of the weak acid. It is not necessary in this experiment to determine the exact volume necessary for the titration.

Mix the neutralized solution from the titration with the remaining portion of weak acid solution. Stir well.

Check the calibration of the pH meter with the standard reference buffers provided to make sure that the meter has not drifted in its readings.

Rinse off the electrode of the pH meter with distilled water, and transfer the electrode to the mixture of unknown weak acid and salt. Stir the solution with a stirring rod. Allow the electrode to stand for 1–2 minutes in the sample, and record the pH of the solution.

From the observed pH, calculate $[H^+]$ in the solution and K_a for the weak acid unknown. Save the solution for use later.

5. Properties of a Buffered Solution

The half-neutralized solution prepared in Part 4 is a buffered system because it contains approximately equal amounts of a free weak acid (HA) and of a salt of the weak acid (Na^+A^-) from the titration. This buffered system should be very resistant to changes in pH, compared to an unbuffered system, when small quantities of strong acid or strong base are added from an external source.

Check the calibration of the pH meter with the standard reference buffers provided, to make sure that the meter has not drifted in its readings.

Place 25 mL of the unknown acid buffered mixture in a small beaker and measure its pH with the meter. Add 5 drops of 0.1 M strong acid HCl to the buffered mixture and stir. Allow the electrode to stand for 1–2 minutes and record the pH. The pH should drop only by a small fraction of a pH unit (if at all).

Place another 25-mL sample of the buffered mixture in a small beaker. Add 5 drops of 0.1 M strong base NaOH to the buffered mixture and stir. Record the pH after allowing the electrode to stand for 2–3 minutes in the solution. The pH should have risen only by a small fraction of a pH unit (if at all).

For comparison, repeat the addition of small amounts of 0.1 M HCl and 0.1 M NaOH to 25-mL portions of one of the standard reference buffered solutions available. Determine the pH of the buffered solution itself, as well as the pH after the addition of strong acid or strong base.

To contrast a buffered system with an unbuffered solution, place 25 mL of distilled water in a small beaker, immerse the electrode, and record the pH. (Do not be surprised if the pH is not exactly 7.0— only pure water that is not in contact with air or glass containers gives pH exactly 7.00.)

Add 5 drops of 0.1 M HCl to the distilled water, and record the pH. It should decrease by several pH units.

Place another 25 mL of distilled water in a beaker. Add 5 drops of 0.1 M NaOH solution, and record the pH. It should increase by several pH units.

Write net ionic equations demonstrating why an HA/A^- buffered system resists changes in its pH when strong acid (H^+) or strong base (OH^-) is added.

6.   **Conductivities of Acid, Base, and Salt Solutions**

Your instructor has set up a "light bulb" conductivity tester and will demonstrate the relative conductivities of the following solutions:

0.1 *M* HCl (strong acid), 0.1 *M* NaOH (strong base), 0.1 *M* acetic acid (weak acid), 0.1 *M* ammonia (weak base).

Why does the light bulb glow more brightly for the HCl and NaOH solutions than for the acetic acid and ammonia solutions?

You instructor will then *combine* samples of the acetic acid and ammonia solutions and test the conductivity of the mixture. Why does the mixture cause the light bulb to glow more brightly than either component of the mixture did?

Name: ______________________________ **Section:** __________

Lab Instructor: ______________________________ **Date:** __________

EXPERIMENT 31

Acids, Bases, and Buffered Systems

Pre-Laboratory Questions

1. Using your textbook, a handbook of chemical data, or an online reference, list three strong acids and three strong bases. Write an equation for each acid and each base showing its ionization in water.

2. Using a handbook of chemical data or online reference, find the ionization equilibrium constants (K_a or K_b) for the following weak acids and weak bases. Write an equation for each acid or base showing its ionization in water, and write the expression in terms of chemical symbols for the ionization equilibrium constant (K_a or K_b) for the weak acid or base.

   a. Boric acid, H_3BO_3

   ____

   ____

   b. Hypochlorous acid, HOCl

   ____

   ____

   c. Propanoic acid, CH_3CH_2COOH

   ____

   ____

    d.   Ethylamine, $C_2H_5NH_2$

_____

_____

    e.   Aniline, $C_6H_5NH_2$

_____

_____

3.   Why are pH-sensing electrodes that are used with the typical laboratory pH meter so fragile?

4.   What is an acid–base *indicator*? How may indicators be used in determining the pH of a solution?

Name: ______________________ Section: __________

Lab Instructor: ______________________ Date: __________

EXPERIMENT 31

Acids, Bases, and Buffered Systems

Results/Observations

1. **pH Using Indicators**

Identification number of unknown sample ______________

Approximate pH of unknown ______________

Which indicators bracketed the pH of your unknown?

Lower pH ______________ Higher pH ______________

Best estimate of unknown's pH ______________

2. **Calibration of pH meter**

Describe how you calibrated the pH meter for subsequent determinations of pH. Which standard reference buffers did you use?

____

____

____

____

____

3. **pH of Salt Solutions**

Salt	*pH*	*Equation*
NaCl		
KBr		
$NaNO_3$		
K_2SO_4		
$NaC_2H_3O_2$		
NH_4Cl		
K_2CO_3		
$(NH_4)_2SO_4$		
$CuSO_4$		
$FeCl_3$		

4. **Determination of K_a for a Weak Acid**

    Identification number of unknown acid _____

    pH of half-neutralized solution of unknown _____

    $[H^+]$ = _____ K_a = _____

5. **Properties of a Buffered Solution**

    pH of buffer (half-neutralized
    acid unknown) before addition of other reagents _____

    pH of buffer with 5 drops HCl _____

    pH of buffer with 5 drops NaOH _____

    pH of buffer (standard reference)
    before addition of other reagents _____

    pH of buffer with 5 drops HCl _____

    pH of buffer with 5 drops NaOH _____

    pH of distilled water _____

    pH of water with 5 drops HCl _____

    pH of water with 5 drops NaOH _____

    Balanced equations demonstrating buffering action:

    _____

    _____

6. **Conductivities of Acid, Base, and Salt Solutions**

    HCl/NaOH observations

    _____

    _____

    Acetic acid/ammonia observations

    _____

    _____

    Mixture observations

    _____

    _____

    Explanation

    _____

    _____

    _____

Questions

1. Although in theory distilled water should have a pH of 7.00, this is only observed in systems where the water has been purged of all dissolved gases and is kept isolated from the surroundings. Carbon dioxide gas in particular—perhaps as exhaled by the person using a pH meter—will lower the pH of distilled water to less than pH 7. Write a chemical equation showing why solutions of carbon dioxide would be expected to be acidic.

2. Choose four salts from the list below and indicate whether an aqueous solution of the salt would be expected to be acidic or basic. Write an equation for each salt showing how you reached your conclusion.

   Sodium carbonate, potassium acetate, ammonium nitrate, potassium fluoride, sodium borate, methylammonium chloride (CH_3NH_3Cl), potassium cyanide, sodium benzoate ($NaC_6H_5CO_2$)

3.   In addition to pH, another important aspect of a buffered system is called the *buffer capacity*. Use your textbook or a chemical dictionary to define this term.

4.   Give 4 examples below of systems that would behave as buffered solutions. For each of your choices, indicate with chemical equations how the components of each buffered solution would consume additional hydronium or hydroxide ion.

(System 1)

(System 2)

(System 3)

(System 4)

Acid–Base Titrations 1:
Analysis of an Unknown Acid Sample

Objective

A sodium hydroxide solution will be prepared and standardized against a known acid, and is then used in titration analyses of unknown acid samples

Introduction

Titration involves measuring the exact volume of a solution of *known* concentration that is required to react with a measured volume of a solution of *unknown* concentration or with a *weighed sample* of unknown solid. A solution of accurately known concentration is called a **standard solution**. Typically, for a solution to be considered a standard solution, the concentration of the solute in the solution must be known to four significant figures.

In many cases (especially with solid solutes) it is possible to prepare a standard solution by accurate weighing of the solute, followed by precise dilution to an exactly known volume in a volumetric flask. Such a standard is said to have been prepared *determinately*. One of the most common standard solutions used in acid–base titration analyses, however, cannot be prepared in this manner.

Solutions of sodium hydroxide are commonly used in titration analyses of samples containing an acidic solute. Although sodium hydroxide is a solid, it is *not* possible to prepare standard sodium hydroxide solutions by mass. Solid sodium hydroxide is usually of questionable purity. Sodium hydroxide reacts with carbon dioxide from the atmosphere and is also capable of reacting with the glass of the container in which it is provided. For these reasons, sodium hydroxide solutions are generally prepared to be *approximately* a given concentration. They are then standardized by titration of a weighed sample of a primary standard acidic substance. By measuring how many milliliters of the approximately prepared sodium hydroxide are necessary to react completely with a weighed sample of a known primary standard acidic substance, it is possible to calculate the concentration of the sodium hydroxide solution. Once a sodium hydroxide solution is prepared, however, its concentration will change with time (for the same reasons outlined earlier). As a consequence, sodium hydroxide solutions must be used relatively quickly.

In titration analyses, there must be some means of knowing when enough titrant has been added to react exactly and completely with the sample being titrated. In an acid/base titration analysis, there should be an abrupt change in pH when the reaction is complete. For example, if the sample being titrated is an acid, then the titrant to be used will be basic (probably sodium hydroxide). When one excess drop of titrant is added (beyond that needed to react with the sample), the solution being titrated will suddenly become basic. There are various dyes, called *indicators* that exist in different colored forms at different pH values. A suitable indicator can be chosen that will change color at a pH value consistent with the point at which the titration reaction is complete. The indicator to be used in this experiment is phenolphthalein, which is colorless in acidic solutions, but changes to a pink form at basic pH.

Safety Precautions	• Protective eyewear approved by your institution must be worn at all times while you are in the laboratory.
	• The primary standard acidic substance potassium hydrogen phthalate (KHP) will be stored in an oven at 110°C to prevent moisture from being absorbed by the crystals. Use tongs or an oven mitt to remove the KHP from the oven and to handle it while it is hot.
	• Sodium hydroxide is extremely caustic, and sodium hydroxide dust is very irritating to the respiratory system. Do not handle the pellets with the fingers. Wash hands after weighing the pellets. Work in a ventilated area and avoid breathing NaOH dust.
	• Use a rubber safety bulb when pipetting. *Never pipet by mouth.*
	• The unknowns to be used are *acidic* and may be irritating and/or damaging to the skin. Avoid contact, and wash after handling them.

Apparatus/Reagents Required

50 mL buret and clamp, buret brush, 5-mL pipet and safety bulb, soap, 1-L glass or plastic bottle with stopper, sodium hydroxide pellets, primary standard grade potassium hydrogen phthalate (KHP), phenolphthalein indicator solution, unknown vinegar or acid salt sample

Procedure

Record all data and observations directly in your notebook in ink.

1. **Preparation of the Burets and Pipet**

   For precise quantitative work, volumetric glassware must be scrupulously clean. Water should run down the inside of burets and pipets in *sheets* and should *not* bead up anywhere on the interior of the glassware. Rinse the burets and the pipet with distilled water to see whether they are clean.

   If not, partially fill the buret with a few milliliters of soap solution, and rotate the buret/pipet so that all surfaces come in contact with the soap.

   Rinse with tap water, followed by several portions of distilled water. If the burets are still not clean, they should be scrubbed with a buret brush. If the pipet cannot be cleaned, it should be exchanged.

   In the subsequent procedure, it is important that water from rinsing a pipet/buret does not contaminate the solutions to be used in the glassware. This rinse water would change the concentration of the glassware's contents. Before using a pipet/buret in the following procedures, *rinse* the pipet/buret with several small portions of the solution that is to be *used* in the pipet/buret. Discard the rinsings.

2. **Preparation of the Sodium Hydroxide Solution**

   Clean and rinse the 1-L bottle and stopper. Label the bottle "Approx. 0.1 *M* NaOH." Put about 500 mL of distilled water into the bottle.

   Weigh out approximately 4 g (0.1 mol) of sodium hydroxide pellets (*Caution!*) and transfer to the 1-L bottle. Stopper and shake the bottle to dissolve the sodium hydroxide.

When the sodium hydroxide pellets have dissolved, add additional distilled water to the bottle until the water level is approximately 1 inch from the top. Stopper and shake thoroughly to mix.

This sodium hydroxide solution is the **titrant** for the analyses to follow. Keep the bottle tightly stoppered when not actually in use (to avoid exposure of the NaOH to the air).

Set up a buret in a buret clamp. See Figure 32-1. Rinse the buret and fill it with the sodium hydroxide solution just prepared.

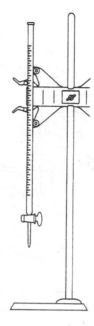

Figure 32-1. Set-up for titration

The buret should be below eye level during filling. Use a funnel when adding liquid to the buret. Make sure there are no air bubbles in the tip before taking a reading of the titrant level. Most buret scales can be read to the nearest 0.01 or 0.02 mL.

3. **Standardization of the Sodium Hydroxide Solution**

Clean and dry a small beaker. Take the beaker to the oven that contains the primary standard grade potassium hydrogen phthalate (KHP).

Using tongs or an oven mitt to protect your hands, remove the bottle of KHP from the oven, and pour a few grams into the beaker. If you pour too much, do *not* return the KHP to the bottle. Return the bottle of KHP to the oven, and take the beaker containing KHP back to your bench. Cover the beaker of KHP with a watch glass.

Allow the KHP to cool to room temperature. While the KHP is cooling, clean three 250-mL Erlenmeyer flasks with soap and water. Rinse the Erlenmeyer flasks with 5–10-mL portions of distilled water. Label the Erlenmeyer flasks as 1, 2, and 3.

When the KHP is completely cool, weigh three samples of KHP between 0.6 and 0.8 g, one for each of the Erlenmeyer flasks. Record the exact weight of each KHP sample *at least* to the nearest milligram, preferably to the nearest 0.1 mg (if an analytical balance is available). Be certain not to confuse the samples while determining their masses.

At your lab bench, add 100 mL of water to KHP sample 1. Add 2–3 drops of phenolphthalein indicator solution. Swirl to dissolve the KHP sample completely.

319

Record the initial reading of the NaOH solution in the buret to the nearest 0.02 mL, remembering to read across the bottom of the curved solution surface (meniscus).

Begin adding NaOH solution from the buret to the sample in the Erlenmeyer flask, swirling the flask constantly during the addition. (See Figure 32-2.) If your solution was prepared correctly, and if your KHP samples are of the correct size, the titration should require at least 20 mL of NaOH solution. As the NaOH solution enters the solution in the Erlenmeyer flask, streaks of red or pink will be visible. They will fade as the flask is swirled.

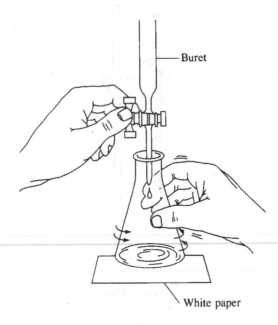

Buret

White paper

Figure 32-2. Titration technique

A right-handed person should operate the stopcock of the buret with the left hand, swirling the flask with the right hand. The tip of the buret should be well inside the flask. The titration is complete when one single drop of titrant causes a permanent pale color change.

Eventually the red streaks will persist for longer and longer periods of time. This indicates the approach of the endpoint of the titration.

Begin adding NaOH 1 drop at a time, with constant swirling, until a single drop of NaOH causes a permanent *pale* pink color that does not fade on swirling. The pink color should last at least 30 seconds, but may eventually fade over time. Record the reading of the buret to the nearest 0.02 mL.

Repeat the titration of the remaining KHP samples. Record both initial and final readings of the buret to the nearest 0.02 mL.

Given that the molar mass of potassium hydrogen phthalate is 204.22 g/mol, calculate the number of moles of KHP in samples 1, 2, and 3.

From the number of moles of KHP present in each sample, and from the volume of NaOH solution used to titrate the sample, calculate the concentration of NaOH in the titrant solution in moles per liter (molarity of NaOH, M). The reaction between NaOH and KHP is of 1:1 stoichiometry.

Example:

Suppose a 0.651 g sample of KHP required 26.25 mL of a sodium hydroxide solution to reach the phenolphthalein endpoint.

The concentration of the NaOH solution can be calculated as follows

$$0.651 \text{ g KHP} \times \frac{1 \text{ mol KHP}}{204.22 \text{ g KHP}} = 0.003188 \text{ mol KHP}$$

At the equivalence point of the titration, mol NaOH = mol KHP

$$\text{Molarity of NaOH} = \frac{0.003188 \text{ mol NaOH}}{0.02625 \text{ L NaOH}} = 0.121 \ M$$

Determine the mean (average) molarity of NaOH based on your three titrations. If your three values for the concentration differ by more than 1% from the mean, weigh out an additional sample of KHP and repeat the titration. Use the average concentration of the NaOH solution for subsequent calculations for the unknown.

4. **Analysis of the Unknown Acid Sample**

Two types of unknown acid samples may be provided. Your instructor may ask you to analyze either or both of these.

a. *Analysis of a Vinegar Solution*

Vinegar is a dilute solution of acetic acid and can be effectively titrated with NaOH using the phenolphthalein endpoint.

Clean and dry a small beaker, and obtain 25–30 mL of the unknown vinegar solution. Cover the vinegar solution with a watch glass to prevent evaporation. Record the code number of the sample. If the vinegar is a commercial product, record its brand name. Clean three Erlenmeyer flasks, and label as samples 1, 2, and 3. Rinse the flasks with small portions of distilled water.

Using the rubber safety bulb to provide suction, rinse the 5-mL pipet with small portions of the vinegar solution, and discard the rinsings.

Using the rubber safety bulb, pipet a 5-mL sample of the vinegar solution into each of the Erlenmeyer flasks. Add approximately 100 mL of distilled water to each flask, as well as 2–3 drops of phenolphthalein indicator solution.

Refill the buret with the NaOH solution, and record the initial reading of the buret to the nearest 0.02 mL. Titrate sample 1 of vinegar in the same manner as in the standardization until a single drop of NaOH causes the pale pink color to appear and persist.

Record the final reading of the buret to the nearest 0.02 mL.

Repeat the titration for the other two vinegar samples.

From the volume of vinegar sample taken, and from the volume and average concentration of NaOH titrant used, calculate the concentration of the vinegar solution in moles per liter.

Example: Suppose a 5.00 mL sample of vinegar requires 19.31 mL of 0.121 M NaOH to reach the phenolphthalein endpoint

$$\text{mol NaOH} = 0.01931 \text{ L NaOH sol.} \times \frac{0.121 \text{ mol NaOH}}{1 \text{ L NaOH sol.}} = 0.002337 \text{ mol NaOH}$$

At the endpoint, mol acetic acid = mol NaOH, so

$$\text{Molarity of acetic acid} = \frac{0.002337 \text{ mol acetic acid}}{0.00500 \text{ L acetic acid used}} = 0.0467 \ M$$

Given that the molar mass of acetic acid is 60.0 g and that the density of the vinegar solution is 1.01 g/mL, calculate the percent by weight of acetic acid in the vinegar solution.

Example:

Using the data in the previous example, and abbreviating acetic acid as "HOAc,"

$$0.467 \ M \text{ HOAc} = \frac{0.467 \text{ mol HOAc}}{1 \text{ L}} \times \frac{60.0 \text{ g HOAc}}{1 \text{ mol HOAc}} \times \frac{1 \text{ L}}{1010 \text{ g}} \times 100\% = 2.77\%$$

b. *Analysis of a Solid Acid*

As you saw with the KHP used in the standardization of NaOH, some solid substances are quite acidic. Your instructor will provide you with a solid acidic unknown substance and will tell you approximately what weight of the substance to use in your analysis. Record the code number of the sample. In this analysis, we assume that each molecule of the unknown acid releases just one proton when it ionizes.

Clean three Erlenmeyer flasks and label them as samples 1, 2, and 3.

Following the instructor's directions, weigh out three samples of the solid unknown, one into each Erlenmeyer flask. Make the weight determination at least to the nearest milligram or, preferably, to the nearest 0.1 mg (if an analytical balance is available).

Dissolve the unknown samples in approximately 100 mL of distilled water, and add 2–4 drops of phenolphthalein indicator solution.

Fill the buret with the NaOH titrant, and record the initial volume to the nearest 0.02 mL. Titrate sample 1 to the pale pink endpoint as described in the standardization of NaOH above. Record the final volume to the nearest 0.02 mL. Repeat the titration for samples 2 and 3.

From the mass of unknown sample taken, and from the volume and concentration of the NaOH used to titrate the sample, calculate the molar mass of the solid unknown acid.

Example: Suppose a 1.325 g sample of unknown acid required 24.95 mL of 0.121 M NaOH to reach the phenolphthalein endpoint.

$$\text{mol NaOH} = 0.02495 \text{ L NaOH sol.} \times \frac{0.121 \text{ mol NaOH}}{1 \text{ L NaOH sol.}} = 0.003019 \text{ mol NaOH}$$

At the equivalence point, mol unknown acid = mol NaOH

$$\text{Molar (or equivalent) mass of acid} = \frac{1.325 \text{ g unknown acid}}{0.003019 \text{ mol unknown acid}} = 439 \text{ g/mol}$$

Because you have no way of knowing whether or not the acid is monoprotic (one replaceable hydrogen atom), this is technically the equivalent mass. If the acid is indeed monoprotic, it is also the molar mass of the acid.

322

Name: ______________________________ **Section:** __________

Lab Instructor: ______________________________ **Date:** __________

EXPERIMENT 32

Acid–Base Titrations 1: Analysis of an Unknown Acid Sample

Pre-Laboratory Questions

1. Using your textbook or an online reference, define each of the following terms:

   Standard solution

   Neutralization

   Titrant

2. Using your textbook or a handbook, look up the formula and structure of potassium hydrogen phthalate (KHP) used to standardize the solution of NaOH in this experiment. Calculate the molar mass of KHP.

3. Using your textbook, a chemical dictionary, or an online reference, write the definition of an acid–base *indicator*. What indicator(s) will be used in this experiment?

4. Suppose a sodium hydroxide solution was standardized against pure solid primary standard grade KHP (see Question 1). If 0.4798 g of KHP required 45.22 mL of the sodium hydroxide to reach a phenolphthalein endpoint, what is the *molarity* of the NaOH solution?

EXPERIMENT 32

Acid–Base Titrations 1:
Analysis of an Unknown Acid Sample

Results/Observations

Standardization of NaOH Titrant (Part 3)

	Sample 1	Sample 2	Sample 3
Mass of KHP taken	_____	_____	_____
Initial NaOH buret reading	_____	_____	_____
Final NaOH buret reading	_____	_____	_____
Volume of NaOH used	_____	_____	_____
Moles of KHP present	_____	_____	_____
Molarity of NaOH solution	_____	_____	_____
Mean molarity of NaOH solution and average deviation	_____		

Analysis of Vinegar Solution (Part 4a)

Identification number (or brand) of vinegar sample used   _____

	Sample 1	Sample 2	Sample 3
Quantity of vinegar taken	_____	_____	_____
Initial NaOH buret reading	_____	_____	_____
Final NaOH buret reading	_____	_____	_____
Volume of NaOH used	_____	_____	_____
Molarity of vinegar	_____	_____	_____
Mean molarity of vinegar and average deviation	_____		
% by mass acetic acid present	_____		

Analysis of a Solid Acid (Part 4b)

Identification number of solid acid used _____

	Sample 1	Sample 2	Sample 3
Weight of unknown taken	_____	_____	_____
Initial NaOH buret reading	_____	_____	_____
Final NaOH buret reading	_____	_____	_____
Volume of NaOH used	_____	_____	_____
Moles of NaOH used	_____	_____	
Molar mass of unknown solid	_____	_____	
Mean molar mass and average deviation	_____		

Questions

1. The solid acids chosen for the analysis were typically monoprotic acidic salts such as $NaHSO_4$, $KHSO_4$, etc. Explain why such salts behave as strong enough acids to be titrated with NaOH using phenolphthalein as indicator.

2. Commercial vinegar is generally $5.0 \pm 0.5\%$ acetic acid by weight. Assuming this to be the true value for your unknown, by how much were you in error in your analysis?

3. Use a chemical dictionary or encyclopedia to explain the difference between an indicator *endpoint* for a titration analysis and the true *equivalence point* for the titration.

EXPERIMENT 33

Acid–Base Titrations 2:
Evaluation of Commercial Antacid Tablets

Objective

Antacid tablets consist of weakly basic substances that are capable of reacting with the hydrochloric acid found in the stomach, converting the stomach acid into neutral or nearly neutral salts. In this experiment, you will determine the amount of stomach acid that some commonly used antacids are able to neutralize.

Introduction

Many commercial products are sold for the relief of 'gastric hyperacidity.' Such products are commonly called *antacids*. In this experiment you will determine the relative effectiveness of several commercial preparations in neutralizing stomach acid, in terms of the amount of acid consumed per antacid tablet.

The human stomach contains cells in its walls that secrete hydrochloric acid (HCl) during the digestive process. The acid is needed during the digestive process to activate the stomach enzyme **pepsin**, which begins the digestion of proteins. A strong acid like HCl also often causes a change in the shapes of food molecules, which opens them to attack by enzymes found in the small intestine. The hydrochloric acid in the stomach is typically found at a concentration of 0.1 M (moles per liter). Occasionally, if a person overeats, or eats certain types of foods, the stomach may secrete more hydrochloric acid than is needed. This may result in an uncomfortable condition of "heartburn".

Commercial antacids come in a variety of formulations. Among those substances found in various formulations are: sodium hydrogen carbonate ($NaHCO_3$, sodium bicarbonate, "bicarb"); calcium carbonate ($CaCO_3$); magnesium hydroxide [$Mg(OH)_2$, milk of magnesia]; and a mixture of magnesium hydroxide and aluminum hydroxide [$Al(OH)_3$]. The reactions of these substances with HCl are shown below:

$$NaHCO_3 + HCl \rightarrow NaCl + H_2O + CO_2$$

$$CaCO_3 + 2HCl \rightarrow CaCl_2 + H_2O + CO_2$$

$$Mg(OH)_2 + 2HCl \rightarrow MgCl_2 + 2H_2O$$

$$Al(OH)_3 + 3HCl \rightarrow AlCl_3 + 3H_2O$$

Notice that the products of these reactions include an inert salt (such as NaCl, $CaCl_2$) and water. Two of the antacid substances (sodium hydrogen carbonate and calcium carbonate) also produce carbon dioxide gas, which may provide added relief to the heartburn (as a belch).

In this experiment, you will attempt to evaluate the effectiveness in neutralizing hydrochloric acid of various commercial antacids and will compare these products to the effectiveness of simple baking soda (sodium bicarbonate). The antacids will be dissolved in an excess of 0.5 M hydrochloric acid, and then the remaining acid (i.e., the portion of the acid that did not react with the antacid) will be titrated with standard sodium hydroxide solution. We use 0.5 M HCl (rather than 0.1 M HCl as found in the stomach) to speed up the dissolving of the antacid tablet.

It is not possible to titrate antacid tablets directly for several reasons. First, many of the weak bases used in antacid tablets [e.g., $CaCO_3$, $Mg(OH)_2$] are not very soluble in water. Second, the bases found in most antacids are weak and become buffered as they are titrated, often leading to an indistinct indicator endpoint. Finally, commercial antacid tablets frequently contain binders, fillers, flavorings, and coloring agents that may interfere with the titration.

Your instructor may ask you to make a cost-effectiveness comparison of the various antacid brands if the pricing information is available.

Safety Precautions	• **Protective eyewear approved by your institution must be worn at all times while you are in the laboratory.** • **The hydrochloric acid and sodium hydroxide solutions may be irritating to skin. Wash if these are spilled.**

Apparatus/Reagents Required

Buret and clamp, 100-mL graduated cylinder, plastic wrap, two brands of commercial antacid tablets, sodium bicarbonate, bromphenol blue indicator solution, standard 0.5 M hydrochloric acid solution (record the exact concentration), standard 0.5 M sodium hydroxide solution (record the exact concentration)

Procedure

Record all data and observations directly in your notebook in ink.

Clean and rinse a buret with water. Then rinse and fill the buret with the available standard 0.5 M sodium hydroxide solution (record the exact concentration from the label on the stock bottle).

Obtain an antacid tablet: record the brand, the active ingredient, the amount of active ingredient, and the price and number of tablets per bottle (if available). Wrap the tablet in a piece of plastic wrap or notebook paper. Crush the tablet (use the bottom of a beaker or a heavy object). Transfer the crushed tablet to a clean 250-mL Erlenmeyer flask.

Obtain about 300 mL of the standard 0.5 M hydrochloric acid (record the exact concentration). With a graduated cylinder, measure exactly 100 mL of the hydrochloric acid and add it to the Erlenmeyer flask containing the crushed antacid tablet. Swirl the flask to dissolve the tablet as much as possible. *Note:* As mentioned earlier, commercial antacid tablets contain various other ingredients, which may not completely dissolve.

After the tablet has dissolved as completely as possible, add 2–5 drops of bromphenol blue indicator solution to the sample. The indicator should be bright yellow at this point, indicating that the solution is acidic, due to the presence of excess HCl.

If the indicator is blue, this means that not enough hydrochloric acid was added to consume the antacid tablet completely. If the solution is blue, add 0.5 M HCl in 10-mL increments (record the amount used) until the sample is yellow.

Record the initial reading of the buret. Titrate the antacid sample with standard NaOH solution until the solution just barely turns blue. Record the final reading of the buret at the color change.

Repeat the procedure two more times using the *same brand* of antacid tablet.

Repeat the procedure three times either with another brand of commercial antacid tablet or with samples of baking soda (sodium bicarbonate) on the order of 0.7 g (record the exact mass used to at least 0.01g). If sodium bicarbonate is used, add the standard 0.5 M HCl to it very slowly to prevent excessive frothing as carbon dioxide is liberated.

For each of your titrations, calculate the mass of HCl consumed by the tablet. Recall that molarity, M, can be expressed either in moles per liter (mol/L), or millimoles per milliliter (mmol/mL). Similarly, just as molar masses may be expressed either in grams/mole (g/mol), or in milligrams/millimole (mg/mmol). For this experiment, mmol/mL and mg/mol will be used, as in the following example.

Example: Suppose an antacid tablet was dissolved as described above in 100. mL of 0.501 M HCl, and then was titrated with standard 0.498 M NaOH to a bromphenol blue endpoint, requiring 26.29 mL. Calculate the number of millimoles of HCl neutralized by the antacid tablet in each titration.

The volume of hydrochloric acid used to dissolve the sample, multiplied by the concentration of the HCl, represents the total number of millimoles of HCl taken.

The volume of sodium hydroxide used in the titration, multiplied by the concentration of the NaOH, represents the number of millimoles of HCl *not* consumed by the tablet.

The difference between these two quantities represents the number of millimoles of HCl that was neutralized by the tablet.

$$\text{mmol HCl used} = (100.\,\text{mL}) \times \left(\frac{0.501\ \text{mmol}}{1\ \text{mL}}\right) = 50.1\ \text{mmol HCl}$$

$$\text{mmol NaOH (not consumed by HCl)} = (26.29\ \text{mL}) \times \left(\frac{0.498\ \text{mmol NaOH}}{1\ \text{mL}}\right) = 13.1\ \text{mmol}$$

$$\text{mmol HCl neutralized} = (50.1\ \text{mmol}) - (13.1\ \text{mmol}) = 37.0\ \text{mmol HCl}$$

Given that the molar mass of HCl is 36.46 g, calculate the mass of HCl consumed by the tablet.

$$(37.0\ \text{mmol HCl}) \times \left(\frac{36.46\ \text{mg HCl}}{1\ \text{mmol HCl}}\right) = 1349\ \text{mg} = 1.35\ \text{g HCl consumed}$$

Cost Effectiveness

There often is a large difference in price among the various "brand name" and generic or store brand ant-acids. Your instructor may ask you to make a comparison among the different brands of antacid used for this experiment if the pricing information is available. To compare the cost effectiveness of one antacid versus another, you need to determine the cost of the antacid per gram of HCl consumed. Consider these two hypothetical antacids:

Brand	No. of tablets in bottle	Price	g HCl consumed/tablet
Brand A	100	$3.99	0.789 g
Brand B	30	$4.99	1.523 g

$$\text{For Brand A: Cost} = \frac{\$3.99}{(100 \text{ tablets}) \times (0.789 \text{ g/tablet})} = \$0.0506 \text{ per g HCl consumed}$$

$$\text{For Brand B: Cost} = \frac{\$4.99}{(30 \text{ tablets}) \times (1.523 \text{ g/tablet})} = \$0.109 \text{ per gram of HCl consumed}$$

So although Brand B consumes much more HCl per tablet, it costs nearly twice as much per gram of HCl consumed because there are fewer tablets in the bottle. If Brand A and Brand B were found to contain the same active ingredient, then Brand A is a much better buy.

EXPERIMENT 33

Acid–Base Titrations 2:
Evaluation of Commercial Antacid Tablets

Pre-Laboratory Questions

1. What is a *standard solution* of acid or base?

2. Rather than titrating the active ingredient in an antacid tablet directly with a standard solution of a strong acid in our buret, in this experiment we first dissolve the antacid tablet in a measured excess of standard acid, and then titrate (with standard sodium hydroxide solution) the portion of the acid that was *not* consumed by the tablet. Explain.

3.   Some of the common bases used as the active ingredient in commercial antacid tablets are listed below. Calculate the number of milliliters of 0.100 M HCl solution that could be neutralized by 1.00 g of each of these substances. Show your calculations at the bottom of this page.

    a. $Al(OH)_3$ _____

    b. $Mg(OH)_2$ _____

    c. $CaCO_3$ _____

    d. $NaHCO_3$ _____

4.   Rather than using traditional antacid tablets containing weakly basic materials that *neutralize* excess stomach acid, a class of drugs called *proton pump inhibitors* has been synthesized which prevent excess acid *from being produced* in the first place. Use a scientific encyclopedia or online reference to discuss briefly how proton pump inhibitors prevent the release of acid in the digestive tract.

Name: _____ Section: _____

Lab Instructor: _____ Date: _____

EXPERIMENT 33

Acid–Base Titrations 2:
Evaluation of Commercial Antacid Tablets

Results/Observations

Concentrations of standard solutions:

HCl _____   NaOH _____

First antacid: Brand _____

	Sample 1	*Sample 2*	*Sample 3*
Mass of tablet used	_____	_____	_____
Volume of 0.1 M HCl used	_____	_____	_____
Initial reading of buret	_____	_____	_____
Final reading of buret	_____	_____	_____
mL of 0.1 M NaOH used	_____	_____	_____
Mass of HCl consumed per g	_____	_____	_____
Mean mass of HCl consumed per gram of tablet	_____		

Second antacid: Brand _____

	Sample 1	*Sample 2*	*Sample 3*
Mass of tablet used	_____	_____	_____
Volume of 0.1 M HCl used	_____	_____	_____
Initial reading of buret	_____	_____	_____
Final reading of buret	_____	_____	_____
mL of 0.1 M NaOH used	_____	_____	_____
Mass of HCl consumed per g	_____	_____	_____
Mean mass of HCl consumed per gram of tablet	_____		

Questions

1. Which of the two antacids tested consumed more HCl per gram? Which consumed more HCl per tablet? If the prices of the antacids are available, which antacid is a better buy?

2. Generally, the substances used as antacids are either weak bases or very insoluble bases. Why is a *strong* soluble base like NaOH not used in antacid tablets?

3. Read the label(s) on the commercial antacid tablets you used. List the brand name(s) and the active antacid ingredients.

4. Some people overuse antacids. Consult a chemical encyclopedia to find out what side effects may occur if antacids are used too frequently.

EXPERIMENT 34

The Determination of Calcium
in Calcium Supplements

Objective

The amount of calcium ion in a dietary calcium supplement will be determined by titration with the important complexing agent ethylenediaminetetraacetic acid (EDTA).

Introduction

The concentration of calcium ion in the blood is very important for the development and preservation of strong bones and teeth. Absorption of too much calcium may result in the build-up of calcium deposits in the joints. More commonly, a low concentration of calcium in the blood may lead to the leaching of calcium from bones and teeth. The absorption of calcium is controlled, in part, by vitamin D.

Growing children in particular need an ample supply of calcium for the development of strong bones. Milk, which is a good natural source of calcium and which is consumed primarily by children, is usually fortified with vitamin D to promote calcium absorption. Similarly, outdoor recess for elementary school children is mandated by law in many states because exposure to sunshine allows the body to synthesize vitamin D.

In later adulthood, particularly among women, the level of calcium in the blood may decrease to the point where calcium is removed from bones and teeth to replace the blood calcium, making the bones and teeth much weaker and more susceptible to fracture. This condition is called osteoporosis and can become very serious, causing shrinking of the skeleton and severe arthritis. For this reason, calcium supplements are often prescribed in an effort to maintain the proper concentration of calcium ion in the blood. Typically, such calcium supplements consist of calcium carbonate, $CaCO_3$.

A standard analysis for calcium ion involves *titration* with a standard solution of the disodium salt of ethylenediaminetetraacetic acid (EDTA). In a titration experiment, a standard reagent of known concentration is added slowly to a measured volume of a sample of unknown concentration until the reaction is complete. From the concentration of the standard reagent, and from the volumes of standard and unknown taken, the concentration of the unknown sample may be calculated. In titration experiments, typically the sample of unknown concentration is measured with a pipet, and the volume of standard solution required is measured with a buret.

$$HOOC–CH_2 \qquad\qquad CH_2–COOH$$
$$\diagdown \qquad\qquad \diagup$$
$$N–CH_2–CH_2–N$$
$$\diagup \qquad\qquad \diagdown$$
$$HOOC–CH_2 \qquad\qquad CH_2–COOH$$

EDTA: Ethylenediaminetetraacetic acid

EDTA is a type of molecule called a complexing agent and is able to form stable, stoichiometric (usually 1:1) compounds with many metal ions. Reactions of EDTA with metal ions are especially sensitive to pH, and typically a concentrated buffer solution is added to the sample being titrated to maintain a relatively constant pH. A suitable indicator, which will change color when the reaction is complete, is also necessary for EDTA titrations. EDTA is usually provided as the disodium dihydrogen salt, Na_2H_2EDTA.

Safety Precautions	• Protective eyewear approved by your institution must be worn at all times while you are in the laboratory.
	• Hydrochloric acid solution is corrosive to eyes, skin, and clothing. Wash immediately if spilled, and clean up all spills on the benchtop.
	• Ammonia is a respiratory irritant and cardiac stimulant. The ammonia buffer solution is very concentrated *and must be kept in the fume exhaust hood at all times.* Pour out the ammonia buffer and transfer it to your calcium sample *while still under the fume exhaust hood.*
	• Use a rubber safety bulb when pipetting samples. Do *not* pipet by mouth.
	• Eriochrome Black indicator is toxic and will stain skin and clothing.

Apparatus/Reagents Required

Buret and clamp, 25-mL pipet, 250-mL volumetric flask, 3 *M* HCl, standard 0.0500 *M* Na_2H_2EDTA solution, Eriochrome Black T indicator, calcium supplement tablet

Procedure

Record all data and observations directly in your notebook in ink.

1. **Dissolving of the Calcium Supplement**

   Obtain a calcium supplement tablet. Record the calcium content of the tablet as listed on the label by the manufacturer. Place the tablet in a clean 250-mL beaker.

   Obtain 25 mL of 3 *M* HCl solution in a graduated cylinder. Over a 5-minute period, add the HCl to the beaker containing the calcium tablet in 5-mL portions, waiting between additions until all frothing of the tablet has subsided.

   After the last portion of HCl has been added, allow the calcium mixture to stand for at least an additional 5 minutes to complete the dissolving of the tablet. Since such tablets usually contain various binders, flavoring agents, and other inert material, the solution may *not* be entirely homogeneous.

   While the tablet is dissolving, clean a 250-mL volumetric flask with soap and tap water. Then rinse the flask with two 10-mL portions of distilled water. Fill your plastic wash bottle with distilled water.

   Taking care not to lose any of the mixture, transfer the tablet solution to the volumetric flask using a small funnel. To make sure that the transfer of the calcium solution has been complete, use distilled water from the plastic wash bottle to rinse the beaker that contained the calcium sample into the volumetric flask. Repeat the rinsing of the beaker twice more. Use a stream of distilled water from the

336

wash bottle to rinse any adhering calcium solution thoroughly from the funnel into the volumetric flask, and then remove the funnel.

Add distilled water to the volumetric flask until the water level is approximately 1-inch below the calibration mark on the neck of the volumetric flask. Then use a medicine dropper to add distilled water until the bottom of the solution meniscus is aligned exactly with the volumetric flask's calibration mark.

Stopper the volumetric flask securely (hold your thumb over the stopper to keep it from being displaced), and then invert the flask and shake it 10–12 times to mix the contents.

2. Preparation for Titration

Set up a 50-mL buret and buret clamp. Check the buret for cleanliness. If the buret is clean, water should run down the inside walls in sheets and should not bead up anywhere. If the buret is not clean enough for use, place approximately 10 mL of soap solution in the buret and scrub with a buret brush for several minutes.

Rinse the buret several times with tap water, and then check again for cleanliness by allowing water to run from the buret. If the buret is still not clean, repeat the scrubbing with soap solution. Once the buret is clean, rinse it with small portions of distilled water, allowing the water to drain through the stopcock.

Obtain a 25-mL volumetric pipet and rubber bulb. Using the bulb to provide suction, fill the pipet with tap water and allow the water to drain out. During the draining, check the pipet for cleanliness. If the pipet is clean, water will not bead up anywhere on the interior during draining.

If the pipet is not clean, use the rubber bulb to pipet 10–15 mL of soap solution. Holding your fingers over both ends of the pipet, rotate and tilt the pipet to rinse the interior with soap for 2–3 minutes. Allow the soap to drain from the pipet, rinse several times with tap water, and check again for cleanliness. Repeat the cleaning with soap if needed. Finally, rinse the pipet with several 5–10-mL portions of distilled water.

Obtain 200 mL of standard 0.0500 M Na$_2$H$_2$EDTA solution in a 400-mL beaker. Keep the EDTA solution covered with a watch glass when not in use.

Transfer 5–10 mL of the Na$_2$H$_2$EDTA solution to the buret. Rotate and tilt the buret to rinse and coat the walls of the buret with the Na$_2$H$_2$EDTA solution. Allow the Na$_2$H$_2$EDTA solution to drain through the stopcock to rinse the tip of the buret.

Rinse the buret twice more with 5–10-mL portions of the standard Na$_2$H$_2$EDTA solution.

After the buret has been thoroughly rinsed with Na$_2$H$_2$EDTA, fill the buret to slightly above the zero mark with the standard Na$_2$H$_2$EDTA solution.

Allow the buret to drain until the solution level is slightly below the zero mark. Read the volume of the buret (to two decimal places), estimating between the smallest scale divisions. Record this volume as the initial volume for the first titration to be performed.

Rinse four clean 250-mL Erlenmeyer flasks with distilled water. Label the flasks as samples 1, 2, 3, and 4.

Transfer the calcium tablet solution from the volumetric flask to a clean, *dry* 400-mL beaker. Using the rubber bulb for suction, rinse the pipet with several 5–10-mL portions of the calcium tablet solution to remove any distilled water still in the pipet.

Pipet *exactly* 25 mL of the calcium solution into each of the four Erlenmeyer flasks. With a graduated cylinder, add approximately 50 mL of distilled water to each sample.

3.   Titration of the Calcium Samples

Each sample in this titration must be treated *one at a time*. Do not add the necessary reagents to a particular sample until you are *ready to titrate*. Because the buffer solution used to control the pH of the calcium samples during the titration contains concentrated ammonia, it will be kept stored in the exhaust hood. Take calcium sample 1 to the exhaust hood and add 10 mL of the ammonia buffer (*Caution!*).

At your bench, add 2–3 drops of Eriochrome T indicator solution to calcium sample 1. The sample should be wine-red at this point; if the sample is blue, consult with the instructor.

Add Na_2H_2EDTA solution from the buret to the sample a few milliliters at a time, with swirling after each addition, and carefully watch the color of the sample. The color change of Eriochrome T is from red to blue, but the sample will pass through a gray transitional color as the endpoint is approached. The endpoint is the *disappearance* of the red color, not the appearance of the blue color.

As the sample becomes gray in color, begin adding Na_2H_2EDTA *dropwise* until the pure blue color of the endpoint is reached. Record to two decimal places the final volume used to titrate sample 1, estimating between the smallest scale divisions on the buret.

Refill the buret, and repeat the titration procedure for the three remaining calcium samples. Remember not to add ammonia buffer or indicator until you are actually ready to titrate a particular sample.

4.   Calculations

For each of the four titrations, use the volume of Na_2H_2EDTA required to reach the endpoint and the concentration of the standard Na_2H_2EDTA solution to calculate how many moles of Na_2H_2EDTA were used. Record.

moles Na_2H_2EDTA = (volume used to titrate in liters) × (molarity)

On the basis of the stoichiometry of the Na_2H_2EDTA /calcium reaction, calculate how many moles of calcium ion were present in each sample titrated. Record.

Using the atomic mass of calcium, calculate the mass of calcium ion present in each of the four samples titrated and the average mass of calcium present.

Given that the average mass of calcium calculated above *represents a 25-mL sample*, taken from a *total volume of 250 mL used to dissolve the tablet*, calculate the mass of calcium present in the original tablet.

Compare the mass of calcium present in the tablet calculated from the titration results with the nominal mass reported on the label by the manufacturer. Calculate the percent difference between these values.

$$\% \text{ difference} = \frac{\text{experimental mass} - \text{label mass}}{\text{label mass}} \times 100$$

Name: _____ Section: _____

Lab Instructor: _____ Date: _____

EXPERIMENT 34

The Determination of Calcium in Calcium Supplements

Pre-Laboratory Questions

1. EDTA is referred to as a *chelating agent*. Use a scientific encyclopedia or online reference to explain this term.

2. Sketch the structural formula for EDTA. What about EDTA makes it able to react with many metal ions?

3. Suppose a 25.0 mL sample of a solution containing calcium ion required 26.5 mL of 0.2071 *M* EDTA solution to reach an Eriochrome Black T endpoint. Calculate the *molarity* of calcium ion in the sample and the *number of milligrams* of calcium contained in the sample.

4. EDTA has many uses in addition to the analysis of metal ions in solution. Use a scientific encyclopedia or online reference to find three additional uses for this substance.

Name: _____ Section: _____

Lab Instructor: _____ Date: _____

EXPERIMENT 34

The Determination of Calcium in Calcium Supplements

Results/Observations

Observation

_____

_____

_____

Did the tablet dissolve *completely*? _____

Was the solution of the tablet *homogeneous*? _____

Titrations

Concentration of *standard* EDTA solution used, M _____

Volume of calcium solution taken for titration, mL _____

Sample	*1*	*2*	*3*	*4*
Final volume of EDTA, mL				
Initial volume of EDTA, mL				
Volume of EDTA used, mL				
Moles of EDTA used				
Moles of Ca^{2+} ion present				
Mass of Ca^{2+} present, g				
Average mass of Ca^{2+} , g				

Average mass of Ca^{2+} present in original tablet, g _____

Listed mass of Ca^{2+} present in tablet (from label) _____

Percent difference between experimental mass and listed mass _____

Questions

1. Suggest at least two reasons why your experimentally determined amount of calcium might differ from that listed by the manufacturer of the calcium supplement you used.

2. Most calcium supplements consist of *calcium carbonate*, $CaCO_3$, since this substance is readily available and relatively cheap. On the basis of your experimentally determined average mass of calcium ion, calculate the mass of calcium carbonate that would be equivalent to this amount of calcium ion.

3. Use your textbook or an encyclopedia of chemistry to list at least three *natural* sources of calcium ion that should be included in the diet.

EXPERIMENT 35

Determination of Iron by Redox Titration

Objective

In this experiment, one of the main types of chemical reactions, oxidation–reduction (or redox) reactions, will be used in the titration analysis of an iron compound.

Introduction

Oxidation–reduction processes form one of the major classes of chemical reactions. In **redox reactions**, electrons are transferred from one species to another. For example, in the following simple reaction

$$Zn(s) + Cu^{2+}(aq) \longrightarrow Zn^{2+}(aq) + Cu(s)$$

electrons are transferred from elemental metallic zinc to aqueous copper(II) ions. This is most easily seen if the overall reaction is written instead as two *half-reactions*, one for the oxidation and one for the reduction:

$$Zn(s) \longrightarrow Zn^{2+}(aq) + 2e^- \qquad oxidation$$
$$Cu^{2+}(aq) + 2e^- \longrightarrow Cu(s) \qquad reduction$$

In this experiment you will use potassium permanganate, $KMnO_4$, as the titrant in the analysis of an unknown sample containing **iron.** Although this method can easily be applied to the analysis of realistic iron *ores*—the native state in which an element is found in nature—for simplicity your unknown sample will contain only iron in the +2 oxidation state. In acidic solution, potassium permanganate rapidly and quantitatively *oxidizes* iron(II) to iron(III), while itself being *reduced* to manganese(II). The half-reactions for the process are:

$$MnO_4^- + 8H^+ + 5e^- \longrightarrow Mn^{2+} + 4H_2O \qquad reduction$$
$$Fe^{2+} \longrightarrow Fe^{3+} + e^- \qquad oxidation$$

When these half-reactions are combined to give the overall balanced chemical reaction equation, a factor of *five* has to be used with the iron half-reaction so that the number of electrons lost in the overall oxidation will equal the number of electrons gained in the reduction:

$$MnO_4^- + 8H^+ + 5Fe^{2+} \longrightarrow Mn^{2+} + 4H_2O + 5Fe^{3+} \quad overall$$

Potassium permanganate has been one of the most commonly used oxidizing agents because it is extremely powerful, as well as inexpensive and readily available. It does have some drawbacks, however. Because $KMnO_4$ is such a strong oxidizing agent, it reacts with practically *anything* that can be oxidized: this tends to make solutions of $KMnO_4$ difficult to store without decomposition or a change in concentration. Because of this limitation, it is common to prepare, standardize, and then use $KMnO_4$ solutions for an analysis all on the same day. It is *not* possible to prepare directly $KMnO_4$ standard solutions determinately by mass: solid potassium permanganate cannot be obtained in a completely pure state because of the high reactivity mentioned above. Rather, potassium permanganate solutions are prepared to be an *approximate* concentration and are then standardized against a known primary standard sample of the same substance that is to be analyzed in the unknown sample.

Potassium permanganate is especially useful among titrants since it requires no indicator to signal the endpoint of a titration. Potassium permanganate solutions—even at fairly dilute concentrations—are intensely colored purple. The product of the permanganate reduction half-reaction, manganese(II), in dilute solution shows almost *no* color. Therefore, during a titration using $KMnO_4$, when 1 *excess drop* of potassium permanganate has been added to the sample, the sample will take on a pale red/pink color (since there are no more sample molecules left to convert the purple MnO_4^- ions to the very faintly colored Mn^{2+} ions).

Safety Precautions	• **Protective eyewear approved by your institution must be worn at all times while you are in the laboratory.**
	• **Potassium permanganate is a strong oxidizing agent and can be damaging to skin, eyes, and clothing. Potassium permanganate solutions will stain skin and clothing if spilled.**
	• **Sulfuric acid solutions are damaging to the skin, eyes, and clothing, especially if they are allowed to concentrate through the evaporation of water. If the sulfuric acid is spilled on the skin, wash immediately and inform the instructor. Clean up all spills on the benchtop.**
	• **Iron salts may be irritating to the skin. Wash after handling.**

Apparatus/Reagents Required

Potassium permanganate, ferrous ammonium sulfate (FAS), 3 *M* sulfuric acid, buret and clamp, 250-mL Erlenmeyer flasks, 1-L glass bottle with cap, iron unknown sample

Procedure

1. **Preparation of Potassium Permanganate Solution**

   Clean out a 1-L glass bottle and its cap. Wash first with soap solution then rinse with tap water to remove all the soap. Finally, rinse with several small portions of distilled water. Then place approximately 600 mL of distilled water in the bottle.

   In a small beaker, weigh out 2.0 ± 0.1 g of potassium permanganate crystals (*Caution!*).

   Using a funnel, transfer the potassium permanganate crystals to the 1-L bottle. Cap the bottle *securely* and shake the mixture to dissolve the potassium permanganate.

   Potassium permanganate is often very *slow* in dissolving; continue shaking the solution until you are absolutely certain that all the potassium permanganate crystals have dissolved completely.

2. Preparation of the Iron Standards

Clean out three 250-mL Erlenmeyer flasks with soap and water, giving a final rinse with several small portions of distilled water. Label the flasks as samples 1, 2, and 3.

The primary standard iron(II) compound to be used for the standardization of the potassium permanganate solution is the salt ferrous ammonium sulfate hexahydrate (Mohr's salt), $FeSO_4 \cdot (NH_4)_2SO_4 \cdot 6H_2O$, which is almost universally abbreviated as "FAS".

Place approximately 3 g of primary standard FAS into a small, clean, dry beaker. Determine the mass of the beaker and FAS to at least the nearest milligram (0.001 g), or better, if an analytical balance is available, make the mass determination to the nearest 0.1 milligram (0.0001 g).

Using a funnel, carefully transfer approximately *one-third* of the FAS in the beaker (approximately 1 g) to the Erlenmeyer flask labeled sample 1. Distilled water from a wash bottle can be used to rinse the salt into the flask if the funnel should clog.

Weigh the beaker containing the remainder of the FAS. The difference in mass from the previous weighing represents the mass of FAS transferred to the sample 1 Erlenmeyer flask.

Using a funnel, carefully transfer approximately *half* the FAS remaining in the beaker (one-third of the original FAS sample) to the Erlenmeyer flask labeled sample 2. Again, rinse the FAS from the funnel into the flask with distilled water from a wash bottle.

Weigh the beaker containing the remaining FAS. The difference in mass from the previous weighing represents the mass of sample 2 that was transferred to the Erlenmeyer flask.

Using a funnel as above, transfer as much as possible of the FAS remaining in the beaker to the Sample 3 Erlenmeyer flask. Finally, weigh the empty beaker and any residual FAS that may not have been transferred completely. The difference in mass from the previous weighing represents the mass of sample 3.

Add 25 mL of distilled water to each sample, and swirl the flasks to dissolve the FAS.

Add 15 mL of 3 M sulfuric acid, H_2SO_4, to each sample (*Caution!*). Sulfuric acid is added to the samples to provide the hydrogen ions, H^+, required for the reduction of the permanganate ion.

3. Standardization of the Potassium Permanganate Solution

Clean out a buret with soap and water, and rinse with several portions of tap water, followed by several small rinsings with distilled water.

Rinse the buret with several 4–6-mL portions of the potassium permanganate solution. Tilt and rotate the buret so that the inside walls are completely rinsed with the permanganate, and allow the solution to run out of the tip of the buret to rinse the tip also. Discard the rinsings.

Finally, using a funnel *and with the buret below eye level,* fill the buret to slightly *above* the zero mark with the potassium permanganate solution. Allow a few milliliters of permanganate to run out of the tip of the buret to remove any air bubbles from the tip.

Since potassium permanganate solutions are so intensely colored, it is generally impossible to see the curved meniscus that the solution surface forms in the buret. In this case, it is acceptable to make your liquid level readings at the point where the *top* surface of the permanganate solution comes in contact with the wall of the buret. Remove the funnel from the buret and take the initial volume reading, to the nearest 0.02 mL. Record.

Place the Erlenmeyer flask containing sample 1 under the tip of the buret, and begin adding potassium permanganate to the sample a few milliliters at a time, swirling the flask after each addition of permanganate. As the permanganate solution is added to the sample, red streaks may be visible in the sample until the permanganate has had a chance to mix with and react with the iron(II) present.

Continue adding permanganate a few milliliters at a time, with swirling, until the red streaks begin to become more persistent. At this point, begin adding the permanganate *1 drop* at a time, swirling the flask constantly to mix. The endpoint is the first appearance of a *permanent*, pale pink color.

After the endpoint has been reached, record the final buret reading (to the nearest 0.02 mL). Calculate the volume of potassium permanganate solution used to titrate FAS sample 1.

Titrate FAS samples 2 and 3 in a similar manner, recording the initial and final volumes of potassium permanganate solution used to the nearest 0.02 mL.

From the mass of each FAS sample, and from the volume of $KMnO_4$ used to titrate the respective samples, calculate three values for the molarity of your potassium permanganate solution, as well as the average molarity. If any one of your individual molarities seems out-of-line from the other two, you should consult with the instructor about performing a fourth standardization titration (using an additional 1-g sample of FAS).

4.   Analysis of the Iron Unknown

Your instructor will provide you with an iron unknown and will inform you of the approximate sample size to weigh out. Prepare three samples of the iron unknown just as you did in Part 2 for the primary standard FAS, being sure to add the sulfuric acid required for the reaction, and being sure to keep the masses of the samples to the amount indicated by the instructor.

Titrate the iron unknown samples as described in Part 3 above, recording initial and final volumes to the nearest 0.02 mL. Calculate the volume of permanganate solution required to reach the pale pink endpoint for each sample.

5.   Calculations

From the *volume* required to titrate each unknown sample, and from the *average molarity* of the $KMnO_4$ solution, calculate the *number of moles of potassium permanganate* required for each iron sample.

From the *number of moles of potassium permanganate required* for titration of each iron sample, calculate the *number of moles of iron present* in each unknown sample.

From the *number of moles of iron present* in each sample, and from the *molar mass of iron*, calculate the *mass of iron present* in each unknown sample.

From the *mass of iron present* in each sample, and from each sample's *mass*, calculate the *percent iron* determined for each of your samples.

Calculate the *average* percent iron by mass present in your unknown sample.

Name: _____ Section: _____

Lab Instructor: _____ Date: _____

EXPERIMENT 35

The Determination of Iron by Redox Titration

Pre-Laboratory Questions

1. Balance the following half-reactions (all of which take place in acidic solution):

   a. $HClO(aq) \longrightarrow Cl^-(aq)$ _____

   b. $NO(aq) \longrightarrow N_2O(g)$ _____

   c. $N_2O(aq) \longrightarrow N_2(g)$ _____

   d. $ClO_3^-(aq) \longrightarrow HClO_2(aq)$ _____

   e. $O_2(g) \longrightarrow H_2O(l)$ _____

   f. $SO_4^{2-}(aq) \longrightarrow H_2SO_3(aq)$ _____

   g. $H_2O_2(aq) \longrightarrow H_2O(l)$ _____

   h. $NO_2^-(aq) \longrightarrow NO_3^-(aq)$ _____

2. Write the balanced redox equation for the reaction of iron(II) ion with permanganate ion in acidic solution.

3. Use a chemical dictionary, handbook, or encyclopedia to list several *other* uses of potassium permanganate (other than in the analysis of iron samples as in this experiment).

4. If 27.31 mL of potassium permanganate solution is required to titrate 1.0521 g of ferrous ammonium sulfate hexahydrate, $FeSO_4(NH_4)_2SO_4 \cdot 6H_2O$, calculate the molarity of the $KMnO_4$ solution.

5. If a 2.743-g sample of an unknown containing iron requires 26.35 mL of the permanganate solution described in Pre-Laboratory Question 4 to reach the endpoint, calculate the % Fe in the unknown.

EXPERIMENT 35

Determination of Iron by Redox Titration

Results/Observations

Standardization of KMnO₄ Solution

	Sample 1	*Sample 2*	*Sample 3*
Mass of FAS taken	_____	_____	_____
Initial KMnO₄ volume	_____	_____	_____
Final KMnO₄ volume	_____	_____	_____
Volume of KMnO₄ used	_____	_____	_____
Moles of iron present	_____	_____	_____
Moles of KMnO₄ present	_____	_____	_____
Molarity of KMnO₄ solution	_____	_____	_____
Mean molarity and average deviation			_____

Analysis of Unknown

	Sample 1	*Sample 2*	*Sample 3*
Mass of unknown taken	_____	_____	_____
Initial KMnO₄ volume	_____	_____	_____
Final KMnO₄ volume	_____	_____	_____
Volume of KMnO₄ used	_____	_____	_____
Moles of KMnO₄ present	_____	_____	_____
Moles of iron present	_____	_____	_____
Mass of iron present	_____	_____	_____
% of iron present	_____	_____	_____
Mean % iron present and average deviation			_____

Questions

1. For what purpose was sulfuric acid added to the iron samples before titrating?

2. Typically, a solid iron unknown for titration is dried in an oven to remove adsorbed water before analysis (the unknowns used in this experiment were dried before dispensing). How would the % Fe determined be affected if the unknowns had not been dried and had contained adsorbed water?

3. Potassium permanganate solutions can be used directly in the titration of samples containing iron in the 2+ oxidation state. Consult a textbook of analytical chemistry or a chemical encyclopedia to find *two other methods* for the determination of iron in unknown samples, and describe those methods here.

EXPERIMENT 36

Determination of Vitamin C in Fruit Juices

Objective

The concentration of Vitamin C in commercially available fruit juices will be determined.

Introduction

Vitamin C is an important essential anti-oxidant found in fresh fruits and vegetables.

Vitamin C (L-ascorbic acid)

Vitamin C is necessary in humans for the production of the protein collagen and as cofactor in several enzyme processes. Although many animals can synthesize Vitamin C, humans lack a necessary enzyme for the synthesis and have to consume Vitamin C in the diet. Lack of Vitamin C in the diet results in the disease condition called *scurvy*, which results in formation of spots on the skin, gum disease and loss of teeth, and bleeding. Before the advent of steamships, sailors on long ocean voyages in sailing ships frequently developed scurvy because of a lack of fresh fruit and vegetables in their diet when far from land. Vitamin C is destroyed by oxygen, light, and heat and so only *fresh* fruits and vegetables contain significant amounts.

In this experiment, you will determine the amount of Vitamin C present in a fruit juice by titration with an iodine/potassium iodide solution. Iodine in the presence of potassium iodide forms the triiodide ion, I_3^-. The triiodide ion is able to oxidize the Vitamin C molecule quantitatively, forming dehydroascorbic acid.

$$\underset{\textit{ascorbic acid}}{C_6H_8O_6} + I_3^- + H_2O \rightarrow \underset{\textit{dehydroascorbic acid}}{C_6H_6O_6} + 3I^- + 2H^+$$

You will determine the concentration of the iodine solution by reacting it first with a *standard* solution containing a known amount of Vitamin C per milliliter. The concentration of the iodine solution will be determined in terms of what we will call the "Vitamin C equivalency factor." This factor will represent how many mg of Vitamin C each milliliter of the iodine solution is able to oxidize. Use of this factor means that we can titrate multiple samples of fruit juices with the iodine, and then just multiply the number of milliliters of iodine that are required in the titration by the equivalency factor to get the Vitamin C content of the fruit juice.

Safety Precautions	• Safety eyewear approved by your institution must be worn at all times while you are in the laboratory, whether or not you are working on an experiment. • The iodine solution will stain skin and clothing if spilled. Transfer the buret and stand to the floor before filling the buret with iodine solution. Use a funnel. • The fruit juices used in this experiment are to be treated as "chemicals" and may not under any circumstances be consumed or removed from the laboratory.

Apparatus/Reagents Required

Buret and clamp; 25-mL pipet and safety bulb; 250-mL volumetric flask; 250-mL Erlenmeyer flasks; assorted beakers; funnel; Vitamin C tablets (250. mg Vitamin C per tablet); 1% starch solution; iodine solution; mortar and pestle

Procedure

1. **Preparation of the Vitamin C Standard Solution**

   Obtain a 250 mL volumetric flask and clean it out with soap and water. Rinse with several portions of tap water. Rinse with several portions of deionized or distilled water. Examine the flask carefully and note the location of the calibration mark on the neck of the flask. When the flask is filled exactly to the calibration mark, the flask will contain 250.0 mL.

   The neck of the flask above the calibration mark must be especially clean so that water does not bead up above the calibration mark during the preparation of the solution. If water beads up in this region, scrub the neck of the flask with a brush, rinse, and check again to see if the flask is clean. Any water retained above the calibration mark will run into the solution being prepared and change the volume of the solution. Make sure the volumetric flask is rinsed several times with distilled or deionized water prior to use.

   Obtain a 250-mg Vitamin C tablet. In this experiment, we will assume that manufacturing specifications at the company that manufactures the Vitamin C are very rigid, and that the tablet contains exactly 250. mg of Vitamin C.

   Vitamin C tablets are typically fairly hard because of binding agents that have been added, and dissolve slowly unless they are first crushed. Use a mortar and pestle to grind the tablet into a uniform powder.

   Place a funnel into the neck of the volumetric flask, and quantitatively transfer the crushed Vitamin C tablet from the mortar, through the funnel, into the volumetric flask. If any of the powdered tablet adheres to the mortar, use a stream of water from your plastic wash bottle to try to wash the tablet particles into the funnel. Once the entire tablet has been transferred to the volumetric flask, add water to the flask about halfway up the long neck of the flask: do not add water all the way to the calibration mark at this point.

   Stopper the volumetric flask and hold the stopper in place while you invert and shake the flask to dissolve the Vitamin C tablet. Continue shaking until all pieces of the tablet have dissolved. Then use a plastic dropper to add water to exactly the calibration mark of the flask.

Given that you dissolved a 250. mg Vitamin C tablet to a final volume of 250.0 mL, calculate the concentration of the standard Vitamin C solution in terms of how many milligrams of Vitamin C are contained per milliliter of the solution.

2. **Preparation of the Buret**

Obtain about 150 mL of iodine solution in a beaker. Keep the beaker covered with a watch glass to prevent evaporation.

The fruit juice analysis will be performed as a titration. In a titration, we prepare a sample in an Erlenmeyer flask of the material to be analyzed, and then determine with a buret the exact volume of a reagent required to react completely with the sample.

Obtain a 50-mL buret and clean it with soap and water until water runs in sheets down the inside of the buret without beading up anywhere on the walls of the buret.

Set up the buret in a clamp on a ring stand. See Figure 36-1.

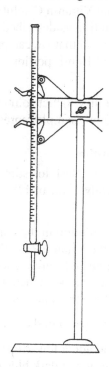

Figure 36-1. Set up for titration.

Move the buret/stand to the floor and place a clean, dry funnel in the stem of the buret. The buret will be filled on the floor because the iodine solution to be placed into the buret can stain skin and clothing if spilled. Add iodine solution to the buret until the liquid level is at least an inch above the top of the scale of the buret.

Remove the funnel and transfer the buret and stand back to the lab bench. Place a beaker under the tip of the buret, open the stopcock of the buret, and run iodine solution through the tip of the buret until there are no air bubbles in the tip. If the liquid level in the buret at this point is not below the top of the scale of the buret, run out a little more iodine solution until the liquid level is just below the top of the scale.

3. **Preparation of the Vitamin C Standard Samples**

    Label three 250-mL Erlenmeyer flasks as samples 1, 2, and 3, then clean out the flasks with soap solution, and finally rinse them with several portions of water. Perform a final rinse with several portions of deionized or distilled water.

    Obtain a 25-mL volumetric pipet and rubber safety bulb. Using the bulb, draw a small amount of soap solution into the pipet (5–10 mL). Then put your fingers over both ends of the pipet, hold the pipet horizontally, and rotate the barrel of the pipet so as to wash all inside surfaces of the pipet with the soap solution. Allow the soap solution to drain from the pipet.

    Rinse the pipet at least three times to remove all traces of soap solution, by drawing small portions of water into the pipet, and then rinsing the inside surfaces of the pipet as described above with deionized or distilled water. Allow the water to drain from the pipet before starting a new rinsing each time.

    Transfer about 100 mL of your standard Vitamin C solution from the volumetric flask to a 250-mL beaker. Using the bulb, draw 5–10 mL of the Vitamin C solution into the pipet, then rinse the inside surfaces of the pipet with the Vitamin C solution as described above. This step is to make sure all rinse water has been removed from the pipet, so that we can pipet an exact amount of Vitamin C without diluting it. Repeat the rinsing with 5–10 mL portions of standard Vitamin C solution twice more.

    Finally, pipet exactly 25.00 mL of the Vitamin C tablet solution into each of the three sample flasks. How many mg of Vitamin C does the 25.00 mL sample represent?

4. **Titration of the Vitamin C Standard Samples**

    Add approximately 75 mL (use the lines on the flask to estimate the quantity) of deionized or distilled water to Sample 1, and then add approximately 1 mL of 1% starch solution also. Swirl the flask to mix.

    As the titration takes place, iodine from the solution in the buret will react with Vitamin C in the solution in the Erlenmeyer flask. When *all* the Vitamin C in the sample has reacted, the next drop of iodine solution added will be *in excess*. Iodine and starch react with each other to form an intensely colored blue/black complex. The appearance of a permanent blue-black color in the flask that does not fade on swirling is taken as the point where the Vitamin C in the sample has completely reacted.

    Take a reading of the initial level of iodine solution in the buret and record.

    Place Sample 1 under the tip of the buret and add a few drops of iodine solution from the buret. As the iodine solution enters the Vitamin C solution, a dark blue (nearly black) cloud will form in the solution: a momentary excess of iodine occurs as the iodine solution enters the solution in the flask. Swirl the flask and note that the dark blue cloud disappears as the iodine mixes with more of the Vitamin C sample. The endpoint of the titration is when adding one additional drop of iodine causes the dark blue color to *remain* even after the solution is swirled, signaling that all the Vitamin C has reacted.

    Continue adding iodine slowly, with swirling of the flask. As you get closer and closer to the endpoint of the titration, it will take longer and longer for the dark blue color to dissipate. Begin adding iodine one drop at a time, with swirling, until one drop of iodine causes the blue color to remain even after the solution is swirled. If you are very careful, you can find the exact drop of iodine that causes the blue color to remain (the solution will only be pale blue, because there will only be a very slight excess of iodine).

    Record the final liquid level in the buret.

    Repeat the procedure for Samples 2 and 3.

Calculate the "Vitamin C equivalency factor" for your three samples and the average value of this factor.

Example:

A 20.00 mL sample of 2.00 mg/mL vitamin C solution required 31.75 mL of iodine solution to reach the blue starch endpoint. Calculate the Vitamin C equivalency factor for the iodine solution in terms of how many mg of Vitamin C each mL of the iodine solution is able to react with.

$$\text{mg Vit. C} = 20.00 \text{ mL Vit. C sol.} \times \frac{2.00 \text{ mg Vit. C}}{1 \text{ mL Vit. C sol.}} = 40.0 \text{ mg Vit. C}$$

$$\frac{40.0 \text{ mg Vit. C}}{31.75 \text{ mL iodine sol.}} = 1.26 \text{ mg Vit. C per mL iodine sol.}$$

Example:

Suppose a fruit juice sample was titrated with the iodine solution described above. If 25.00 mL of the fruit juice required 19.25 mL of iodine solution to reach the blue starch endpoint, how many milligrams of Vitamin C are contained in each milliliter of the fruit juice?

$$19.25 \text{ mL iodine sol.} \times \frac{1.26 \text{ mg Vit. C}}{1 \text{ mL iodine sol.}} = 24.3 \text{ mg Vit. C}$$

This mass of Vitamin C was contained in 25.00 mL of the fruit juice. Therefore the number of milligrams of Vitamin C per milliliter of the fruit juice is

$$\frac{24.3 \text{ mg Vit. C}}{25.00 \text{ mL juice}} = 0.972 \text{ mg Vit. C per mL of juice}$$

5. **Titration of Fruit Juice Samples**

Your instructor will have available for you one or more freshly opened samples of whole fruit juice. Take about 100 mL of one of the juices in a small beaker and record which juice you took. If your juice sample has "pulp" in it, filter the juice either through filter paper or through a double-thickness of cheesecloth held in your gravity funnel.

Clean out the three Erlenmeyer flasks and pipet 25.00 mL of juice into each.

Refill your buret with iodine solution if needed and take a reading of the initial liquid level.

Add approximately 75 mL of deionized or distilled water to fruit juice Sample 1, and then add approximately 1 mL of 1% starch solution also. Swirl the flask to mix.

Titrate Sample 1 using the same technique as used for titrating the standard Vitamin C samples. Because some fruit juices are cloudy, the appearance of the endpoint may not be as sharp as for the standard Vitamin C samples. Try to find the point where one additional drop of iodine from the buret will cause a permanent blue color that does not fade on swirling to appear in the fruit juice sample.

Record the final liquid level in the buret.

Repeat the process for fruit juice Samples 2 and 3.

Using the average "Vitamin C equivalency factor" you determined for your standard Vitamin C samples, calculate the number of milligrams of Vitamin C present per milliliter of fruit juice.

Compare your results with those obtained by other students. Of the fruit juices available in the lab, which contains the most Vitamin C per milliliter?

6. **Additional/Optional Titrations**

Your instructor may ask you to titrate additional fruit juice samples so that you can make your own comparison of Vitamin C content.

You may also be asked to investigate how heating or exposure to air affects the Vitamin C level in fruit juices. For example, you might titrate a fruit juice that was left in an open container overnight, or which had been heated strongly such as might happen during the canning process.

Regardless of the type of juice, or how the juice has been treated, the titration process is the same as you have performed for the fresh juice.

Name: _____ **Section:** _____

Lab Instructor: _____ **Date:** _____

EXPERIMENT 36

Determination of Vitamin C in Fruit Juices

Pre-Laboratory Questions

1. Given the equation for the oxidation of ascorbic acid (Vitamin C) to dehydroascorbic acid given in the Introduction, write the half-reactions for the oxidation and the reduction processes. How many electrons does each ascorbic acid molecule lose when it is oxidized?

2. Why is it important that the interior surface above the calibration mark of a volumetric flask be absolutely clean? What error would be introduced in the concentration of a solution if water beaded up above the calibration mark? Would the solution be more concentrated or less concentrated than expected?

3. Suppose a 25.00 mL pipet sample of 1.50 mg/mL ascorbic acid solution was titrated with an iodine solution, requiring 31.75 mL of the iodine solution to reach the blue-black starch endpoint. Suppose this same iodine solution was then used to titrate 25.00 mL of fruit juice, requiring 29.34 mL to reach the endpoint. What is the concentration of ascorbic acid in the fruit juice sample in mg/mL?

4. Use an online source or a chemical encyclopedia to list five foods that are especially good sources of Vitamin C.

5. Vitamin C is a very popular vitamin! Many claims are made about it in the popular press. Use an online source or a chemical encyclopedia to list five conditions or processes in which Vitamin C is proposed to be beneficial.

Name: ______________________ Section: __________

Lab Instructor: ______________________ Date: __________

EXPERIMENT 36

Determination of Vitamin C in Fruit Juices

Results/Observations

Concentration of standard Vitamin C solutions ______________________

Titration of the Vitamin C Standard Samples (Part 4)

	Sample 1	*Sample 2*	*Sample 3*
Initial reading of buret			
Final reading of buret			
mL of iodine solution used			
Vitamin C equivalency factor			

Mean Vitamin C equivalency factor ______________________

Titration of Fruit Juice Samples (Part 5)

Which fruit juice did you use? ______________________

	Sample 1	*Sample 2*	*Sample 3*
Initial reading of buret			
Final reading of buret			
mL of iodine solution used			
mg Vitamin C/mL juice			

Mean mg Vitamin C/mL juice ______________________

Titration of Additional Samples (Part 6)

Which additional juice sample did you use? ______________________

____

	Sample 1	*Sample 2*	*Sample 3*
Initial reading of buret			
Final reading of buret			
mL of iodine solution used			
mg Vitamin C/mL juice			

Mean mg Vitamin C/mL juice ______________________

Questions

1. Show how you calculated the Vitamin C equivalency factor for your Sample 1 standard.

2. Of the fruit juices available in the lab, which had the highest Vitamin C content per milliliter? Which contained the least Vitamin C per milliliter?

3. If you performed an additional titration involving a juice that had been left open to the air or which had been heated, how did the Vitamin C content of that sample compare to the Vitamin C content in the fresh juices? Explain.

4. In addition to protecting against scurvy, Vitamin C is recommended in the diet because it is an *anti-oxidant*. Use an online source or a chemical encyclopedia to define *anti-oxidant*.

EXPERIMENT 37

Electrochemistry 1: Chemical Cells

Objective

Oxidation–reduction reactions find their most important use in the construction of voltaic cells (chemical batteries). In this experiment, several such cells will be constructed and their properties studied.

Introduction

Electrochemistry is the detailed study of *electron transfer* (or oxidation–reduction) reactions. Electrochemistry is a very wide field of endeavor, covering such subjects as batteries, corrosion and reactivities of metals, and electroplating. This experiment will briefly examine some of these topics.

Consider the following reaction equation:

$$Zn(s) + Cu^{2+}(aq) \longrightarrow Zn^{2+}(aq) + Cu(s)$$

In this process, metallic elemental zinc has been added to a solution of dissolved copper(II) ion. Reaction occurs, and the metallic zinc dissolves, producing a solution of zinc ion. Concurrently, metallic elemental copper forms from the copper(II) ion that had been present in solution. To see what really is happening in this reaction equation, it is helpful to split the given equation into two *half-reactions:*

$$Zn(s) \longrightarrow Zn^{2+}(aq) + 2e^- \quad \textit{oxidation}$$

$$Cu^{2+}(aq) + 2e^- \longrightarrow Cu(s) \quad \textit{reduction}$$

$$Zn(s) + Cu^{2+}(aq) \longrightarrow Zn^{2+}(aq) + Cu(s) \quad \textit{overall}$$

The zinc half-reaction is called an *oxidation* half-reaction. **Oxidation** is a process in which a species *loses electrons* to some other species. In the zinc half-reaction, metallic zinc loses electrons in becoming zinc(II) ions. The copper half-reaction is called a *reduction* half-reaction. **Reduction** is a process in which a species *gains electrons* from some other species. In the copper half-reaction, copper(II) ions gain electrons (i.e., the electrons that had been lost by the zinc atoms) and become metallic copper.

In the zinc/copper reaction, metallic zinc has *replaced* copper(II) ion from a solution. This has happened because metallic zinc is *more reactive* than metallic copper, and zinc is more likely to be found *combined* in a compound than as the free elemental metal. The common metallic elements have been investigated for their relative reactivities and have been arranged into what is called the **electromotive series**. A portion of this series follows:

K, Na, Ba, Ca, Mg, Al, Mn, Zn, Cr, Cd, Fe, Co, Ni, Sn, Pb, H, Sb, Bi, As, Cu, Hg, Ag, Pt, Au

The more reactive elements are at the *left* of this series, and the elements become progressively less reactive moving toward the right of the series. For example, you will notice that zinc comes considerably *before* copper in the series, showing that zinc is more reactive than copper.

Notice that H (hydrogen) appears in the series. Metals to the left of hydrogen are capable of *replacing* hydrogen ion (H^+) from acids, with evolution of gaseous elemental hydrogen (H_2). In fact, the first four elements of the series are so reactive that they will even replace hydrogen from pure cold water. Elements in the series that come to the *right* of H will *not* replace hydrogen from acids and will consequently generally *not* dissolve in mineral acids. You will note that the elements at the far right side of the series include the so-called noble metals: silver, platinum, and gold. These metals are used in jewelry because they have such low reactivities and can maintain a shiny, attractive appearance. In addition to hydrogen, an element at the left of the electromotive series can replace any element to its right.

The fact that a reactive metal can replace a less reactive metal from its compounds might be interesting in itself, but there is much more implied by the electromotive series. In the zinc/copper reaction discussed at the beginning of this Introduction, we considered putting a piece of metallic zinc into *direct contact* with a solution of copper(II) ion so that electrons could flow directly from zinc atoms to copper ions. A far more useful version of this same experiment would be to set up the reaction so that the zinc metal and copper(II) ion solution are *physically separate* from one another (in separate beakers, for example) but are connected *electrically* by a conducting wire. (See Figure 37-1.) The reaction that occurs involves a transfer of electrons, which can now occur through the wire, thereby producing an electrical **current.** We could place a motor or light bulb along the wire joining the zinc/copper beakers and make use of the electrical current produced by the reaction. We have constructed a **battery** (or voltaic cell) consisting of a zinc half-cell and a copper half-cell. A second connection will have to be made between the two beakers to complete the electrical circuit, however.

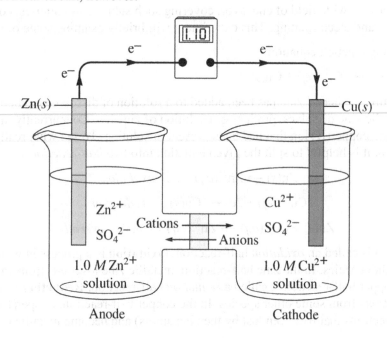

Figure 37-1. Schematic of a zinc/copper voltaic cell

Electrons flow spontaneously through the wire from the zinc half-cell to the copper half-cell when the switch is closed. Ions move to/from the salt bridge to compensate for the change in positive charge in the half-cells.

In common practice, a glass tube containing a nonreactive salt solution (a **salt bridge**) is used to do this, or, alternatively, a porous porcelain cup is used to contain one half-reaction, and is then placed in a beaker containing the second half-reaction. Reactions of voltaic cells are exergonic. They take place with the *release* of energy. This energy can be put to use if the cell is set up correctly.

It was indicated earlier that gold is the least reactive metal in the electromotive series. This would mean that metallic gold could not be produced by some more active metal replacing gold ions from solution. Yet gold is quite commonly electroplated from solutions of Au^{3+} ions onto more common, cheaper metals. The reduction of Au^{3+} ions to metallic gold is an endergonic process, requiring the *input* of energy from an external source to overcome the reluctance of Au^{3+} ions to undergo reduction. If an electrical current of sufficient voltage from an outside source is passed through a solution of Au^{3+} ions, it will provide the required energy, and metallic gold will be produced. When an electrical current is used to force a reaction to occur that would ordinarily not be capable of spontaneously occurring, **electrolysis** is said to be taking place. You will perform several electrolysis reactions in this experiment.

You will examine the relative reactivity of some metals and verify a small portion of the electromotive series. You will also set up several batteries and measure the voltage delivered by the cells. You will finally study the effect of *concentration* on the potential exhibited by a cell.

Safety Precautions	• **Protective eyewear approved by your institution must be worn at all times while you are in the laboratory.** • **Salts of metal ions may be toxic. Wash hands after use.** • **Dispose of all solutions as directed by the instructor** • **Wash hands after using sulfuric acid. Sulfuric acid *concentrates* as water evaporates from it and becomes more dangerous.**

Apparatus/Reagents Required

Equipment for voltaic cells (beakers, porous porcelain cup, voltmeter), 1 *M* sulfuric acid, 1 *M* magnesium sulfate, 1 *M* copper(II) sulfate, 1 *M* sodium sulfate, 1 *M* zinc sulfate, 0.1 *M* copper sulfate, 0.1 *M* zinc sulfate, small strips of metallic zinc and copper, magnesium turnings, 24-well plate, 4-inch strips of magnesium, copper, and zinc metals

Procedure

Record all data and observations directly in your notebook in ink.

1. **The Electromotive Series**

   In separate wells of the 24-well test plate, add 10 drops of solutions of each of the following: 1 *M* sulfuric acid, 1 *M* magnesium sulfate, 1 *M* copper(II) sulfate, 1 *M* sodium sulfate, and 1 *M* zinc sulfate. Place a small strip of metallic zinc in each well so that the metal is partially covered by the solution in the test plate.

   Allow the solutions to stand for about 15 minutes. Examine the zinc strips for evidence of reaction, both during the 15-minute waiting period and after removing them from the test plate wells. Determine which ionic species zinc is capable of *replacing* from solution, and write equations for the reactions that take place.

   Repeat the process using new 10-drop samples of the same solutions, but substituting first copper and then magnesium in place of the zinc metal.

   On the basis of your results, arrange the following elements in order of their activity: H, Mg, Cu, Zn, Na.

2. **Voltaic Cells**

   Using strips of copper, zinc, and magnesium metals as electrodes, and solutions of the sulfates of these metals, you will set up three voltaic cells and will measure the cell potentials (voltages). The following procedure is described in terms of a copper/zinc voltaic cell. You will also set up copper/magnesium and zinc/magnesium voltaic cells.

Obtain a porous porcelain cup from your instructor and place it in a 400-mL beaker of distilled water for 5 minutes to wet the cup. The porcelain cup is very fragile and expensive: be careful with it.

Add 15–20 mL of 1 M CuSO$_4$ to a 100-mL beaker. Obtain a 4-inch strip of copper metal and clean it with sandpaper. Place the copper metal strip into the beaker containing the copper sulfate solution to serve as an electrode.

Remove the porcelain cup from the water bath, and add 10–15 mL of 1 M ZnSO4 to the cup. Obtain a 4-inch strip of zinc metal and clean it with sandpaper. Place the zinc metal strip into the porous cup containing the zinc sulfate solution to serve as an electrode.

Connect one lead of the voltmeter to the copper strip, and connect the other lead of the voltmeter to the zinc strip.

Place the porous cup containing the Zn|Zn^{2+} half-cell into the beaker containing the Cu|Cu^{2+} half-cell. Allow the cell to stand until the voltage reading on the voltmeter has stabilized; then record the *highest* voltage obtained.

Using the table of standard reduction potentials in your textbook, calculate the *standard potential* for the copper/zinc voltaic cell. Calculate the % *difference* between your experimentally determined voltage and the standard voltage.

Using the same method as discussed for the copper/zinc cell, construct copper/magnesium and zinc/magnesium cells, and measure their potentials. Calculate the standard cell potential for both of these cells, and calculate the % *difference* between your experimental voltage and the standard voltage.

3.   Effect of Concentration on Cell Potential

Prepare a copper/zinc voltaic cell as in Part 2, using 1 M ZnSO$_4$ solution as before, but replace the 1 M CuSO$_4$ solution with 0.1 M CuSO$_4$ solution. Measure the voltage of the cell. How does the decrease in concentration of copper ion affect the voltage of the cell?

Prepare a copper/zinc voltaic cell as in Part 2, using 1 M CuSO$_4$ solution as before, but replace the 1 M ZnSO$_4$ solution with 0.1 M ZnSO$_4$ solution. Measure the voltage of the cell. Does the decrease in concentration of Zn^{2+} ion affect the voltage measured? Why?

Prepare a voltaic cell as in Part 2, only using Cu|Cu^{2+}(1 M) as one half cell and Cu|Cu^{2+}(0.1 M) as the other half-cell. Measure the voltage of this cell. Why does the cell show a potential difference? What is the "driving force" in this cell? This type of cell is sometimes called a "concentration cell."

EXPERIMENT 37

Electrochemistry 1: Chemical Cells

Pre-Laboratory Questions

1. Explain to a friend who has not taken any chemistry courses what a *voltaic cell* consists of, and how it may be used as a source of electricity.

2. What type of reaction takes place at the cathode in a voltaic cell? At the anode? Give an example of a voltaic cell, and indicate the balanced half-reaction illustrating each process

3. How is the voltage (potential) developed by a voltaic cell dependent on the *concentrations* of the ionic species involved in the cell reaction?

4. Given the following two standard half-cells and their associated standard reduction potentials, what cell reaction occurs when the half-cells are connected to form a battery? What is the standard potential of the battery? What voltage would have to be applied to this cell to *reverse* the spontaneous process?

$$Mn^{2+} + 2e^- \rightarrow Mn(s) \qquad E° = -1.185 \text{ V}$$

$$Ni^{2+} + 2\,e^- \rightarrow Ni(s) \qquad E° = -0.23 \text{ V}$$

Name: ______________________ **Section:** __________

Lab Instructor: ______________________ **Date:** __________

EXPERIMENT 37

Electrochemistry 1: Chemical Cells

Results/Observations

1. The Electromotive Series

Observations for reactions of metallic zinc

with 0.1 *M* sulfuric acid ______________________

with 1 *M* $MgSO_4$ ______________________

with 1 *M* $CuSO_4$ ______________________

with 1 *M* Na_2SO_4 ______________________

Observations for reactions of metallic copper

with 1 *M* sulfuric acid ______________________

with 1 *M* $MgSO_4$ ______________________

with 1 *M* $ZnSO_4$ ______________________

with 1 *M* Na_2SO_4 ______________________

Observations for reactions of metallic magnesium

with 1 *M* sulfuric acid ______________________

with 1 *M* $CuSO_4$ ______________________

with 1 *M* $ZnSO_4$ ______________________

with 1 *M* Na_2SO_4 ______________________

Write balanced equations for any reactions that occurred.

____

____

____

____

____

Order of activity

____

2. Voltaic Cells

Copper/zinc cell

Observation of zinc electrode _____

Observation of copper electrode _____

Voltage measured for the cell _____ % difference from $E°$ _____

Balanced chemical equation for the cell reaction: _____

Copper/magnesium cell

Observation of magnesium electrode _____

Observation of copper electrode _____

Voltage measured for the cell _____ % difference from $E°$ _____

Balanced chemical equation for the cell reaction: _____

Zinc/magnesium cell

Observation of magnesium electrode _____

Observation of zinc electrode _____

Voltage measured for the cell _____ % difference from $E°$ _____

Balanced chemical equation for the cell reaction: _____

3. Effect of Concentration on Cell Potential

Voltage measured with 0.1 M Cu(II) _____

Why is the voltage measured lower when [Cu(II)] is decreased?

Voltage measured with 0.1 M Zn(II) _____

How was the measured voltage affected by the decrease in Zn(II) concentration? Why?

Voltage of Cu concentration cell _____

Why does this cell show a positive potential?

What is the driving force in the concentration cell? Hint: $\Delta G = \Delta H - T\Delta S$

Questions

1. Sketch a schematic representation of a typical voltaic cell, using any reaction that is *not* discussed in this experiment. Indicate the direction of electron flow in your cell. Write balanced half-reactions for the oxidation and reduction processes in your cell, and calculate the standard potential for your cell.

2. Voltages listed in references for voltaic cells are given in terms of standard cell potentials (voltages). What is a *standard cell*? Was your initial voltaic cell a standard cell? Why or why not?

3. As a standard voltaic cell runs, the voltage delivered by the cell drops with time. Why does this happen?

4. In Part 3 of this experiment, you prepared a *concentration cell* (in which copper was used as the electrode in both half cells, but in which the concentration of copper ion differed between the half cells). Explain briefly how *entropy* is the driving force in such a cell.

Electrochemistry 2: Electrolysis

Objective

Electrolysis is the use of an electrical current to *force* a chemical reaction to occur that would ordinarily *not proceed spontaneously*. When an electrical current is passed through a molten or dissolved electrolyte between two physically separated electrodes, two chemical changes take place. At the positive electrode (also called the **anode**), an oxidation half-reaction takes place. At the negative electrode (referred to as the **cathode**), a reduction half-reaction takes place. Exactly what half-reaction occurs at each electrode is determined by the relative ease of oxidation–reduction of all the species present in the electrolysis cell. In this experiment you will study the electrolysis of water itself and also the electrolysis of a salt solution.

Introduction

When an electrical current is passed through water, two electrochemical processes take place. At the anode, water molecules are *oxidized:*

$$2H_2O \longrightarrow O_2 + 4H^+ + 4e^-$$

Gaseous elemental oxygen is produced at the anode and can be collected and tested. The solution in the immediate vicinity of the electrode becomes acidic as H^+ ions are released. At the cathode during the electrolysis of water, water molecules are *reduced*:

$$4e^- + 4H_2O \longrightarrow 2H_2 + 4OH^-$$

Gaseous elemental hydrogen is produced at the cathode and may be collected and tested. The solution in the immediate vicinity of the electrode becomes basic as OH^- ions are released.

The two processes above are called **half-reactions**. It is the *combination* of the two half-reactions, taking place at the same time but in different locations, that constitutes the overall reaction in the electrolysis cell. The overall cell reaction that takes place is obtained by adding together the two half-reactions and canceling species common to both sides:

$$6H_2O \longrightarrow O_2 + 2H_2 + 4H^+ + 4OH^-$$

However, when the hydrogen ions and hydroxide ions migrate toward each other in the cell, they will react with each other as follows:

$$H^+ + OH^- \longrightarrow H_2O$$

resulting in the production of four water molecules. This leaves the final overall equation for what occurs in the cell as simply

$$2H_2O \longrightarrow O_2 + 2H_2$$

Note the *coefficients* in this final overall equation. Twice as many moles of elemental hydrogen gas are produced as moles of elemental oxygen gas. If the gases produced are collected, then according to Avogadro's law, the *volume* of hydrogen collected should be twice the volume of oxygen collected.

The reactions that take place in an electrolysis cell are always those that require the *least expenditure of energy*. In the above discussion, we considered the electrolysis of water itself. Since water was the only reagent present in any quantity, water was both oxidized at the anode and reduced at the cathode. Now let's consider what will happen if we electrolyze a solution containing a dissolved salt. As an example, we will consider electrolyzing a solution of the salt potassium iodide, KI.

Two possible oxidation half-reactions must be considered. Depending on which species present in the solution is more easily oxidized, one of these half-reactions will represent what actually occurs in the cell:

$$2H_2O \longrightarrow O_2 + 4H^+ + 4e^-$$

$$2I^- \longrightarrow I_2(s) + 2e^-$$

In the first half-reaction, *water* is being oxidized. This half-reaction would generate elemental oxygen gas, whose presence can be detected with a glowing splint. In addition, the pH of the solution in the region of the anode would be expected to *decrease* as hydrogen ion is generated by the electrode process. An indicator might be added to determine whether the pH changes in the region of the anode. In the second possible half-reaction, elemental *iodine* is generated. Elemental iodine is slightly soluble in water, producing a brown solution. If this were to be the oxidation half-reaction, you would notice the solution surrounding the anode in the cell becoming progressively more brown as the electrolysis occurs.

The reduction half-reaction that takes place at the cathode in this experiment could also be either of two processes, again depending on which reduction requires a lower expenditure of energy:

$$2H_2O + 2e^- \longrightarrow H_2 + 2OH^-$$

$$K^+ + e^- \longrightarrow K(s)$$

If the reduction of *water* is the actual half-reaction, as in the first example, hydrogen gas will be generated at the cathode and could be collected and tested for its flammability. Notice that hydroxide ion is also produced. Hydroxide ion will make the solution basic in the region of the cathode. As before, an indicator might be added to detect this change in pH in the region of the cathode. If the actual reduction, on the other hand, were that of *potassium ion*, the cathode would be expected to increase in size (and mass) as potassium metal was plated out on the surface of the cathode. It actually is more complicated than that. As you may recall from the previous experiment (37: *Chemical Cells*), potassium metal is an extremely active metal and would immediately react with the water that was present according to

$$2\,K(s) + H_2O(l) \longrightarrow 2\,K^+(aq) + 2\,OH^-(aq) + H_2(g)$$

Note that the ultimate result for the cathode would still be formation of hydroxide ion and hydrogen gas!

By careful observation in this experiment, you should be able to determine which oxidation and which reduction of those suggested actually take place in the cell.

Safety Precautions	• Protective eyewear approved by your institution must be worn at all times while you are in the laboratory. • A 9-volt battery will be used as the source of electrical current for the electrolysis. Be aware that even a small batter can cause an electrical shock if care is not exercised. • Hydrogen gas is produced in the reaction. Hydrogen gas is extremely flammable. Use caution during its generation and testing. • Potassium iodide may be irritating to the skin. Avoid contact. • Elemental iodine will stain the skin and clothing.

Apparatus/Reagents Required

Electrolysis apparatus (9-volt battery and leads, graphite electrodes, two test tubes for collecting gases evolved), two rubber stoppers to fit the test tubes tightly, wood splints, ruler, 1 M sodium sulfate solution, potassium iodide, sodium thiosulfate, universal indicator solution and color chart, starch solution

Procedure

Record all data and observations directly in your notebook in ink.

1. **Electrolysis of Water**

   Place approximately 200 mL of distilled water in a 400-mL beaker. Add 2–3 mL of 1 M sodium sulfate to the water and stir. The sodium sulfate is added to help the electrical current pass more easily through the cell.

   Arrange the dc power supply (9-volt battery) and graphite electrodes as indicated in Figure 38-1, using connecting wires terminating in alligator clips to make the connections. *Do not connect the battery at this point, however.* Make sure that the electrodes do not touch each other, and be certain that the electrodes are arranged in such a way that the test tubes can be inverted over them easily.

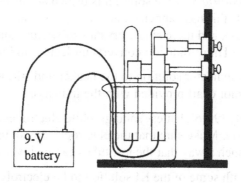

Figure 38-1. Apparatus for the electrolysis of water with collection of the evolved gases
Beware of the electrical shock hazard.

Fill each of the test tubes to be used for collecting gas with some of the water to be electrolyzed. Take one of the test tubes, and place your finger over the mouth of the test tube to prevent loss of water.

Invert the test tube, and lower the test tube into the water in the beaker. Remove your finger, and place the test tube over one of the electrodes so that the gas evolved at the electrode surface will be directed into the test tube.

If the liquid in the test tube is lost during this procedure, remove the test tube, refill with water, and repeat the transfer. Repeat the procedure with the other test tube and the remaining electrode.

Have the instructor check your set-up before continuing.

If the instructor approves, connect the battery to begin the electrolysis. Allow the electrolysis to continue until one of the test tubes is just *filled* with gas (hydrogen).

Disconnect the battery.

Stopper the test tubes while they are still *under the surface* of the water in the beaker, and remove. One test tube should be filled with gas (hydrogen), whereas the other test tube should be only about half-filled with gas (oxygen), with the remainder of the test tube filled with water.

2. **Testing of the Gases Evolved**

With a ruler, measure the approximate height of gas contained in each test tube as an index of the volume of gas that was generated. Do the relative amounts of hydrogen and oxygen generated seem to correspond to the stoichiometry of the reaction?

Ignite a wooden splint. Using a clamp or test tube holder to protect your hands, hold the test tube containing the hydrogen gas upside down (hydrogen is lighter than air) and remove the stopper. Bring the flame near the open mouth of the test tube. Describe what happens to the hydrogen when ignited.

Ignite a second wooden splint in a burner flame or match; then *blow out* the splint quickly so that the wood is still glowing. Remove the stopper from the oxygen test tube and insert the splint. Describe what happens to the splint.

3. **Electrolysis of Potassium Iodide Solution**

Weigh out approximately 2.5 g of potassium iodide and dissolve in 150 mL of distilled water. This results in an approximately 0.1 M KI solution.

The KI solution should be *colorless.* If the solution is brown at this point, some of the iodide ion present has been oxidized. If this has happened, add a *single* crystal of sodium thiosulfate and stir. If the brown color does not fade, add more single crystals of sodium thiosulfate until the potassium iodide solution is colorless. (Thiosulfate ion converts molecular iodine, I_2, to iodide ion, I^-.)

Add 4–5 drops of universal indicator solution to the beaker and stir. Record the color and pH of the solution. Keep handy the color chart provided with the indicator.

Arrange the dc power supply (9-volt battery) and graphite electrodes as indicated in Figure 38-1, but *do not connect the battery yet.* Make sure that the electrodes do not touch each other and that the electrodes are arranged in such a way that the test tubes can be inverted over them easily.

Fill each of the test tubes with some of the KI solution to be electrolyzed. Take one of the test tubes and place your finger over its mouth to prevent loss of solution.

Invert the test tube, and lower the test tube into the solution in the beaker. Remove your finger, and place the test tube over one of the electrodes so that the substances evolved at the electrode surface will be directed into the test tube.

If the liquid in the test tube is lost during this procedure, remove the test tube, refill with solution, and repeat the transfer. Repeat the procedure with the other test tube and the remaining electrode.

Wash your hands at this point to remove potassium iodide.

Have the instructor check your set-up before continuing.

If the instructor approves, connect the 9-volt battery to begin the electrolysis. Examine the electrodes for evolution of gas or deposition of a solid. Allow the electrolysis to continue for several minutes. (If a gas is generated in the cell reaction, stop the electrolysis when the test tube above the electrode is filled with the gas.)

Observe and record the color changes that take place in the solution in the region of the electrodes. Be careful to distinguish between color changes associated with the indicator and the possible production of elemental iodine (brown color). By reference to the color chart provided with the indicator, determine what pH changes (if any) have occurred near the electrodes.

While the test tubes are still under the surface of the solution in the beaker, stopper them, and remove them from the solution in the beaker.

The possible oxidation and reduction half-reactions for this system were listed in the introduction to this choice and are repeated below. By testing the contents of the two test tubes, determine which half-reactions actually occurred.

$$\text{Oxidations:} \quad 2H_2O \longrightarrow O_2 + 4H^+ + 4e^-$$

$$2I^- \longrightarrow I_2(s) + 2e^-$$

$$\text{Reductions:} \quad 2H_2O + 2e^- \longrightarrow H_2 + 2OH^-$$

$$K^+ + e^- \longrightarrow K(s)$$

If a gas is present in either test tube, use a clamp to hold the test tube, and test the gas with a glowing wood splint. If you suspect the gas is hydrogen, invert the test tube (hydrogen is lighter than air), remove the stopper, and bring the wood splint near the mouth of the test tube. Hydrogen will explode with a loud pop. If you suspect the gas is oxygen, hold the test tube upright with the clamp, remove the stopper, and insert the glowing splint. Oxygen will cause the splint to burst into full flame.

If elemental iodine were produced, one of the test tubes would contain a brown solution. Confirm the presence of iodine by addition of a few drops of starch (iodine forms an intensely-colored blue-black complex with starch). If iodine had been produced, the electrode at which the oxidation occurred would probably be coated with a thin layer of gray-black iodine crystals.

If metallic potassium were produced, it would have plated out as a thin gray-white coating on one of the electrodes, then reacted rapidly with the water.

After determining what oxidation and what reduction have actually occurred in the electrolysis cell, combine the appropriate half-reactions into the overall cell reaction for the electrolysis.

EXPERIMENT 38

Electrochemistry 2: Electrolysis

Pre-Laboratory Questions

1. In this experiment we collect the gases produced by the electrolysis of water and measure the volumes of the gases to see whether the volumes correspond to the stoichiometry of the reaction. Look up the solubilities of gaseous hydrogen and gaseous oxygen in a handbook to see whether there will be any problem with one gas dissolving more than the other as it is generated, thereby affecting the volumes of gases measured. Write what you find.

2. Suppose 25 mL of gaseous hydrogen is collected through the electrolysis of water. What volume of gaseous oxygen should also be collected? Why?

3. Describe the qualitative tests for oxygen gas and for hydrogen gas.

4. When the electrolysis of an aqueous solution of an ionic salt is undertaken, very often the half-reactions that actually occur involve *water* molecules rather than the ions of the salt. Consult your textbook to learn how you can predict what half-reactions are most likely to occur in an electrolysis experiment. Summarize your findings.

5. Suppose aqueous solutions of each of the following substances were electrolyzed. Write the half-reactions that would be most likely to occur and the overall equation for the electrolysis.

    a. $NaCl(aq)$ reduction   _____

                         oxidation   _____

                         overall   _____

    b. $KBr(aq)$ reduction   _____

                         oxidation   _____

                         overall   _____

    c. $CuCl_2(aq)$ reduction   _____

                         oxidation   _____

                         overall   _____

EXPERIMENT 38

Electrochemistry 2: Electrolysis

Results/Observations

1. **Electrolysis of Water**

   Observations on electrolysis

   Approximately how long did it take to fill the test tube with H_2? _____

   Height of gas in O_2 tube _____ in H_2 tube _____

   Ratio of H_2/O_2 heights _____ error _____

2. **Testing of the Evolved Gases**

   Observation on testing H_2 with flame

   Observation on testing O_2 with glowing wood splint

3. **Electrolysis of Potassium Iodide**

   Was it necessary to add crystals of sodium thiosulfate to the potassium iodide solution? If so, how many crystals did you add?

   _____

   Color and pH of KI solution before electrolysis _____

   Color and pH of KI solution in region of the anode _____

   Color and pH of KI solution in region of the cathode _____

   What gas(es) were evolved during the electrolysis? At which electrode? How did the gas(es) respond when tested with the glowing wood splint?

Was elemental iodine produced? How was this confirmed?

_____

_____

Was elemental potassium produced? How was this indicated?

_____

_____

Appearance of cathode after electrolysis

_____

_____

Appearance of anode after electrolysis

_____

_____

What half-reactions occurred in the electrolysis of aqueous KI?

_____

What is the overall cell reaction?

_____

_____

Questions

1. How could the *speed* of an electrolysis such as you performed be increased?

2. Why was a very small amount of sodium sulfate added to the water to be electrolyzed in Part 1?

3. In order for electrolysis to take place, the current used must be of sufficient voltage. Use your textbook to determine the *minimum voltage* needed to electrolyze pure water.

EXPERIMENT 39

Gravimetric Analysis 1: Determination of Chloride Ion

Objective

An unknown compound will be analyzed for the amount of chloride ion it contains by precipitation and weighing of the resultant silver chloride produced when the unknown is treated with silver nitrate solution.

Introduction

Gravimetric analysis is a standard classical method for determining the amount of a given component present in many solution- or solid-unknown samples. The method involves *precipitating* the component of interest from the unknown by means of some added reagent. From the *mass* of the precipitate, the percentage of the unknown component in the original sample may be calculated.

Generally, the reagent causing precipitation is chosen to be as specific as possible for the component of interest. The process is intended to remove and weigh *only* the particular component of interest from the unknown sample. The precipitating agent must also be chosen carefully so that it *completely* precipitates the component under study. The resulting precipitate must have an extremely low solubility, must be of known composition, and must be chemically stable. Handbooks of chemical data list suggested precipitating agents for routine gravimetric analysis of many unknowns.

A complete gravimetric analysis includes a series of distinct steps. First, a precise, known amount of original unknown sample must be taken for the analysis. If the unknown is a solid, an appropriately sized portion is taken and weighed as precisely as permitted by the balances available (typically to the nearest 0.001 or 0.0001 g). If the unknown is a solution, an appropriately sized aliquot (a known fraction) is taken with a pipet. The sample of unknown taken for the analysis must be large enough that precision may be maintained at a high level during the analysis, but not so large that the amount of precipitate generated cannot be handled easily.

Next, the unknown sample must be brought into solution (if it is not already dissolved). For some solid samples, such as metal ores, this may involve heating with acid to effect the dissolving. Frequently, the pH of the solution of unknown must be adjusted before precipitation can take place; this is usually done with concentrated buffer systems to maintain the pH constant throughout the analysis.

The reagent that causes the precipitation is then added to the sample. The precipitating reagent is added slowly, and in fairly dilute concentration, to allow large, easily filterable crystals of precipitate to form. The precipitate formed is allowed to stand for an extended period, perhaps at an elevated temperature, to allow the crystals of precipitate to grow as large as possible. This waiting period is called **digestion** of the precipitate.

The precipitate must then be filtered to remove it from the liquid. Although filtration could be accomplished with an ordinary gravity funnel and filter paper, this would probably be very slow. The precipitates produced in gravimetric analysis are often very finely divided and would tend to clog the pores of the filter paper. Specialized *sintered glass* filtering funnels have been prepared for routine gravimetric analyses.

Rather than filter paper, such funnels have a *frit plate* constructed of several layers of very fine compressed glass fibers that act to hold back the particles of precipitate. Such glass funnels can use *suction* to speed up the removal of liquid from a precipitate, can be cleaned easily before and after use, and are not affected by reagents in the solution. A typical sintered glass funnel is shown in Figure 43-1 and the set-up for suction filtration using the funnel is shown in Figure 43-2.

30mL-30C

Figure 39-1. A sintered glass filter funnel (crucible)

Be careful not to scratch the fritted glass plate during cleaning: the glass plate takes the place of filter paper in this kind of funnel.

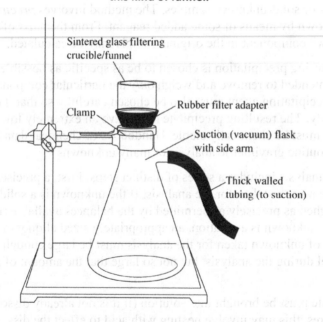

Sintered glass filtering crucible/funnel

Clamp

Rubber filter adapter

Suction (vacuum) flask with side arm

Thick walled tubing (to suction)

Figure 39-2. Set-up for suction filtration

The sintered-glass filtering crucible is resting on a soft rubber filter adapter, which makes a tight seal between the crucible and the suction flask.

Once the precipitate has been transferred to the sintered glass funnel, it must be washed to remove any adhering ions. As the precipitate forms, certain of the other species present in the mixture may have been *adsorbed* on the surface of the crystals. The liquid chosen for washing the crystals of precipitate is usually chosen so as to remove chemically such adhering ions. The washing of the precipitate must be performed carefully to prevent *peptization* of the precipitate. The wash liquid must not redissolve the precipitate or break up the crystals to the point where they might be lost through the pores of the filter.

After the precipitate has been filtered and washed, it must be dried. Drying is usually accomplished in an oven whose temperature is rigorously controlled at 110°C. The oven must be hot enough to boil off water adhering to the crystals but cannot be so hot that it might decompose the crystals. Some precipitating reagents are organic in nature and cannot stand very strong heating.

Finally, the dried precipitate is weighed. From the mass and composition of the precipitate, the mass of the component of interest in the original sample is determined. Generally, gravimetric analyses are done in triplicate (or even quadruplicate) as a check on the determination. The precision expected in a good gravimetric analysis is very high, and if there is any major deviation in the analyses, it is probably due to some source of error either in the procedure or by the operator.

One of the most common gravimetric analyses is that of the chloride ion, Cl^-. Chloride samples determined by this method typically include simple inorganic salts, samples of brine, or even body fluids.

In the analysis, a weighed sample containing chloride ion is first dissolved in dilute nitric acid. Then a dilute solution of silver nitrate is added, which causes the formation of a precipitate of silver chloride. The silver chloride is then digested, filtered, dried, and weighed. The net ionic reaction for the process is simply

$$Ag^+(aq) + Cl^-(aq) \longrightarrow AgCl(s)$$

The mass of chloride ion present in the original sample is given by

$$\text{Mass of AgCl precipitated} \times \frac{\text{molar mass of } Cl^-}{\text{molar mass of AgCl}}$$

and the percent of chloride ion in the original sample is given by

$$\frac{\text{mass of } Cl^-}{\text{mass of sample}} \times 100\%$$

The sample is dissolved originally in dilute nitric acid (rather than distilled water) to prevent precipitation of any *other* negative ions that might be present in the sample. For example, if carbonate ion (or bicarbonate ion) were present in a sample, they too would be precipitated by silver ion. Nitric acid destroys or masks such interferences by converting the competing ions to soluble species.

As initially formed, the particles of AgCl are too small to be filtered easily. Digestion of the AgCl allows smaller particles of AgCl to combine to form larger crystals that can be filtered more quantitatively. Digestion of the precipitate may be effected by either of two methods. If it is desired to complete the analysis during a single laboratory period, then the precipitate and supernatant solution can be heated almost to boiling for approximately an hour. If the analysis is to be finished at a later time, the precipitate and supernatant liquid can be allowed to stand in a dark place for a few days. Silver ion can be reduced by light to metallic silver, with loss of chloride ion to the atmosphere. Loss of chloride ion to this side reaction would affect the mass of precipitate obtained.

Safety Precautions	• **Protective eyewear approved by your institution must be worn at all time while you are in the laboratory.**
	• **Silver nitrate solution will stain the skin. The stain will wear off in 2-3 days, but you may wish to wear gloves. Nitrates are strong oxidizing agents and are toxic.**
	• **Nitric acid solutions are corrosive. Exercise caution in their use.**
	• **Handle hot glassware with tongs or an oven mitt.**

Apparatus/Reagents Required

Medium-porosity sintered glass funnels; suction filtration set-up; oven set reliably to 110°C; several clean, dry beakers for storage of funnels/samples; stirring rods with "rubber policeman"; polyethylene squeeze bottle; chloride unknown; 6 *M* nitric acid; 5% silver nitrate solution

Procedure

1. **Preparation of the Sintered-Glass Crucibles**

   Obtain two medium-porosity sintered glass funnels or crucibles. If the funnels appear to be dirty, wash them with soap and water, but *do not scrub the fritted plate*.

   Rinse thoroughly with tap water and then with small portions of distilled water. Set up a suction filtration apparatus, and place one of the funnels into position for filtration.

   Add 20 to 30 mL of distilled water to the funnel and apply suction to pull the water through the fritted plate of the funnel. Add another portion of distilled water and repeat. Repeat the process with the other funnel.

   If the funnels do not appear clean at this point, consult the instructor.

   Obtain two small beakers that can accommodate the funnels and label the beakers as 1 and 2; also label the beakers with your name. Transfer one funnel to each beaker; then place the beakers into the oven for at least 1 hour to dry. The oven should be kept at a temperature of at least 110°C.

   While the funnels are drying, go on to the next part of the experiment.

2. **Preparation and Precipitation of the Chloride Samples**

   Obtain an unknown chloride ion sample, and record the identification number in your notebook.

   Clean two 400- or 600-mL beakers and label them as 1 and 2.

   Weigh out two samples of approximately 0.4 g of the chloride unknown (one into each beaker), making the weight determinations to the precision of the balance (at least to the nearest 0.001 g).

   Add 150–200 mL of distilled water to each beaker, followed by 5 mL of 6 *M* nitric acid (*Caution!*). Stir to dissolve the samples, using separate stirring rods for each beaker (do not remove the stirring rods from the beaker from this point on, to prevent loss of the sample).

   Obtain approximately 50 mL of 5% silver nitrate solution. Each of your samples should require no more than 20–25 mL of silver nitrate to precipitate the chloride ion completely (assuming your chloride samples are close to 0.4 g in mass).

The two chloride samples may be treated one at a time, or, if two burners are available, they may be treated in parallel.

Heat the dissolved chloride sample almost to the boiling point (but do not let it actually boil) and add 5 mL of 5% silver nitrate solution. A white precipitate of silver chloride will form immediately.

While stirring the solution, continue to add silver nitrate solution slowly until a total of 15–18 mL has been added. Continue to heat the solution without boiling to allow the silver chloride precipitate to digest.

When it appears that the silver chloride has settled completely to the bottom of the beaker, and the solution above the precipitate has become clear, slowly add another 1 mL of silver nitrate. Observe the silver nitrate as it enters the chloride sample beaker to see whether additional precipitate forms.

If no additional precipitate forms at this point, continue to heat the solution for 10 minutes. Then transfer to a cool, dark place for approximately 1 hour.

If additional precipitate has formed during this period, add another 3–4 mL of silver nitrate, stir, and allow the sample to settle again. After the precipitate has settled completely, test the solution above the sample with 1 mL of silver nitrate to see whether the precipitation is complete yet. Continue in this way until it is certain that all of the chloride ion has been precipitated. Finally, transfer the sample to a cool, dark place for approximately 1 hour.

3. **Preparation for Filtration**

While the silver chloride sample is settling, remove the funnels or crucibles from the oven, and allow them to cool for 10 minutes. Cover the beakers containing the funnels to avoid any possible contamination.

When the funnels are cool, weigh them to the precision of the balance (at least to the nearest 0.001 g). Be sure not to mix up the funnels during the weighing process. Handle the funnels with tongs or strips of paper during the weighing to avoid getting finger marks on the funnels.

Clean out the suction filtration apparatus in preparation for the filtration of the silver chloride.

Prepare a washing solution consisting of approximately 2 mL of 6 *M* nitric acid in 200 mL of distilled water. Place this solution into a squeeze wash bottle.

4. **Filtration of the Silver Chloride**

Set up funnel 1 on the suction filtration apparatus.

Remove chloride sample 1 from storage, and slowly pour some of the supernatant liquid into the funnel under suction. Continue to pour off the supernatant liquid into the funnel until only a small amount of liquid remains in the chloride sample.

Carefully begin the transfer of the precipitated silver chloride, a little at a time, waiting for the liquid to be drawn off by the suction before transferring more solid. Use the stirring rod to help transfer the precipitate. Continue this process until most of the AgCl has been transferred to the funnel.

Using the rubber policeman on your stirring rod, scrape any remaining silver chloride from the beaker into the funnel. Using 5–10-mL portions of the dilute nitric acid solution in the squeeze bottle, rinse the beaker two or three times into the funnel. When it is certain that all the silver chloride has been transferred from the beaker to the funnel, scrape any silver chloride adhering to the stirring rod or rubber policeman into the funnel, and put the funnel into its numbered beaker. Transfer the beaker to the oven and dry the funnel for 1 hour.

Repeat the procedure for sample 2, using funnel 2.

When the funnels have dried for 1 hour, remove them from the oven and allow them to cool in the covered beakers for 10 minutes.

When the funnels have cooled, weigh them to the precision of the balance (at least to the nearest 0.001 g).

5. Calculations

From the *difference in mass* for each funnel, calculate the quantity of silver chloride precipitated from sample 1 and sample 2.

Calculate the mass of chloride ion present in each sample.

Calculate the *percent* chloride present in each sample.

Calculate the *mean* percent chloride for your two determinations.

EXPERIMENT 39

Gravimetric Analysis 1:
The Determination of Chloride Ion

Pre-Laboratory Questions

1. Gravimetric analysis is a traditional method of chemical analysis based on *mass*. For this reason, if performed correctly, a gravimetric analysis tends to be very precise. Use a scientific encyclopedia or online reference to find some advantages and disadvantages of gravimetric analysis over other analysis methods (titration, instrumental methods, etc.).

2. Given that the solubility product, K_{sp}, for AgCl is 1.8×10^{-10}, calculate the solubility of AgCl in moles per liter and in grams per liter.

3. A chloride unknown weighing 0.4272 g is dissolved in acidic solution and is treated with silver nitrate. The silver chloride precipitate that forms is filtered, dried, and weighed. The weight of silver chloride obtained is 0.7519 g. Calculate the percentage of chloride ion in the unknown chloride sample.

4. What is meant by *digestion* of a precipitate? Why is this performed before a precipitate is filtered?

EXPERIMENT 39

Gravimetric Analysis 1:
Determination of Chloride Ion

Results/Observations

Identification number of sample _____

	Sample 1	*Sample 2*
Mass of sample taken, g	_____	_____
Mass of empty funnel, g	_____	_____
Mass of funnel with AgCl, g	_____	_____
Mass of AgCl collected, g	_____	_____
Mass of Cl in the AgCl collected, g	_____	_____
% chloride in the sample	_____	_____
Mean % chloride	_____	

Observation on the silver chloride precipitate

Questions

1. You were told to let the precipitated chloride samples stand in a cool, *dark* place for an hour before filtering them. Why was storage in a *dark* place important? If you had stored your samples in a brightly lighted area, what error might have been introduced into your analysis?

2. Why is it customary to add a slight *excess* of silver nitrate solution when precipitating chloride ion?

3. You were told to add the silver nitrate reagent to your chloride sample slowly, and to heat the samples for several minutes to allow the precipitate of silver chloride to digest. These procedures were intended to prevent the silver chloride precipitate from *occluding* other ions. What error would be introduced if other ions were trapped within the crystals of silver chloride as they formed?

Gravimetric Analysis 2: Determination of Sulfate Ion

Objective

An unknown compound will be analyzed for the amount of sulfate ion it contains by precipitation and weighing of the resultant barium sulfate produced when the unknown is treated with barium chloride solution.

Introduction

Gravimetric analysis is a standard classical method for determining the amount of a given component present in many solid and solution unknown samples. The method involves *precipitating* the component of interest from the unknown by means of some added reagent. From the *mass* of the precipitate, the percentage of the unknown component in the original sample may be calculated.

Generally, the reagent causing precipitation is chosen to be as specific as possible for the component of interest. The process is intended to remove and weigh *only* the particular component of interest from the unknown sample. The precipitating agent must also be chosen carefully so that it *completely* precipitates the component under study. The resulting precipitate must have an extremely low solubility, must be of known composition, and must be chemically stable. Handbooks of chemical data list suggested precipitating agents for routine gravimetric analysis of many unknowns.

A complete gravimetric analysis includes a series of distinct steps. First, a precise, known amount of original unknown sample must be taken for the analysis. If the unknown is a solid, an appropriately sized portion is taken and weighed as precisely as permitted by the balances available (typically to the nearest 0.001 or 0.0001 g). If the unknown is a solution, an appropriately sized aliquot (a known fraction) is taken with a pipet. The sample of unknown taken for the analysis must be large enough that precision may be maintained at a high level during the analysis, but not so large that the amount of precipitate generated cannot be handled easily.

Next, the unknown sample must be brought into solution (if it is not already dissolved). For some solid samples, such as metal ores, this may involve heating with acid to effect the dissolving. Frequently, the pH of the solution of unknown must be adjusted before precipitation can take place; this is usually done with concentrated buffer systems to maintain the pH constant throughout the analysis.

The reagent that causes the precipitation is then added to the sample. The precipitating reagent is added slowly, and in fairly dilute concentration, to allow large, easily filterable crystals of precipitate to form. The precipitate formed is allowed to stand for an extended period, perhaps at an elevated temperature, to allow the crystals of precipitate to grow as large as possible. This waiting period is called **digestion** of the precipitate.

The precipitate must then be filtered to remove it from the liquid. Although filtration could be accomplished with an ordinary gravity funnel and filter paper, this would probably be very slow. The precipitates produced in gravimetric analysis are often very finely divided and would tend to clog the pores of the filter paper. Specialized *sintered glass* filtering funnels have been prepared for routine gravimetric analyses.

Rather than filter paper, such funnels have a *frit plate* constructed of several layers of very fine compressed glass fibers that act to hold back the particles of precipitate. Such glass funnels can use *suction* to speed up the removal of liquid from a precipitate, can be cleaned easily before and after use, and are not affected by reagents in the solution. A typical sintered glass funnel is shown in Figure 44-1 and the set-up for suction filtration using the funnel is shown in Figure 44-2.

30mL-30C

Figure 40-1. A sintered glass filter funnel (crucible)

Be careful not to scratch the fritted glass plate during cleaning: the glass plate takes the place of filter paper in this kind of funnel.

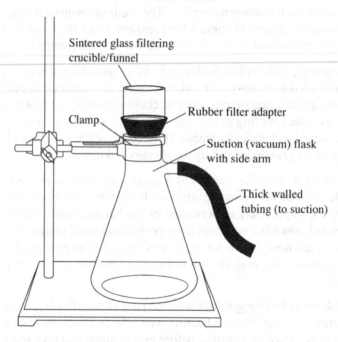

Sintered glass filtering
crucible/funnel

Rubber filter adapter

Clamp

Suction (vacuum) flask
with side arm

Thick walled
tubing (to suction)

Figure 40-2. Set-up for suction filtration

The sintered-glass filtering crucible is resting on a soft rubber filter adapter, which makes a tight seal between the crucible and the suction flask.

Once the precipitate has been transferred to the sintered glass funnel, it must be washed to remove any adhering ions. As the precipitate forms, certain of the other species present in the mixture may have been *adsorbed* on the surface of the crystals. The liquid chosen for washing the crystals of precipitate is usually chosen so as to remove chemically such adhering ions. The washing of the precipitate must be performed carefully to prevent *peptization* of the precipitate. The wash liquid must not redissolve the precipitate or break up the crystals to the point where they might be lost through the pores of the filter.

After the precipitate has been filtered and washed, it must be dried. This is usually accomplished in an oven whose temperature is rigorously controlled at 110°C. The oven must be hot enough to boil off water adhering to the crystals but cannot be so hot that it might decompose the crystals. Some precipitating reagents are organic in nature and cannot stand very strong heating.

Finally, the dried precipitate is weighed. From the mass and composition of the precipitate, the mass of the component of interest in the original sample is determined. Generally, gravimetric analyses are done in triplicate (or even quadruplicate) as a check on the determination. The precision expected in a good gravimetric analysis is very high, and if there is any major deviation in the analyses, it is probably due to some source of error either in the procedure or by the operator.

Sulfate ion is removed from acidic solution by barium ion, producing a very finely divided crystalline precipitate of barium sulfate:

$$Ba^{2+}(aq) + SO_4^{2-}(aq) \longrightarrow BaSO_4(s)$$

Although the analysis of sulfate ion itself is important, this analysis may be extended to nearly any compound containing the element sulfur by first oxidizing the sulfur present to sulfate ion. For example, sulfur in metallic sulfides or sulfite compounds can be oxidized to sulfate ion by reaction with potassium permanganate, or the sulfur compounds may be heated to a high temperature in the presence of sodium peroxide to cause the oxidation. Similarly, organic compounds containing sulfur (such as some of the amino acids) can also be oxidized with sodium peroxide.

The solution from which the sulfate is precipitated must be *acidic* for two reasons. First, barium sulfate ordinarily crystallizes in very tiny particles that often occlude impurities as they form: the presence of acid helps larger crystals of $BaSO_4$ to form. Second, in a real sample, other anions (such as carbonate) would also be precipitated by barium ion. Adding acid minimizes the precipitation of other anions that might be present by converting those ions to soluble species.

The mass of sulfate ion present in the original sample is given by

$$\text{Mass of } BaSO_4 \text{ precipitated} \times \frac{\text{molar mass of } SO_4^{2-}}{\text{molar mass of } BaSO_4}$$

and the percent of sulfate ion in the original sample is given by

$$\frac{\text{mass of } SO_4^{2-}}{\text{mass of sample}} \times 100\%$$

Safety Precautions	• **Protective eyewear approved by your institution must be worn at all time while you are in the laboratory.** • **Barium compounds are toxic. Wear disposable gloves during their use.** • **Hydrochloric acid can burn clothing and skin. Wash immediately if spilled.** • **Use tongs or an oven mitt to handle hot glassware.**

Apparatus/Reagents Required

Two sintered glass funnels or crucibles, suction filtration set-up, unknown sulfate ion sample, 5% barium chloride solution, 6 M hydrochloric acid, 0.1 M silver nitrate solution, stirring rods with rubber policeman, hotplate, distilled water, disposable gloves

Procedure

1.   Preparation of the Sintered Glass Crucibles

Obtain two medium-porosity sintered glass funnels or crucibles. If the funnels appear to be dirty, wash them with soap and water, but *do not scrub the fritted plate*.

Rinse thoroughly with tap water and then with small portions of distilled water. Set up a suction filtration apparatus, and place one of the funnels into position for filtration. See Figure 44-2.

Add 20 to 30 mL of distilled water to the funnel and apply suction to pull the water through the fritted plate of the funnel. Add another portion of distilled water and repeat. Repeat the process with the other funnel.

If the funnels do not appear clean at this point, consult the instructor.

Obtain two small beakers that can accommodate the funnels and label the beakers as 1 and 2; also label the beakers with your name. Transfer one funnel to each beaker; then place the beakers into the oven for at least 1 hour to dry. The oven should be kept at a temperature of at least 110°C.

While the funnels are drying, go on to the next part of the experiment.

2.   Preparation and Precipitation of the Sulfate Samples

Obtain an unknown sulfate sample, and record the identification number in your notebook.

Clean two 600-mL beakers, rinse with distilled water, and label them as 1 and 2. Weigh out two 0.6-g samples of the unknown sulfate sample, one into each beaker. Make the weight determinations to at least the nearest milligram (0.001 g), and do not go much above 0.65 g total mass of sample.

Add approximately 150 mL of distilled water to each beaker, followed by 2 mL of 6 M hydrochloric acid (*Caution!*). Wearing disposable gloves, obtain about 100 mL of 5% barium chloride solution, and divide into two 50-mL portions in separate clean beakers.

The precipitation of barium sulfate is carried out at an elevated temperature to assist in the formation of larger crystals. On a hotplate, heat sulfate ion sample 1 and one of the beakers of barium chloride solution nearly to boiling.

Using paper towels or an oven mitt to protect your hands, slowly pour the hot barium chloride solution in small portions into the sulfate ion sample. Stir vigorously for several minutes while continuing to heat the mixture on the hotplate.

Allow the precipitate of barium sulfate to settle to the bottom of the beaker, leaving a clear layer of supernatant liquid. Test for complete precipitation of sulfate ion by adding 1–2 drops of barium chloride solution to the clear layer.

If more $BaSO_4$ solid forms in the clear layer, add an additional 5 mL of barium chloride, stir, and allow the precipitate to settle again. Repeat the testing process until it is certain that all sulfate ion has been precipitated.

After all of the sulfate ion has been precipitated, cover the beaker with a watch glass, and continue to heat the mixture to 75–90°C on the hotplate for at least 1 hour to digest the precipitate. While the first sample is heating, precipitate the second sulfate ion sample by the same process.

3. Preparation for Filtration

Remove the two filtering funnels from the oven using tongs or an oven mitt to protect your hands, and allow them to cool covered for 10 minutes. When the funnels have reached room temperature, weigh them to the nearest milligram (0.001 g). Be careful not to confuse funnels 1 and 2 during the weighing. Return the funnels to their storage beaker for later use. Handle the funnels with tongs or paper strips during the weighing procedure to avoid getting finger marks on them.

Clean out the suction filtration apparatus in preparation for the filtration of the barium sulfate.

4. Filtration of the Barium Sulfate

The barium sulfate samples must be filtered while still hot and must be washed (to remove contaminating ions) with hot water. Heat 200–250 mL of water to 80–90°C before beginning the filtration.

Set up funnel 1 on the suction filtration apparatus. Using paper towels or an oven mitt to protect your hands from the heat, remove sulfate sample 1 from the hotplate, and slowly pour some of the supernatant liquid into the funnel under suction. Continue to pour off the supernatant liquid into the funnel until only a small amount of liquid remains in the sulfate sample.

Carefully begin the transfer of the precipitated barium sulfate, a little at a time, waiting for the liquid to be drawn off by the suction before transferring more solid. Use the stirring rod to help transfer the precipitate. Continue this process until most of the $BaSO_4$ has been transferred to the funnel.

Using the rubber policeman on your stirring rod, scrape any remaining barium sulfate from the beaker into the funnel. Using 5–10-mL portions of the hot distilled water, rinse the beaker two or three times into the funnel. When it is certain that all the barium sulfate has been transferred from the beaker to the funnel, scrape any barium sulfate adhering to the stirring rod or rubber policeman into the funnel.

Wash the precipitate with three 10-mL portions of hot distilled water. Wash the precipitate with a fourth 10-mL portion of hot water, but catch the wash liquid (after it passes through the funnel) in a test tube.

To the wash liquid in the test tube, add 2–3 drops of 0.1 M silver nitrate solution to check for the presence of chloride ion (from the $BaCl_2$ precipitating reagent). If a precipitate (of AgCl) forms in the wash liquid, wash the $BaSO_4$ precipitate in the funnel three more times with small portions of hot water. After the repeat washings, test again for chloride ion and wash the precipitate again if necessary.

After it is certain that all chloride ion has been washed from the $BaSO_4$ precipitate, continue suction for 2–3 minutes. Finally, disconnect the suction and transfer the funnel into its numbered beaker. Transfer the beaker to the oven and dry the sample for at least 1 hour.

Repeat the procedure for sample 2, using funnel 2.

When the funnels have dried for 1 hour, remove them from the oven and allow them to cool in the covered beaker for 10 minutes. When the funnels have cooled, weigh them to the precision of the balance (at least to the nearest 0.001 g).

5.   Calculations

From the difference in mass between the empty filtering funnels and the mass of the funnel containing precipitate, calculate the *mass of BaSO₄* precipitated from each sample.

From the molar masses of the sulfate ion and barium sulfate, calculate the *mass of sulfate ion* contained in each sample.

On the basis of the mass of sulfate ion present and the mass of the original sample, calculate the *percentage* by mass of sulfate in the original unknown.

Calculate the *mean* percentage of sulfate for your two determinations.

Name: _____ Section: _____

Lab Instructor: _____ Date: _____

EXPERIMENT 40

Gravimetric Analysis 2:
The Determination of Sulfate Ion

Pre-Laboratory Questions

1. Given that the solubility product, K_{sp}, for $BaSO_4$ is 1.1×10^{-10}, calculate the solubility of $BaSO_4$ in moles per liter and in grams per liter.

2. A 1.5928-g sample containing sulfate ion was treated with barium chloride reagent, and 0.2471 g of barium sulfate was isolated. Calculate the percentage of sulfate ion in the sample.

3. Why is it necessary to *wash* a precipitate before the precipitate is dried and weighed?

4. You were told to add the barium chloride reagent to your sulfate sample slowly, and to heat the samples for several minutes to allow the precipitate of barium sulfate to digest. These procedures were intended to prevent the barium sulfate precipitate from *occluding* other ions. What error would be introduced if other ions were trapped within the crystals of barium sulfate as they formed?

EXPERIMENT 40

Gravimetric Analysis 2:
Determination of Sulfate Ion

Results/Observations

Identification number of sample   _____

	Sample 1	*Sample 2*
Mass of sample taken, g	_____	_____
Mass of empty funnel, g	_____	_____
Mass of funnel with $BaSO_4$, g	_____	_____
Mass of $BaSO_4$ collected, g	_____	_____
Mass of sulfate in the $BaSO_4$, g	_____	_____
% sulfate in the sample	_____	_____
Mean % sulfate	_____	

Appearance of sulfate solution

Appearance of precipitate

Questions

1. This analysis, performed on a real sample, is subject to error because of the nature of the precipitate formed: the crystals of barium sulfate formed initially tend to occlude foreign ions (which add to the mass of the precipitate). How might this effect be prevented?

2. Barium sulfate is used in medicine as the active ingredient in the "barium cocktails" given before a series of X rays is taken of the upper gastrointestinal tract. Use your textbook, a chemical encyclopedia, or online source to find out what particular property of barium sulfate makes it useful in this connection.

3. $BaCl_2$ would also have precipitated some other ions in addition to the sulfate ion had they been present in the sample. For example, if carbonate ion, CO_3^{2-}, had been present in the sample, it would have been precipitated by the barium ion in barium chloride:

$$Ba^{2+}(aq) + CO_3^{2-}(aq) \longrightarrow BaCO_3(s)$$

How was the precipitation of carbonate ion avoided?

EXPERIMENT 41

Preparation of a Coordination Complex of Copper(II)

Objective

Coordination compounds generally contain a central metal atom, to which a fixed number of molecules or ions (called *ligands*) are coordinately covalently bonded in a characteristic geometry. Coordination complexes are very important, in both inorganic and biological systems. In this experiment, you will prepare an intensely colored coordination compound of copper(II). You also will perform some simple tests to characterize the compound.

Introduction

Coordination complex compounds play a vital role in our everyday lives. For example, the molecule *heme* in the oxygen-bearing protein hemoglobin contains coordinated iron atoms. *Chlorophyll,* the molecule that enables plants to carry on photosynthesis, is a coordination compound of magnesium.

In general, coordination compounds contain a central metal ion that is bound to several other molecules or ions by coordinate covalent bonds. The metal ion behaves as a Lewis acid (capable of accepting electron pairs in coordinate bonds). The ligand molecules or ions behave as Lewis bases (capable of providing electron pairs in coordinate bonds). For example, the compound to be synthesized in this experiment, tetramminecopper(II) sulfate, consists of a copper(II) ion surrounded by four coordinated ammonia molecules. The unshared pair of electrons on the nitrogen atom of an ammonia molecule is donated in forming the coordinate bond to the copper(II) ion.

One property of most transition metal coordination compounds that is especially striking is their color. Generally the coordinate bond between the metal ion and the ligand is formed using empty low-lying *d*-orbitals of the metal ion. Transitions of electrons within the *d*-orbitals correspond to wavelengths of visible light, and generally these transitions are very intense.

Coordination complexes of metal ions are some of the most beautifully colored chemical substances known; frequently they are used as pigments for paints. For example, the pigment known as Prussian blue is a coordination complex of iron with the cyanide ion, $Fe_4[Fe(CN)_6]_3$.

The copper/ammonia complex you will prepare today has an intensely rich blue color.

Safety Precautions	• Protective eyewear approved by your institution must be worn at all time while you are in the laboratory.
	• Copper compounds are toxic and may be irritating to the skin. Wash after handling. Copper compounds must never be poured down the drain because copper(II) is toxic to the bacteria in sewage treatment systems.
	• Ammonia solutions are caustic and irritating to the skin. Ammonia vapor is a respiratory irritant and cardiac stimulant. Work with ammonia only in the fume exhaust hood.
	• Ethyl alcohol is flammable. Avoid all flames during its use.
	• Concentrated hydrochloric acid is damaging to skin, eyes, and clothing, and its vapor is a respiratory irritant. Wash after use and rinse with a large amount of water if any HCl is spilled on the skin. Use concentrated HCl only in the fume exhaust hood.
	• Dispose of all materials as directed by the instructor.

Apparatus/Reagents Required

Suction filtration apparatus, pH paper, copper(II) sulfate pentahydrate, concentrated ammonia solution, concentrated hydrochloric acid, ethyl alcohol (ethanol)

Procedure

Record all data and observations directly in your notebook in ink.

1. **Preparation of the Complex**

   Weigh out approximately 1 g of copper(II) sulfate pentahydrate to the nearest 0.01 g (record).

   Dissolve the copper salt in approximately 10 mL of distilled water in a beaker or flask. Stir thoroughly to make certain that all the copper salt has dissolved before proceeding. Record the color of the solution at this point, that is, the color of the tetraaquo copper(II) complex, $[Cu(H_2O)_4]^{2+}$.

   Transfer the copper solution to the *exhaust hood*, and, with constant stirring, slowly add 5 mL of concentrated ammonia solution (*Caution!*). The first portion of ammonia added will cause a light blue precipitate of copper(II) hydroxide to form. As you continue adding ammonia, this precipitate will redissolve as the ammonia complex forms. Record the color of the mixture after all of the ammonia has been added.

   To decrease the solubility of the tetramminecopper(II) complex, add approximately 10 mL of ethyl alcohol with stirring. A deep blue solid should precipitate.

   Allow the solid precipitate to stand for 5–10 minutes, and then filter the precipitate under suction. See Figure 41-1.

   While it is on the filtering funnel, wash the precipitate with two 5-mL portions of ethyl alcohol, stir, and apply suction to the precipitate until it appears dry.

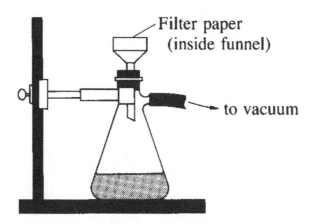

Figure 41-1. Set-up for suction filtration using a Buchner funnel

Wet the filter paper so that it adheres to the funnel and will not be displaced when the solution to be filtered is added.

Allow the precipitate to dry in air for at least 1 hour. Weigh the dried precipitate and record the yield.

On the basis of the weight of copper(II) sulfate pentahydrate taken, calculate the theoretical and percentage yields of the product, tetramminecopper(II) sulfate monohydrate, $[Cu(NH_3)_4]SO_4 \cdot H_2O$.

2.   Tests on the Product

Dissolve a small amount of the product in a few milliliters of water. In the exhaust hood, add concentrated HCl dropwise (*Caution!*) until a color change is evident.

Ammonia molecules have a stronger affinity for protons than for copper ions. When HCl is added to the copper complex, the ammonia molecules are converted to ammonium ions, which no longer have an unshared pair of electrons and so cannot bond to the copper(II) ion.

Heat a very small amount of the copper complex in a test tube in a burner flame. Hold a piece of moist pH test paper near the mouth of the test tube while it is being heated.

When the copper complex is heated, the ammonia molecules are driven out of it. Ammonia is a base and should change the color of the pH test paper.

Cautiously waft toward your nose some of the vapor produced from the complex as it is heated. The odor of ammonia should be detected.

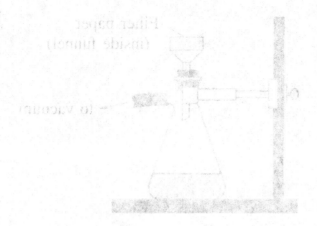

Filter paper
(inside funnel)

(to vacuum)

EXPERIMENT 41

Preparation of a Coordination Complex of Copper(II)

Pre-Laboratory Questions

1. For the synthesis to be performed in this experiment, on the basis of 1.00 g of copper(II) sulfate pentahydrate taken initially, what is the theoretical yield of product tetramminecopper(II) sulfate hydrate?

2. What is a *ligand*? What properties of a molecule might make the molecule able to act as a ligand in forming coordination complexes with the ions of transition metals?

3. Briefly explain why so many coordination compounds of transition metals are *brightly colored*.

4. In addition to the three examples of coordination compounds mentioned in this experiment (heme, chlorophyll, Prussian blue), use a scientific encyclopedia or online reference to find three other examples of important coordination compounds. Give their formulas and uses.

EXPERIMENT 41

Preparation of a Coordination Complex of Copper(II)

Results/Observations

Mass of copper(II) sulfate pentahydrate taken _____

Theoretical yield of product _____

Mass of copper complex obtained _____

% yield _____

Effect of adding HCl to product

_____

_____

Conclusion

_____

_____

_____

_____

Balanced equation _____

Observation on heating product

_____

_____

_____

Effect on pH paper _____

Odor of vapor from heating product _____

Balanced equation _____

Questions

1. Use your textbook, a chemical encyclopedia, or an online reference to describe the structure, shape, and bonding found in the $[Cu(NH_3)_4]^{2+}$ complex ion. Sketch a representation of the shape of the molecule.

2. Copper(II) ion exists in water as the $[Cu(H_2O)_4]^{2+}$ ion. Ammonia forms stronger bonds to copper than does water and hence was able to *replace* the water molecules in this complex ion to form the tetrammine complex, $[Cu(NH_3)_4]^{2+}$. What properties of a ligand make it a "strong" ligand? Use your textbook, an encyclopedia of chemistry, or an online reference to find an example of a ligand that would be "stronger" than ammonia and thus would be able to replace ammonia from $[Cu(NH_3)_4]^{2+}$.

3. Suppose that rather than reacting copper(II) with *ammonia*, you had reacted it with an *ammonium salt*, such as NH_4Cl. Would the same coordination complex have formed? Why or why not? What product would have formed (if any)?

Inorganic Preparations 1:
Synthesis of Sodium Thiosulfate Pentahydrate

Objective

Synthetic chemistry is one of the most important and practical fields of science. The synthetic chemist designs laboratory procedures for the production of new and useful compounds from natural resources or other simpler synthetic substances. In this experiment you will prepare sodium thiosulfate pentahydrate.

Introduction

Sodium thiosulfate pentahydrate, $Na_2S_2O_3 \cdot 5H_2O$, is a substance that finds several important uses, both in the chemistry laboratory and in commercial situations. The thiosulfate ion, $S_2O_3^{2-}$, has a Lewis structure that is comparable to that of the sulfate ion, SO_4^{2-}, with one of the oxygen atoms of sulfate replaced by an additional sulfur atom:

sulfate   thiosulfate

The presence of the second sulfur atom in thiosulfate ion makes this species a better reducing agent than sulfate ion. This ability of thiosulfate to serve as a reducing agent is at the basis of most of its common uses.

Standard solutions of sodium thiosulfate are used as the titrant in analyses of samples containing elemental iodine, I_2:

$$2S_2O_3^{2-}(aq) + I_2(aq) \longrightarrow S_4O_6^{2-}(aq) + 2I^-(aq)$$

These titrations are monitored by watching for the disappearance of the brown color of elemental iodine as the equivalence point is approached. This is usually enhanced by addition of a small amount of starch near the endpoint of the titration, which results in the production of the characteristic blue-black color of the starch/iodine complex.[1] Although not many real samples would be expected to contain elemental iodine itself, oftentimes iodine can be generated in a sample containing an oxidizing agent by addition of an excess amount of potassium iodide:

$$2I^-(aq) \longrightarrow I_2(aq) + 2e^-$$

Naturally, the quantity of elemental iodine produced is directly related to the quantity of oxidizing agent present in the original sample. By titration of the elemental iodine thus produced with standard thiosulfate solution, the quantity of oxidizing agent in the original sample, which resulted in the production of iodine, may be determined.

---

[1] You saw this in Experiment 30, *The Solubility Product Constant of Calcium Iodate.*

Before digital cameras became the norm, sodium thiosulfate pentahydrate was used in photography, and still is, to an extent. The substance is known more commonly as *hypo* when used for that purpose. Monochrome photographic film contains a thin coating of silver bromide, AgBr, spread on plastic. Silver bromide is sensitive to light: silver ions are reduced to silver atoms when exposed to light through a camera's lens. The production of elemental silver on the film forms the negative image. When exposed photographic film is developed, any unreacted silver bromide must be removed from the film or it will also be reduced when the film is exposed to further light. Thiosulfate ion complexes and dissolves silver ions, but it is not able to dissolve metallic silver:

$$AgBr(s) + 2S_2O_3^{2-}(aq) \longrightarrow Ag(S_2O_3)_2^{3-}(aq) + Br^-(aq)$$

When sodium thiosulfate is used to fix the film, only the negative image (elemental silver) remains on the film. Any AgBr that was not reduced by light in forming the image is washed away.

Sodium thiosulfate may be prepared by reaction between aqueous sodium sulfite and elemental sulfur. Since elemental sulfur is not soluble in water, a small amount of detergent is added to the system to promote wetting of the sulfur.

Safety Precautions	• **Protective eyewear approved by your institution must be worn at all time while you are in the laboratory.** • **Sodium sulfite is toxic and irritating to the skin. Sodium sulfite evolves toxic, irritating sulfur dioxide gas if acidified or heated.** • **Elemental iodine will stain skin and clothing. Silver salts will stain the skin if spilled and not washed off quickly: the stain is elemental silver and will take several days to wear off.** • **As the solution of sodium thiosulfate is concentrated, spattering may occur. Keep the evaporating dish covered with a watch glass while heating.** • **Use tongs or an oven mitt when handling hot glassware**

Apparatus/Reagents Required

Sodium sulfite, powdered sulfur, detergent, 0.1 *M* I$_2$/KI, 0.1 *M* silver nitrate, 0.1 *M* sodium bromide, evaporating dish, crystallizing dish, suction filtration apparatus, gravity funnel, 6 *M* hydrochloric acid.

Procedure

Record all data and observations directly in your notebook in ink.

1. **Synthesis of Sodium Thiosulfate**

    Weigh 15 g of sodium sulfite and 10 g of powdered sulfur (each to the nearest 0.1 g) into a clean 400-mL beaker.

    Add approximately 75 mL of water and 5 drops of concentrated laboratory detergent to the beaker. Stir to dissolve the sodium sulfite. Determine the pH of the mixture with pH paper.

    Transfer the beaker to a ringstand, cover with a watchglass, and heat the mixture to a *gentle* boil for approximately 30 minutes.

    The presence of the detergent makes this mixture tend to bubble and foam excessively. Keep the heat as *low as possible* while still maintaining boiling.

    Stir the mixture frequently during the heating period to promote the mixing of the powdered sulfur. During the heating, watch for a subtle change in the appearance and color of the powdered sulfur: the bright yellow color of the sulfur becomes somewhat *lighter* as the reaction proceeds.

    After boiling for 30 minutes, check the pH of the solution with pH paper. The pH of the solution should have dropped to nearly neutral. If the pH has not dropped to at least pH 8, continue heating for an additional 15 minutes. When the pH has reached the nearly neutral point, remove the heat.

    Allow the mixture to cool until it can be handled easily. Filter the mixture through filter paper on a gravity funnel to remove unreacted sulfur. Collect the filtrate in a clean beaker.

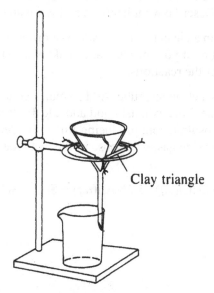

Clay triangle

Figure 42-1. Set up for gravity filtration of the solution

**Gravity filtration is used when we want to
remove solid impurities from a solution of interest.**

Using the beaker calibration marks, note the approximate volume of the solution at this point.

Transfer the filtrate to an evaporating dish. Cover the evaporating dish with a watchglass, and heat the solution on a ringstand using a very small flame. Do not rapidly boil the solution, or a portion of the product may be lost to spattering.

Continue heating with a small flame until the volume of the solution has been reduced approximately 50%.

Protect your hands with a towel while transferring the hot contents of the evaporating dish to a clean, dry crystallizing dish. Allow the solution to cool to room temperature.

Scratch the bottom and sides of the crystallizing dish with a stirring rod during this cooling period to assist nucleation of crystal formation.

If no crystals have formed by the time the solution has reached room temperature, return the solution to the evaporating dish, and reduce its volume by an additional 25%. Cool the solution again to room temperature to allow crystallization.

Filter the crystals through a Büchner funnel under suction. Allow the suction to continue for several minutes to promote drying of the crystals.

Remove a tiny portion of the product for the tests indicated. Allow the remainder of the product to dry in the air for approximately 1 hour. Weigh the product and calculate the percent yield.

2.   Tests on Sodium Thiosulfate

Place 5 mL of distilled water in a clean test tube. Add 5–6 drops of 0.1 M I_2/KI reagent, followed by a small amount of your product. Describe what happens and write an equation for the reaction.

Obtain 2–3 mL of 0.1 M NaBr in a clean test tube. Add 3–4 drops of 0.1 M silver nitrate. Record your observations. Add a small amount of your product and shake the test tube. Record your observations and write balanced equations for the reactions.

Place 5 mL of distilled water in a clean test tube. Add a small amount of your product and stir to dissolve. Take the test tube to the fume exhaust hood and add 10 drops of 6 M HCl. Adding acid causes the breakdown of the thiosulfate ion, producing finely divided elemental sulfur. The reaction also generates noxious sulfur dioxide gas, so dispose of the test tube contents while still under the fume exhaust hood.

$$Na_2S_2O_3(aq) + 2HCl(aq) \rightarrow 2NaCl(aq) + S(s) + SO_2(g) + H_2O(l)$$

EXPERIMENT 42

Inorganic Preparations 1:
Synthesis of Sodium Thiosulfate Pentahydrate

Pre-Laboratory Questions

1. Write a balanced chemical equation for the synthesis of sodium thiosulfate pentahydrate, using aqueous sodium sulfite and elemental sulfur as reactants.

2. Use a chemical encyclopedia or an online reference to find 5 commercial uses of sodium thiosulfate and record your findings.

3. Write balanced chemical equations for the reaction of thiosulfate ion with each of the following:

    a.   Elemental iodine, I_2 _____

    b.   Silver bromide, AgBr _____

    c.   Hydrochloric acid, HCl _____

4. If 11.5 g of sodium sulfite is reacted with 9.95 g of powdered sulfur, which substance is present in excess? By how many grams is this substance in excess? Show your reasoning.

EXPERIMENT 42

Inorganic Preparations 1:
Synthesis of Sodium Thiosulfate Pentahydrate

Results/Observations

Mass of sodium sulfite taken _____

Mass of sulfur taken _____

pH of initial reaction mixture _____

Observation on heating mixture

pH of mixture after heating _____

Explanation of pH change

Theoretical yield of product _____

Actual yield of product _____

% yield _____

Observation on treating I_2/KI solution with product

Observation on mixing NaCl and $AgNO_3$

Equation for reaction _____

Observation on adding product to NaCl/$AgNO_3$ mixture

Questions

1. Why is the initial solution of sodium sulfite basic?

2. What purpose does the detergent serve in the synthesis?

3. Sodium thiosulfate pentahydrate is a deliquescent substance. Use a chemical dictionary or online reference to explain the meaning of the term *deliquescent*.

4. Sodium thiosulfate is the salt of a weak acid ($H_2S_2O_3$). What would happen to a solution of sodium thiosulfate if a quantity of *strong* acid (for example, HCl) were added?

EXPERIMENT 43

Inorganic Preparations 2:
Preparation of Sodium Hydrogen Carbonate

Objective

Synthetic chemistry is one of the most important and practical fields of science. The synthetic chemist designs laboratory procedures for the production of new and useful compounds from natural resources or other simpler synthetic substances. In this experiment you will prepare sodium hydrogen carbonate (baking soda) by a method that mimics the industrial process.

Introduction

Sodium hydrogen carbonate (sodium bicarbonate, $NaHCO_3$), more commonly known as baking soda, finds many uses both in the home and in commercial applications. Sodium hydrogen carbonate, when combined with certain solid acidic substances, is used as a leavening agent in the form of commercial baking powders. When baking powder is mixed with water or milk (in batter, for example), carbon dioxide gas is released:

$$HCO_3^-(aq) + H^+(aq) \rightarrow H_2O(l) + CO_2(g)$$

The release of carbon dioxide causes the batter to rise and become filled with air bubbles, adding to the texture of the baked good.

Sodium hydrogen carbonate has several other uses in the home. Because it is weakly basic, baking soda is used in some over-the-counter antacid tablets ("bicarb"). Finely powdered baking soda can adsorb many molecules on the surface of its crystals and reacts chemically with vapors of acidic substances. For this reason, it is sold as a deodorizer for the home.

The synthesis of sodium hydrogen carbonate in this experiment duplicates some of the steps of the industrial synthesis. Dry ice (solid CO_2) is added to a saturated solution of sodium chloride in concentrated aqueous ammonia

$$NaCl(aq) + NH_3(aq) + CO_2(s) + H_2O(l) \rightarrow NH_4Cl(aq) + NaHCO_3(s)$$

In addition to serving as a source of carbon dioxide, dry ice also provides a means for *separating* the two products of the reaction. Dry ice provides a very low temperature for the reaction. At low temperatures, $NaHCO_3$ is much less soluble in water than is NH_4Cl and crystallizes from the solution much more readily. By filtering of the reaction mixture while it is still very cold, the sodium hydrogen carbonate product can be separated.

The yield of sodium hydrogen carbonate from this synthesis is typically much less than the theoretical amount expected (less than 50%). First, although $NaHCO_3$ is less soluble than NH_4Cl at low temperatures, a fair amount of $NaHCO_3$ does remain in solution. Second, a side reaction involving the same reagents is possible, in which significant amounts of ammonium hydrogen carbonate, NH_4HCO_3, are produced. The solid product isolated from the system is typically a mixture of sodium and ammonium hydrogen carbonates.

Once the solid has been filtered, however, ammonium hydrogen carbonate can be removed by simple heating of the solid over a steam bath:

$$NH_4HCO_3(s) \longrightarrow NH_3(g) + H_2O(g) + CO_2(g)$$

The products of this reaction are all *gases*, which are evolved from the solid mixture when heated, leaving pure sodium hydrogen carbonate.

Safety Precautions	• Protective eyewear approved by your institution must be worn at all time while you are in the laboratory. • Ammonia is a severe respiratory irritant and a strong cardiac stimulant. Confine all use of ammonia solutions to the fume exhaust hood. • Dry ice will cause severe frostbite to the skin if handled with the fingers. Use tongs, a scoop, or insulated gloves to pick up the dry ice. • The mixture of sodium/ammonium hydrogen carbonates will spatter when heated. Use caution when heating. • Hydrochloric acid is damaging to skin, eyes, and clothing. Wash immediately if the acid is spilled.

Apparatus/Reagents Required

Sodium chloride, concentrated ammonia solution, crushed dry ice, tongs, 3 *M* hydrochloric acid

Procedure

Record all data and observations directly in your notebook in ink.

Set up a suction filtration apparatus (See Figure 47-1) for use in filtering the product. The product of the reaction must be filtered before it has a chance to warm up too much, so it is important to have the filtration apparatus set up and ready to use before continuing.

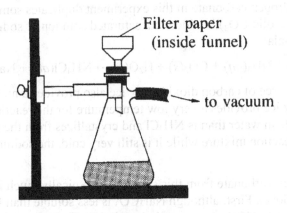

Figure 43-1. Suction filtration apparatus

The filter paper must fit the funnel and must lie flat on the base of the funnel. Wet the filter paper to keep it from being displaced when the solution to be filtered is added.

1. **Preparation of the compound**

    Clean a 250-mL Erlenmeyer flask, and fit with a solid stopper.

    Weigh out 15–16 g of NaCl and transfer to the Erlenmeyer flask. This quantity of NaCl is more than is actually needed for the synthesis, but the excess amount will speed up the saturation of the ammoniacal solution. Bring the flask containing the NaCl to the fume exhaust hood where the ammonia solution is kept.

    In the fume exhaust hood, add 50 mL of concentrated aqueous ammonia solution to the Erlenmeyer flask containing NaCl. Stopper the flask, but keep the flask *in the fume exhaust hood.*

    Swirl and shake the flask *in the fume exhaust hood* for 10–15 minutes to saturate the ammonia solution with NaCl. Do not "cheat" on the time in this step. If the solution is not saturated, no product will be obtained.

    After the ammonia solution has been saturated with NaCl, some undissolved NaCl will remain.

    Allow the solid to settle, and then *decant* the clear saturated ammonia/NaCl solution into a clean 250-mL beaker, avoiding the transfer of the solid as much as possible. Keep the beaker of solution *in the fume exhaust hood.*

    Using tongs or a plastic scoop to protect your hands from the cold, fill a 150-mL beaker approximately half full of crushed dry ice. This represents 60–70 g of CO_2.

    In the fume exhaust hood, over a 15-minute period, add the dry ice in small portions to the ammonia/NaCl solution. Stir the mixture with a glass rod throughout the addition of dry ice. Do not add the dry ice so quickly that the solution freezes. The temperature of the system at this point is 50–60 degrees below 0°C and is beyond the range of the thermometers in your locker.

    As the last portions of dry ice are added, a solid ($NaHCO_3/NH_4HCO_3$) will begin to form in the mixture. Allow the mixture to stand until the temperature has risen to approximately –5°C and all dry ice has dissolved or evaporated.

    Quickly filter the mixture on the suction filtration apparatus before the temperature has a chance to rise much above 0°C. Allow suction to continue for several minutes to remove liquid from the solid.

2. **Purification of the Product**

    Weigh a clean, dry watchglass to the nearest 0.1 g, and transfer the solid mixture to the watch glass.

    Place the watchglass over a beaker of boiling water, and heat for 10–15 minutes to decompose ammonium hydrogen carbonate. As the solid on the watchglass is heated, it will begin to melt and bubble as gases are evolved. Beware of spattering.

    After the heating period, allow the watchglass and solid to cool; then reweigh the watchglass.

    Determine the yield of $NaHCO_3$ in grams.

    Assuming that approximately 10 g of the initial 15–16 grams of NaCl actually dissolved in the ammonia solution, calculate an approximate percent yield for the synthesis.

3. **Testing the Product**

Since the sodium hydrogen carbonate product and the sodium chloride starting material are both white powders, perform the following simple test to confirm that the product of the reaction is not merely recrystallized sodium chloride.

Add a small portion of your product to a clean, dry test tube, and add a small portion of sodium chloride to a second test tube.

Add 5–6 drops of 3 M HCl to each test tube, and record your observations. Write an equation for the reaction that occurs.

EXPERIMENT 43

Inorganic Preparations 3:
Preparation of Sodium Hydrogen Carbonate

Pre-Laboratory Questions

1. Why must dry ice never be handled with the fingers?

2. It is stressed in the experiment, that the set-up for suction filtration should be prepared *before* performing the chemical reaction that produces the product. Explain.

3. Write the balanced chemical equation for the synthesis of sodium hydrogen carbonate using aqueous sodium chloride, ammonia, and dry ice as starting materials.

4. Why must this reaction be performed in the fume exhaust hood until you are ready to collect and filter the product of the reaction?

5. The desired product of this synthesis, sodium hydrogen carbonate, is separated from the other major product, ammonium chloride, on the basis of its *solubility* at low temperatures. However, it is not possible to separate ammonium hydrogen carbonate from sodium hydrogen carbonate on this basis. Use a handbook of chemical data to find and write information about the solubilities of these three substances.

	Solubility at 0°C	**Solubility at 20°C**
$NaHCO_3$	_____	_____
NH_4Cl	_____	_____
NH_4HCO_3	_____	_____

6. The preparation of sodium hydrogen carbonate in this experiment is similar to the industrial process for its preparation. Use a scientific encyclopedia or online reference to describe the key steps in the *Solvay process* for the production of sodium hydrogen carbonate.

EXPERIMENT 43

Inorganic Preparations 3:
Preparation of Sodium Hydrogen Carbonate

Results/Observations

Observation on saturating concentrated ammonia solution with NaCl

Observation on adding dry ice to NaCl/NH$_3$ solution

Approximate temperature at which crystals of product appeared   _____

Mass of empty watchglass   _____

Observation on heating product crystals on watchglass

Mass of watchglass with NaHCO$_3$ _____

Theoretical yield of NaHCO$_3$, based on 10 g of NaCl taken   _____

Actual yield of NaHCO$_3$ _____

% yield   _____

Observation on treating product with HCl

Equation for reaction

Observation on treating NaCl with HCl

Questions

1. Bubbling CO_2 gas through concentrated sodium hydroxide solutions can also produce $NaHCO_3$. Write an equation for this method of synthesis.

2. One of the most important industrial chemical substances is sodium carbonate, Na_2CO_3. Strong heating of sodium hydrogen carbonate, $NaHCO_3$, can produce this substance. Write an equation for this process.

3. Both sodium hydrogen carbonate and sodium carbonate have important commercial uses. Use a chemical dictionary or online source to discuss two such uses of each substance.

EXPERIMENT 44

Qualitative Analysis of Organic Compounds

Objective

The elements present in a typical covalent organic or biological compound can be determined by sodium fusion of the compound, followed by qualitative analysis for the ions produced by the fusion.

Introduction

When a new compound is prepared or isolated from some natural source, the techniques of **qualitative analysis** are frequently applied as a first step in identifying the compound. However, the traditional techniques of qualitative analysis generally apply only to inorganic *ions* in aqueous solution.

When a new compound does not consist of ions, but rather consists of covalently bonded molecules, frequently the molecule can be broken up into *ions* by controlled decomposition of the compound. The reagent(s) used to cause the decomposition of the original unknown compound are chosen in such a way that the ions produced by the decomposition will clearly reflect those elements that were present initially in the original covalently bonded unknown. In this experiment, a small amount of an unknown organic compound will be dropped into a hot vapor of elemental sodium. This technique is called **sodium fusion**.

The organic compound to be analyzed consists basically of a chain of carbon atoms to which various other atoms are attached. Sodium is an extremely strong reducing agent that will cause the breakup of the organic compound's carbon atom chain. It also will convert those other atoms that are covalently bonded to the carbon chain to inorganic *ions*. Once the covalent compound has been fused with sodium, the ionic mixture produced can then be dissolved in water and tested. Sodium itself, of course, is oxidized to Na^+ in this process.

Some elements that occur commonly in organic compounds are nitrogen, sulfur, and the halogens (chlorine, bromine, and iodine). When an organic compound undergoes a sodium fusion, the carbon present in the compound is reduced partially to elemental carbon, which is usually quite visible on the walls of the test tube used for the fusion. Any nitrogen present in the original compound is converted by the fusion to the cyanide ion, CN^-. Any sulfur present is converted to the sulfide ion, S^{2-}, and any halogens present are converted to the halide ions, Cl^-, Br^-, and I^-.

The test for cyanide ion involves forming a precipitate of an iron/cyanide complex of characteristic dark blue color (Prussian blue). The test for nitrogen is very sensitive to conditions, so follow the directions given exactly. Sometimes, if the sodium is not heated to a high enough temperature during the fusion, very little cyanide ion is produced. In this case, the dark blue precipitate may not form, but a small amount of a greenish solid may result. The green solid is sometimes taken as a weakly positive test for nitrogen. It is usually helpful to repeat the sodium fusion to confirm the presence of nitrogen if only the greenish solid is obtained.

The presence of sulfide ion from the original organic compound can be confirmed by several techniques. If the solution prepared from the sodium fusion products is acidified, hydrogen sulfide gas, H_2S, will be generated. The hydrogen sulfide may be detected by its reaction with lead(II) ions impregnated on test paper, causing the test paper to darken as black PbS forms.

$$Pb^{2+} + S^{2-} \longrightarrow PbS(s)$$
colorless   *black*

425

On the other hand, a solution that contains sulfide ions may be treated with sodium nitroprusside test reagent, in which case a purple/violet color will result.

The presence of halide ions from the original organic compound can be confirmed by treating the aqueous solution of the fusion products with silver ion reagent. Halide ions precipitate with silver ion.

$$\underset{colorless}{Ag^+} + Cl^- \longrightarrow \underset{white}{AgCl(s)}$$

$$\underset{colorless}{Ag^+} + (Br^-, I^-) \longrightarrow \underset{yellow}{AgBr(s), AgI(s)}$$

If more than one halide ion was present in the original compound, the mixed precipitate of silver halides can be analyzed for each of the possible halogens by the usual techniques. Before adding silver ions to the fusion products, however, it may be necessary to treat the sample to remove cyanide ions and sulfide ions (if they were present). Otherwise, these ions also will precipitate with the silver ions, masking the presence of the halogens.

Safety Precautions	• **Protective eyewear approved by your institution must be worn at all times while you are in the laboratory.** • *Caution!* **Elemental sodium is dangerous! For this reason, your instructor will perform the sodium fusion for you. Do not handle the sodium yourself! The sodium fusion will be performed by your instructor in the fume exhaust hood: occasionally during the fusion, an amount of noxious vapor may be released by the reaction.** • **Do not add water to the sodium fusion test tube until it has been established with all certainty that all elemental sodium has been destroyed by reaction with methanol. Sodium reacts violently with water.** • **When heating liquids in a test tube, use a small flame, hold the test tube at a 45° angle, and move the test tube quickly through the flame. Only a few seconds are required to heat a liquid in a test tube. Make sure the mouth of the test tube is not aimed at yourself or your neighbors.** • **Hydrogen cyanide and hydrogen sulfide are both extremely toxic gases. Perform those portions of the experiment that generate these gases in the *fume exhaust hood*.** • **Methyl alcohol is poisonous, and its vapor is toxic.** • **Sulfuric, nitric, and acetic acids are dangerous to skin, eyes, and clothing. Potassium hydroxide is caustic and can burn the skin. Wash immediately if these acids are spilled and inform the instructor.**

Apparatus/Reagents Required

Your instructor will have available clean dry test tubes, containing single small pellets of sodium metal (your instructor will perform the sodium fusion for you: do not attempt yourself), unknown organic compound, methyl alcohol, 0.5 M iron(II) sulfate (ferrous sulfate, 0.5 M iron(III) chloride (ferric chloride), 3 M potassium hydroxide, 3 M sulfuric acid, 3 M acetic acid, 3 M nitric acid, sodium nitroprusside reagent, lead acetate test paper, 0.1 M silver nitrate, 3 M aqueous ammonia, 0.1 M iron(III) nitrate (ferric nitrate), 0.1 M potassium permanganate, methylene chloride

Procedure

Record all data and observations directly in your notebook in ink.

1. **Instructor Procedure**

Because of the dangers associated with elemental sodium, your instructor will perform the sodium fusion for you. *Do not attempt the fusion yourself.*

Obtain an unknown organic compound and record its identification number. Bring your unknown sample to the fume exhaust hood where the instructor will be performing the sodium fusions. *The safety shield of the hood should be closed as much as possible during the fusion to protect the observers.*

The instructor will light a burner in the fume hood and will adjust the flame so it will be as hot as possible. The sodium pellet will be heated quickly and strongly, until it melts and then vaporizes (sodium vapor is dark gray or purple).

When the sodium vapor has risen about half an inch in the test tube, the instructor will quickly add your unknown sample from a spatula. A very vigorous reaction will take place, and the test tube should darken as the carbon of the organic compound is deposited on the walls of the test tube. A cloud of dark vapor may be released from the test tube.

The instructor will continue heating the test tube for 15–20 seconds and will then remove the heat and allow the test tube to cool completely to room temperature. Once the test tube containing the fusion product has cooled to room temperature, you may take the tube back to your own lab bench.

2. **Student Procedure**

Before water can be added to the sodium fusion test tube to dissolve the products of the fusion, any excess sodium present must be *destroyed*. Add about 5 mL of methyl alcohol to the test tube, and *allow the test tube to stand for at least 5 minutes.* Methyl alcohol reacts with sodium, but at a much slower speed than does water.

Hydrogen gas will bubble from the solution as the methanol destroys the sodium. Allow the methanol to react until no further bubbling is evident, and then wait an additional 5 minutes before proceeding.

When it is certain that all sodium has been destroyed, add 15 mL of distilled water to the test tube.

In the fume exhaust hood, set up a 600-mL beaker half full of water and begin heating for use as a boiling-water bath. Place the sodium fusion test tube in the boiling-water bath and heat for 5–10 minutes to assist the dissolving of the inorganic ions produced by the fusion.

While the test tube is heating, scrape down the walls of the tube to assist the dissolving. Remember that the black deposit on the test tube is carbon, which does not dissolve in water.

Filter the solution from the sodium fusion through a gravity funnel/filter paper to remove carbon (see Figure 44-1), and divide the filtrate into five portions in separate clean test tubes.

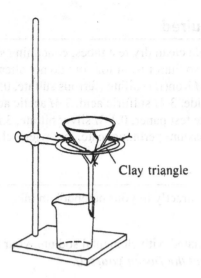

Figure 44-1. Set-up for gravity filtration of the solution of the fusion products

a.   Test for Nitrogen as Cyanide Ion

To one of the five portions of unknown solution filtrate, add 5–6 drops of 0.5 M ferrous sulfate solution (freshly prepared: the solution should be green) and 5–6 drops of 3 M potassium hydroxide. Stir to mix. A precipitate will form.

Heat the test tube in the boiling-water bath for 5 minutes. Cool the solution briefly, and stir in 2–3 drops of 0.5 M ferric chloride solution.

Dissolve the reddish-brown precipitate of iron hydroxides by adding 3 M sulfuric acid dropwise, with stirring, until the brown precipitate *barely* dissolves.

Cool the test tube in an ice bath, and then filter the solution through filter paper. If nitrogen was present in the unknown compound tested, a dark blue precipitate of Prussian blue will remain on the filter paper.

A green precipitate remaining on the filter paper is a weakly positive test for nitrogen. To be certain that nitrogen is present, repeat the sodium fusion on an additional sample of the unknown. A brown precipitate on the filter paper indicates that no nitrogen was present in the unknown.

b.   Tests for Sulfur as Sulfide Ion

Two separate tests for sulfur will be performed.

To a second portion of the filtrate from the sodium fusion, add a few drops of sodium nitroprusside reagent. The appearance of a violet/purple color indicates that sulfur is present, but this color may not persist for more than a few seconds.

To a third portion of the filtrate from the sodium fusion, add acetic acid dropwise until the solution is acidic to pH paper.

Transfer the test tube to the boiling-water bath in the fume exhaust hood. Moisten a piece of lead acetate test paper with a drop of distilled water, and hold the test paper over the open mouth of the test tube heating in the water bath.

If hydrogen sulfide is being evolved from the sample as it is being heated, the lead acetate test paper will darken as lead sulfide forms on the paper.

c.   Test for Halogens as Halide Ions

Before the filtrate from the sodium fusion can be tested for the presence of halide ions, any cyanide ion (nitrogen) or sulfide ion (sulfur) must be removed. Transfer one of the filtrate samples to a small beaker, add 10–15 mL of distilled water, and add 3 M acetic acid until the solution is acidic to pH paper. (Withdraw a drop of the solution with a stirring rod and touch the stirring rod to the pH paper. Do not add the pH paper to the test tube or the color will leach from the pH test paper.)

Transfer the beaker to the exhaust hood, and boil the solution for 5 minutes. If the solution threatens to boil away completely during this heating period, add small portions of distilled water to the test tube.

While the solution is being boiled, test the vapors coming from the beaker with moist lead acetate paper. When the lead acetate paper no longer darkens, you can assume that both cyanide and sulfide ions have been expelled from the solution. Allow the solution to cool to room temperature.

Add 10-15 drops of 0.1 M silver nitrate to the solution. If a precipitate of silver halide forms, at least one of the halogens is present in the unknown. Filter the precipitate through a piece of filter paper on a gravity funnel, and wash it with several drops of distilled water.

Transfer the precipitate to a clean test tube with a spatula, and add 10 drops of 0.1 M silver nitrate, 6–7 drops of 3 M ammonia, and 2–3 mL of distilled water.

If the white precipitate *completely* dissolves in the silver nitrate/ammonia solution, chloride is present and bromide and iodide are absent.

If the precipitate does not dissolve, or appears to only *partially* dissolve, then bromide, iodide, and chloride ions may *all* be present.

If the precipitate did not dissolve completely, filter the precipitate, collecting the clear filtrate.

Add 3 M nitric acid dropwise to the clear filtrate until it is acidic to pH paper. If a precipitate (AgCl) reappears when the solution is acidified, then chloride is present in the unknown.

Take the remaining portion of the *original* filtered solution from the sodium fusion, and transfer it to a small beaker. Add 10–15 mL of distilled water and sufficient 3 M acetic acid to make the solution acidic to pH paper.

Transfer the beaker to the exhaust hood, and boil the solution for 5 minutes. If the solution threatens to boil away completely during this heating period, add small portions of distilled water to the test tube.

Test the vapors coming from the solution as it boils with moist lead acetate paper to make certain that all cyanide and sulfide ion have been boiled away.

Transfer the solution to a small test tube. Add 10 drops of 3 M nitric acid to the solution, followed by 2–3 mL of 0.1 M ferric nitrate solution. Add 10 drops of methylene chloride, stopper, and shake. Ferric nitrate is capable of oxidizing iodide ion but not bromide ion.

The appearance of a violet color in the methylene chloride layer at this point indicates the presence of iodine in the original organic sample. Save this sample for determining whether bromide ion is present.

Use a disposable pipet and bulb to remove the methylene chloride layer from the test tube used in testing for iodide ion.

Add another 1 mL of ferric nitrate and 10 drops of methylene chloride. Stopper and shake the test tube. If the methylene chloride layer still shows the violet color of iodine, remove the methylene chloride layer with the disposable pipet, and add another 10 drops of methylene chloride.

Repeat the addition and pipet removal of methylene chloride until no violet color is present (indicating that all iodine has been removed from the sample).

When all iodine has been removed, add 5 drops of 3 M nitric acid. Then add 0.1 M potassium permanganate until the sample retains the pink color of the permanganate. The $KMnO_4$ will oxidize bromide ions to elemental bromine. Add 10 drops of methylene chloride, stopper, and shake. A yellow-orange color in the methylene chloride layer indicates that bromine was present in the original organic compound.

EXPERIMENT 44

Qualitative Analyis of Organic Compounds

Pre-Laboratory Questions

1. Sodium is used in the elemental analysis of organic compounds because it is a very strong reducing agent. What is a *reducing agent*? Give an example of how sodium acts as a reducing agent.

2. This experiment is concerned with detecting what elements are present in an organic compound. For new compounds, the molecular formula of the compound must be determined, followed by a determination of the structure of the compound. What information about a compound (beyond what elements are present) is needed to determine the molecular formula and structure.

3. Describe briefly what you expect to see in a *positive test* involving each of the following:

    a.   the test for nitrogen as cyanide ion

    b.   the test for sulfur using sodium nitroprusside solution

    c.   the test for sulfur using lead acetate test paper

    d.   the general test for the halides using silver nitrate solution

4. What precautions must be taken when working with sodium metal? Why? Write an equation for the reaction of sodium metal with water. What product(s) of that reaction is/are dangerous? Why?

5. In the test to be performed to detect nitrogen (as cyanide) in your sample, the blue pigment *Prussian blue* will form if nitrogen is present. Use a scientific encyclopedia or online reference to find the composition and formula of Prussian blue.

Name: _____ **Section:** _____

Lab Instructor: _____ **Date:** _____

EXPERIMENT 44

Qualitative Analysis of Organic Compounds

Results/Observations

Identification number of unknown _____

Observation on adding unknown to hot sodium vapor (performed by instructor)

Observation on adding methyl alcohol to sodium fusion products

Results/observations on test for nitrogen

Results/observations on sodium nitroprusside test for sulfide

Results/observations on lead acetate paper test for sulfide

Result on adding silver nitrate to fusion product solution

Results/observations on adding ammonia followed by nitric acid

Results/observations on adding ferric nitrate/methylene chloride

Results/observations on adding KMnO₄/methylene chloride

List those elements present in your unknown: _____

Questions

1. Water and methyl alcohol both react with sodium metal. Why was methyl alcohol used (rather than water) to destroy any remaining sodium present after fusion of the organic sample?

2. Why was the lead acetate test paper *moistened* before it was used to test for hydrogen sulfide gas being evolved from the boiling acidified solution?

3. Describe briefly how the three halide ions (Cl⁻, Br⁻, and I⁻) were separated from one another and then identified.

4. Why is it necessary to remove sulfide and cyanide ion from the sodium fusion products before testing for the presence of halide ions? How specifically would sulfide and cyanide ions interfere with the halide ion tests?

Organic Chemical Compounds

Objective

There are millions of known carbon compounds. In this experiment, several of the major families of carbon compounds will be investigated, which will demonstrate some of the characteristic properties and reactions of each family.

Introduction

Carbon atoms have the correct electronic configuration and size to enable long chains or rings of carbon atoms to be built up. This ability to form long chains is unique among the elements, and is called **catenation**. There are known molecules that contain thousands of carbon atoms, all of which are attached by covalent bonds between the carbon atoms. No other element is able to construct such large molecules with its own kind.

Hydrocarbons

The simplest types of carbon compounds are known as **hydrocarbons**. Such molecules contain only carbon atoms and hydrogen atoms. There are various sorts of molecules among the hydrocarbons, differing in the arrangement of the carbon atoms in the molecule, or in the bonding between the carbon atoms. Hydrocarbons consisting of chain-like arrangements of carbon atoms are called **aliphatic hydrocarbons**, whereas molecules containing the carbon atoms in a closed ring shape are called **cyclic hydrocarbons**. For example, the molecules hexane and cyclohexane represent aliphatic and cyclic molecules, each with six carbon atoms:

$$CH_3-CH_2-CH_2-CH_2-CH_2-CH_3$$

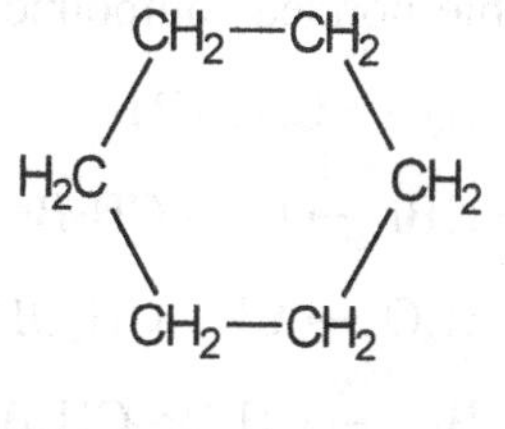

$$
\begin{array}{c}
CH_2-CH_2 \\
H_2C \qquad\qquad CH_2 \\
CH_2-CH_2
\end{array}
$$

Hexane (left) and cyclohexane (right)

Notice that all the bonds between carbon atoms in hexane and cyclohexane are single bonds. Molecules containing only singly bonded carbon atoms are said to be **saturated**.

Other molecules are known in which some of the carbon atoms may be attached by double or even triple bonds and are described as being **unsaturated**. The following are representations of the molecules 1-hexene and cyclohexene, which contain double bonds:

$$CH_2{=}CH{-}CH_2{-}CH_2{-}CH_2{-}CH_3$$

$$\begin{array}{c} CH{=}CH \\ H_2C \qquad CH_2 \\ CH_2{-}CH_2 \end{array}$$

1-hexene (left) and cyclohexene (right)

Hydrocarbons with only single bonds are not very reactive. One reaction all hydrocarbons undergo, however, is combustion. Hydrocarbons react with oxygen, releasing heat and/or light. For example, the simplest hydrocarbon is methane, CH_4, which constitutes the major portion of natural gas. Methane burns in air, producing carbon dioxide, water vapor, and heat:

$$CH_4 + 2O_2 \longrightarrow CO_2 + 2H_2O$$

Single-bonded hydrocarbons also will react slowly with the halogen elements (chlorine and bromine). A halogen atom replaces one (or more) of the hydrogen atoms of the hydrocarbon. For example, if methane is treated with bromine, bromomethane is produced, with concurrent release of hydrogen bromide:

$$CH_4 + Br_2 \longrightarrow CH_3Br + HBr$$

The hydrogen bromide produced is generally visible as a fog evolving from the reaction container, produced as the HBr mixes with water vapor in the atmosphere.

Molecules with double and triple bonds are much more reactive than single-bonded compounds. Unsaturated hydrocarbons undergo reactions at the site of the double (or triple) bond, in which the fragments of some outside reagent attach themselves to the carbon atoms that had been involved in the double (triple) bond. Such reactions are called **addition reactions**. The orbitals of the carbon atoms that had been used for the extra bond instead are used for attaching to the fragments of the new reagent. For example, the double-bonded compound ethylene (ethene), $CH_2{=}CH_2$, will react with many reagents

$$CH_2{=}CH_2 + H_2 \longrightarrow CH_3{-}CH_3$$

$$CH_2{=}CH_2 + HBr \longrightarrow CH_3{-}CH_2Br$$

$$CH_2{=}CH_2 + H_2O \longrightarrow CH_3{-}CH_2OH$$

$$CH_2{=}CH_2 + Br_2 \longrightarrow CH_2Br{-}CH_2Br$$

Notice that in all of these reactions of ethylene, the product is saturated and no longer contains a double bond. Many of the reactions of ethylene indicated only take place in the presence of a catalyst, or under reaction conditions that we will not be able to investigate in this experiment.

Cyclic compounds containing only single or double bonds undergo basically the same reactions indicated earlier for aliphatic hydrocarbons. One class of cyclic compounds, called the **aromatic hydrocarbons**, has unique properties that differ markedly from those we have discussed. The parent compound of this group of hydrocarbons is benzene:

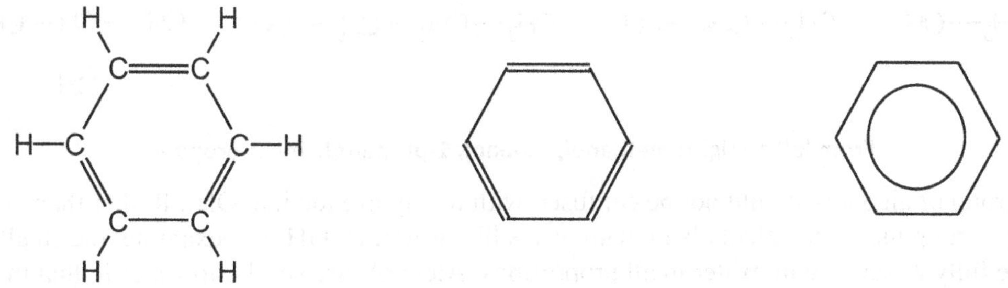

Benzene: three equivalent representations

Although we typically draw the structure of benzene indicating double bonds around the ring of carbon atoms, in terms of its reactions, benzene behaves as though it did not have any double bonds. For example, if elemental bromine is added to benzene (with a suitable catalyst), benzene does not undergo an addition reaction as would be expected for a compound with double bonds. Rather, it undergoes a substitution reaction of the sort that would be expected for single-bonded hydrocarbons:

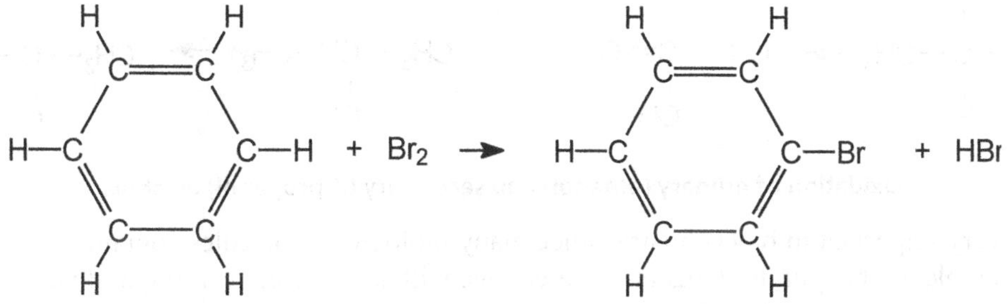

Benzene (left) and bromobenzene (right)

This reaction indicates that perhaps benzene does not have any double bonds in reality. Several Lewis dot structures can be drawn for benzene, with the "double bonds" in several different locations in the ring. Benzene exhibits the phenomenon of **resonance**; that is, the structure of benzene shown, as well as any other structure that might be drawn on paper, does not really represent the true electronic structure of benzene. The electrons of the "double bonds" of benzene are, in fact, **delocalized** around the entire ring of the molecule and are not concentrated in any particular bonds between carbon atoms. The ring structure of benzene is part of many organic molecules, including some very common biological compounds.

Functional Groups

While the hydrocarbons just described are of interest, the chemistry of organic compounds containing atoms other than carbon and hydrogen is far more varied. When another atom, or group of atoms, is added to a hydrocarbon framework, the new atom frequently imparts its own very strong characteristic properties to the resulting molecule. An atom or group of atoms that is able to impart new and characteristic properties to an organic molecule is called a **functional group**. With millions of carbon compounds known, it has been helpful to chemists to divide these compounds into **families**, based on what functional group they contain. Generally it is found that all the members of a family containing a particular functional group will have many similar physical properties and will undergo similar chemical reactions.

Alcohols

Simple hydrocarbon chains or rings that contain a hydroxyl group bonded to a carbon atom are called **alcohols**. The following are several such compounds:

$$CH_3-OH \qquad CH_3-CH_2-OH \qquad CH_3-CH_2-CH_2-OH \qquad CH_3-\underset{\underset{OH}{|}}{CH}-CH_3$$

From left to right: methanol, ethanol, 1-propanol, and 2-propanol

The –OH group of alcohols should not be confused with the hydroxide ion, OH⁻. Rather than being basic, the hydroxyl group makes the alcohols in many ways like water, H–OH. For example, the smaller alcohols are fully miscible with water in all proportions. Alcohols are similar to water in that they also react with sodium, releasing hydrogen. Alcohols, however, react more slowly with sodium than does water.

Many alcohols can be oxidized by reagents such as permanganate ion or dichromate ion, with the product depending on the structure of the alcohol. For example, ethyl alcohol is oxidized to acetic acid: When wine spoils, vinegar is produced. On the other hand, 2-propanol is oxidized to propanone (acetone) by such oxidizing agents:

$$CH_3-\underset{\underset{H}{|}}{\overset{\overset{H}{|}}{C}}-OH \rightarrow CH_3-\underset{\underset{OH}{|}}{C}=O \qquad\qquad CH_3-\underset{\underset{OH}{|}}{CH}-CH_3 \rightarrow CH_3-\underset{\underset{O}{\|}}{C}-CH_3$$

Oxidation of primary (ethanol) and secondary (2-propanol) alcohols

Alcohols are very important in biochemistry, since many biological molecules contain the hydroxyl group. For example, carbohydrates (sugars) are alcohols, with a particular carbohydrate molecule generally containing several hydroxyl groups.

Aldehydes and Ketones

An oxygen atom that is double bonded to a carbon atom of a hydrocarbon chain is referred to as a **carbonyl group**. Two families of organic compounds contain the carbonyl function, with the difference between the families having to do with where along the chain of carbon atoms the oxygen atom is attached. Molecules having the oxygen atom attached to the first (terminal) carbon atom of the chain are called **aldehydes**, whereas molecules having the oxygen atom attached to an interior carbon atom are called **ketones**.

Aldehydes are *intermediates* in the oxidation of alcohols. For example, when ethyl alcohol is oxidized, the product is first acetaldehyde, which is then further oxidized to acetic acid:

$$CH_3-\underset{\underset{H}{|}}{\overset{\overset{H}{|}}{C}}-OH \rightarrow CH_3-\underset{\underset{H}{|}}{C}=O \rightarrow CH_3-\underset{\underset{OH}{|}}{C}=O$$

Oxidation of ethanol to acetaldehyde (mild oxidation) and acetic acid (strong oxidation)

Ketones are the end product of the oxidation of alcohols in which the hydroxyl group is attached to an interior carbon atom (called secondary alcohols), as was indicated earlier for 2-propanol.

Aldehydes and ketones are important biologically because all carbohydrates are also either aldehydes or ketones, as well as being alcohols. The common sugars glucose and fructose are shown next. Notice that glucose is an aldehyde, whereas fructose contains the ketone functional group.

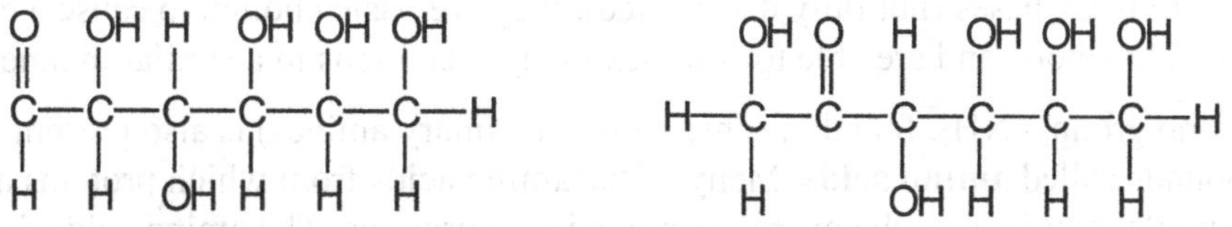

Glucose (left) and fructose (right)

Benedict's reagent easily oxidizes molecules such as glucose and fructose, which contain both the carbonyl and hydroxyl functional groups. Benedict's reagent is a specially prepared solution of copper(II) ion that is reduced when added to certain sugars, producing a precipitate of red copper(I) oxide. Benedict's reagent is the basis for the common test for sugars in the urine of diabetics. A strip of paper or tablet containing Benedict's reagent is added to a urine sample. The presence of sugar is confirmed if the color of the test reagent changes to the color of copper(I) oxide. Benedict's reagent reacts only with aldehydes and ketones that also contain a correctly located hydroxyl group, but will not react with simple aldehydes or ketones.

Organic Acids

Molecules that contain the carboxyl group at the end of a carbon atom chain behave as acids. For example, the hydrogen atom of the carboxyl group of acetic acid ionizes as:

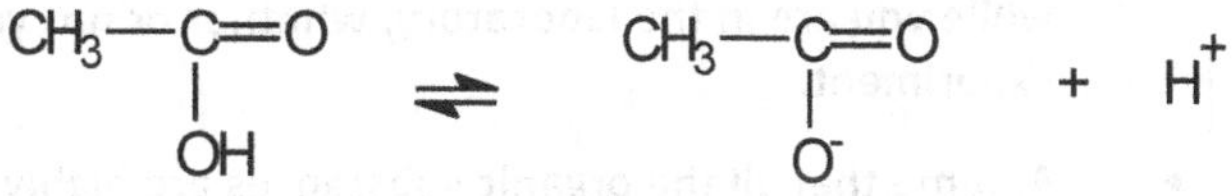

Ionization of acetic acid

On paper, the carboxyl group looks like a cross between the carbonyl group of aldehydes and the hydroxyl group of alcohols, but the properties of the carboxyl group are completely different from either of these. The hydrogen atom of the carboxyl group is lost by organic acids because the remaining ion has more than one possible resonance form, which leads to increased stability for this ion, relative to the unionized molecule.

Organic acids are typically *weak* acids, with only a few of the molecules in a sample being ionized at any given time. However, organic acids are strong enough to react with bicarbonate ion or to cause an acidic response in pH test papers or indicators.

Organic Bases

There are several types of organic compounds that show basic properties, the most notable of such compounds forming the family of **amines**. Amines are considered to be organic derivatives of the weak base ammonia, NH_3, in which one or more of the hydrogen atoms of ammonia is replaced by chains or rings of carbon atoms. The following are some typical amines:

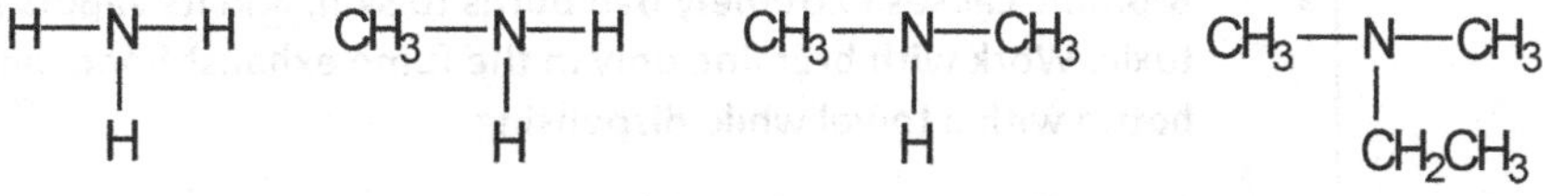

From left to right: ammonia, methylamine, dimethylamine, and ethyldimethyamine

Like ammonia, organic amines react with water, releasing hydroxide ion from the water molecules:

$$NH_3 + H_2O \rightleftarrows NH_4^+ + OH^-$$

$$CH_3-NH_2 + H_2O \rightleftarrows CH_3NH_3^+ + OH^-$$

Although amines are weak bases (not fully dissociated), they are basic enough to cause a color change in pH test papers and indicators, and are able to complex many metal ions in a similar manner to ammonia.

The so-called amino group, –NH_2, found in some amines (primary amines) is also present in the class of biological compounds called **amino acids**. Many of the amino acids from which proteins are constructed are remarkably simple, considering the overall complexity of proteins. The amino acids found in human protein contain a carboxyl group, with an amino group attached to the carbon atom next to the carboxyl group in the molecule's carbon chain. This sort of amino acid is referred to as an alpha amino acid. One simple amino acid found in human protein is shown here:

$$H_2N—C—C{=}O$$

Glycine (2-aminoethanoic acid)

Safety Precautions	• Safety eyewear approved by your institution must be worn at all times while you are in the laboratory, whether or not you are working on an experiment.
	• Assume that all the organic substances are highly flammable. Use only small quantities, and keep them away from open flames. Use a hot water bath if a source of heat is required.
	• Assume that all organic substances are toxic and are capable of being absorbed through the skin. Avoid contact and wash after use. Inform the instructor of any spills.
	• Assume that the vapors of all organic substances are harmful. Use small quantities and work in the fume exhaust hood.
	• Sodium reacts violently with water, liberating hydrogen gas and enough heat to ignite the hydrogen. Do not dispose of sodium in the sinks. Add any excess sodium to a small amount of ethyl alcohol, and allow the sodium to react fully with the alcohol before disposal.
	• Potassium permanganate, potassium dichromate, and Benedict's reagent are toxic. Permanganate and dichromate can harm skin and clothing. Exercise caution and wash after use. Chromium compounds are suspected mutagens/carcinogens.
	• Bromine causes extremely bad burns to skin, and its vapors are extremely toxic. Work with bromine only in the fume exhaust hood, and handle the bottle with a towel while dispensing.
	• Some of the chemicals in this experiment may be environmental hazards. Do not pour down the drain. Dispose of all materials as directed by the instructor.

Apparatus/Reagents Required

hexane; 1-hexene; cyclohexane; cyclohexene; toluene; ethyl alcohol; methyl alcohol; isopropyl alcohol; *n*-butyl alcohol; *n*-pentyl alcohol; *n*-octyl alcohol; 10% formaldehyde solution; 10% acetaldehyde solution; acetone; 10% glucose; 10% fructose; acetic acid; propionic acid; butyric acid; ammonia; 10% butylamine; 1% bromine in methylene chloride; 1% aqueous potassium permanganate; Benedict's reagent; sodium pellets (instructor only); 5% acidified potassium dichromate solution; pH paper; 10% sodium bicarbonate solution; 1 *M* copper sulfate solution; gloves

Procedure

Record all data and observations directly on the report pages in ink.

1. **Hydrocarbons**

   Your instructor will ignite small portions of hexane, 1-hexene, cyclohexane, cyclohexene, and toluene to demonstrate that they burn readily (do *not* attempt this yourself). Notice that toluene (an aromatic compound) burns with a much sootier flame than the other hydrocarbons. This is characteristic of aromatics.

   Obtain small (0.5-mL) portions of hexane, 1-hexene, cyclohexane, cyclohexene, and toluene in separate small test tubes. Take the test tubes to the exhaust hood, and add a few drops of 1% bromine solution (*Caution!*) to each. Those samples containing double bonds will decolorize the bromine immediately (addition reaction).

   Those samples that did not decolorize the bromine immediately will gradually lose the bromine color (substitution reaction).

   Expose those test tubes that did *not* immediately decolorize the bromine to bright sunlight, an ultraviolet lamp, or a high-intensity incandescent lamp. The reaction is sensitive to wavelengths of ultraviolet light.

   Blow across the open mouth of the test tubes to see if a fog of hydrogen bromide becomes visible

   Obtain additional samples of hexane, 1-hexene, cyclohexane, cyclohexene, and toluene in separate small test tubes. Add a few drops of 1% potassium permanganate to each test tube, stopper, and shake for a few minutes. Those samples containing double bonds will cause a change in the purple color of permanganate.

2. **Alcohols**

   Obtain small samples of methyl, ethyl, isopropyl, *n*-butyl, *n*-pentyl, and *n*-octyl alcohols in separate clean test tubes. Add an equal quantity of distilled water to each test tube, stopper, and shake.

   Record the solubility of the alcohols in water. In alcohols with only a few carbon atoms, the hydroxyl group contributes a major portion of the molecule's physical properties, making such alcohols miscible with water. As the carbon atom chain of an alcohol gets larger, the alcohol becomes more hydrocarbon in nature and less soluble in water.

   Obtain 2-drop samples of methyl, ethyl, isopropyl, *n*-butyl, *n*-pentyl, and *n*-octyl alcohols in separate clean test tubes.

   Wearing disposable gloves, add 10 drops of acidified potassium dichromate to each sample, transfer to a hot water bath, and heat until a reaction is evident. Dichromate ion oxidizes primary and secondary alcohols.

441

Obtain small samples of methyl, ethyl, and isopropyl alcohols in separate clean test tubes.

Have the instructor add a tiny pellet of sodium to each test tube. The vigor of the reaction depends on the structure of the alcohol. Smaller alcohols are more reactive than larger alcohols.

Wait until all the sodium in the test tubes has completely reacted. Then add about 5 mL of distilled water to the test tubes, and test the resulting solution with pH paper. The *alkoxide* ions that remain from the alcohols after reaction with sodium are strongly basic in aqueous solution.

3. **Aldehydes and Ketones**

Obtain 2-drop samples of formaldehyde, acetaldehyde, acetone, glucose, and fructose in separate test tubes.

Wearing gloves, add 10 drops of acidified potassium dichromate solution to each sample, and transfer the test tubes to a hot water bath. Aldehydes are oxidized by dichromate, being converted into organic acids.

Record which samples change color.

Obtain small samples of formaldehyde, acetaldehyde, acetone, glucose, and fructose in separate test tubes.

To each test tube add 5 mL of Benedict's reagent, and transfer to a hot water bath. Molecules containing a hydroxyl group next to a carbonyl group on a carbon atom chain will cause a color change in Benedict's reagent. This test is applied to urine for the detection of sugar.

4. **Organic Acids**

Obtain small samples of acetic acid, propionic acid, and butyric acid in separate small test tubes.

Add 5 mL of 10% sodium bicarbonate solution to each test tube: a marked fizzing will result as the acids release carbon dioxide from the bicarbonate.

Obtain small samples of acetic acid, propionic acid, and butyric acid in separate small test tubes.

Add 5 mL of water to each test tube, and test the solution with pH test paper.

5. **Organic Bases**

Obtain small samples of aqueous ammonia and of *n*-butylamine solution in separate small test tubes.

Determine the pH of each sample with pH test paper.

Add a few drops of 1 *M* copper sulfate solution to each test tube and shake. The color of the copper sulfate solution should darken as the metal/amine complex is formed.

EXPERIMENT 45

Organic Chemical Compounds

Pre-Laboratory Questions

Draw structural formulas for each of the following substances. If the common name of the substance is given, write the IUPAC name for the substance.

1. heptane   IUPAC name _____

2. 2-propanol   IUPAC name _____

3. l-heptene   IUPAC name _____

4. formaldehyde   IUPAC name _____

5. cyclohexene   IUPAC name _____

6. acetaldehyde   IUPAC name _____

7. cyclohexane   IUPAC name _____

8. acetone   IUPAC name _____

9. toluene   IUPAC name _____

10. acetic acid   IUPAC name _____

11. ethyl alcohol   IUPAC name _____

12. propanoic acid   IUPAC name _____

13. methyl alcohol   IUPAC name _____

14. *n*-butylamine   IUPAC name _____

Name: ______________________ Section: __________

Lab Instructor: ______________________ Date: __________

EXPERIMENT 45

Organic Chemical Compounds

Results/Observations

1. **Hydrocarbons**

Observation on igniting (instructor demonstration)

____

Which compounds decolorized bromine immediately?

____

Which compounds decolorized bromine on exposure to light?

____

Which compounds produced an HBr fog?

____

Which compounds caused a color change in $KMnO_4$?

____

2. **Alcohols**

Which alcohols were fully soluble in water?

____

Observation on $K_2Cr_2O_7$ test

____

Odors of ethyl alcohol and *n*-butyl alcohol oxidation products

____

Relative vigor of reaction of alcohols with sodium (instructor demonstration)

____

pH of water solution of alcohol/sodium product

____

3. **Aldehydes and Ketones**

Observation on $K_2Cr_2O_7$ test

____

Odor of acetaldehyde/dichromate product

_____

Observation of Benedict's test

_____

4.   **Organic Acids**

Observation on test with bicarbonate

_____

pH of aqueous solution of organic acid samples

_____

5.   **Organic Bases**

pH of aqueous solution of ammonia/_n_-butylamine

_____

Observation on adding $CuSO_4$

_____

Questions

1.   When sodium bicarbonate was added to an organic acid sample, carbon dioxide was evolved. Is this test specific for organic acids, or does it apply to acids in general? Write the balanced equation for the process.

2.   When potassium permanganate was shaken with alkenes, the color of the permanganate changed from the purple color of MnO_4^- to the brown-black color of manganese(IV) oxide. What sort of reaction must this be? What is the original alkene converted to in the reaction?

3.   When ethyl alcohol was treated with potassium dichromate, the alcohol was oxidized to acetic acid, yet it was indicated that acetaldehyde is an _intermediate_ in this oxidation. How might aldehydes be isolated during such an oxidation, preventing them from being further oxidized to the acid?

EXPERIMENT 46

Ester Derivatives of Salicylic Acid

Objective

Esters are an important class of organic chemical compounds. The bonding in esters is analogous to that found in some important classes of biological compounds. In this experiment, two esters of salicylic acid with important medicinal properties will be prepared.

Introduction

Two common esters, acetylsalicylic acid and methyl salicylate, are very important over-the-counter drugs. Since ancient times, it has been known that the barks of certain trees, when chewed or brewed as a tea, had *analgesic* (pain-killing) and *antipyretic* (fever-reducing) properties. The active ingredient in such barks was determined to be **salicylic acid**. When chemists first isolated *pure* salicylic acid, however, it proved to be much too harsh to the linings of the mouth, esophagus, and stomach for direct use as a drug in the pure state. Salicylic acid contains the phenolic (–OH) functional group in addition to the carboxyl (acid) group, and it is the combination of these two groups that leads to the harshness of salicylic acid on the digestive tract.

Because salicylic acid contains both the organic acid group (–COOH) and the phenolic (–OH) group, salicylic acid is capable of undergoing two separate esterification reactions, depending on whether it is behaving as an acid (through the –COOH) or as an alcohol analogue (through the –OH). Research was conducted in an attempt to modify the salicylic acid molecule in such a manner that its desirable analgesic and antipyretic properties would be preserved but its harshness to the digestive system would be decreased. The Bayer Company of Germany, in the late 1800s, patented an ester of salicylic acid that had been produced by reaction of salicylic acid with acetic anhydride (a compound related to acetic acid).

| Salicylic acid | Acetic anhydride | Aspirin | Acetic acid |

The ester, commonly called *acetylsalicylic acid* or by its original trade name (aspirin), no longer has the phenolic functional group. The salicylic acid has acted as an alcohol when reacted with acetic anhydride. Acetylsalicylic acid is much less harsh to the digestive system. When acetylsalicylic acid reaches the intestinal tract, however, the basic environment of the small intestine causes hydrolysis of the ester (the reverse of the esterification reaction) to occur. Acetylsalicylic acid is converted back into salicylic acid in the small intestine and is then absorbed into the bloodstream in that form. Aspirin tablets sold commercially generally contain binders (such as starch) that help keep the tablets dry and prevent the acetylsalicylic acid in the tablets from decomposing into salicylic acid. Since the other component in the production of acetylsalicylic acid is acetic acid, one indication that aspirin tablets have decomposed is an odor of vinegar from the acetic acid released by the hydrolysis.

The second common ester of salicylic acid that is used as a drug is methyl salicylate. When salicylic acid is heated with methyl alcohol, the carboxyl group of salicylic acid is esterified, producing a strong-smelling liquid ester (methyl salicylate).

| Salicylic acid | Methyl alcohol | Methyl salicylate | Water |

The *mint* odor of many common liniments sold for sore muscles and joints is due to this ester. Methyl salicylate is absorbed through the skin when applied topically and may permit the pain-killing properties of salicylic acid to be localized on the irritated area. Methyl salicylate is a skin irritant, however, and causes a sensation of warmth to the area of the skin where it is applied. This is usually considered to be a desirable property of the ester. Methyl salicylate is also used as a flavoring/aroma agent in various products and is referred to commercially as oil of wintergreen.

Safety Precautions	• Protective eyewear approved by your institution must be worn at all time while you are in the laboratory.
	• Most of the organic compounds used or produced in this experiment are highly flammable. All heating will be done using a hotplate, and no flames will be permitted in the laboratory.
	• Sulfuric acid is used as a catalyst for the esterification reaction. Sulfuric acid is dangers and can burn skin, eyes, and clothing. If it is spilled, wash *immediately* before the acid has a chance to cause a burn, and inform the instructor.
	• Acetic anhydride (rather than acetic acid itself) is used in the synthesis of aspirin. Acetic anhydride can seriously burn the skin, and its vapors are harmful to the respiratory tract. If any is spilled, wash *immediately* and inform the instructor. Confine the pouring of acetic anhydride to the fume exhaust hood.
	• The vapors of the esters produced in this experiment may be harmful. When determining the odors of the esters, do not deeply inhale the vapors. Merely waft a small amount of vapor from the ester toward your nose.
	• Dispose of all materials as directed by the instructor.

Apparatus/Reagents Required

Hotplate, suction filtration apparatus, ice, melting point apparatus, salicylic acid, acetic anhydride, methyl alcohol (methanol), 50% sulfuric acid, sodium bicarbonate, 1 M iron(III) chloride

Procedure

Record all data and observations directly in your notebook in ink.

1. **Preparation of Aspirin**

   Set a 400-mL beaker about half full of water to warm on a hotplate in the exhaust hood. A water heating bath at approximately 70°C is desired, so use the lowest setting of the hotplate control. Check the temperature of the water before continuing. See Figure 50-1.

   Weigh out approximately 1.5 g of salicylic acid and transfer to a clean 125-mL Erlenmeyer flask.

   In the exhaust hood, add approximately 4 mL of acetic anhydride (*Caution!*) and 3–4 drops of 50% sulfuric acid (*Caution!*) to the salicylic acid. Stir carefully until the mixture is homogeneous.

   Transfer the Erlenmeyer flask to the beaker of 70°C water (see Figure 46-1) and heat for 15–20 minutes, stirring occasionally. Monitor the temperature of the water bath during this time, and do not let the temperature rise above 70°C.

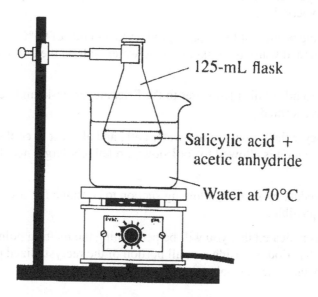

   Figure 46-1. Heating bath for synthesis of aspirin

   At the end of the heating period, cool the Erlenmeyer flask in an ice bath until crystals begin to form. If crystallization does not take place, scratch the walls and bottom of the flask with a stirring rod to promote formation of crystals.

   To destroy any excess acetic anhydride that may be present, add 50 mL of cold water, stir, and allow the mixture to stand for at least 15 minutes to permit the hydrolysis to take place.

   Remove the liquid from the crystals by filtering under suction in a Büchner funnel (see Figure 46-2). Wash the crystals with two 10-mL portions of cold water to remove any excess reagents. Continue suction through the crystals for several minutes to help dry them.

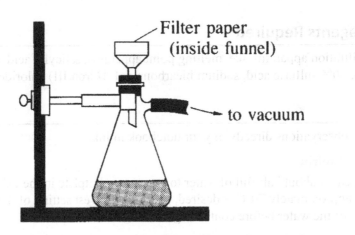

Figure 46-2. Suction filtration apparatus

The filter paper must fit the funnel exactly and must lie flat on the base of the funnel. Wet the filter paper to help it to adhere to the funnel.

You will perform some tests on this *crude* product today, and your instructor may ask you to save a small portion of your crystals for a melting point determination to be performed during the next laboratory period. If your instructor directs, set aside a small portion of your crystals in a beaker to dry until the next lab period.

Dissolve the remaining portion of the aspirin product in 4–5 mL of 95% ethyl alcohol in a 125-mL Erlenmeyer flask. Stir and warm the mixture in the water bath on the hotplate until the crystals have dissolved completely.

Add approximately 15 mL of distilled water to the Erlenmeyer flask, and heat to re-dissolve any crystals that may have formed.

Remove the Erlenmeyer flask from the hot-water bath, and allow it to cool slowly to room temperature. If the solution is allowed to cool slowly, relatively large, needlelike crystals of aspirin should form.

Filter the purified product by suction, allowing suction to continue for several minutes to dry the crystals as much as possible.

If your instructor has indicated that you will be performing the melting point determination for your aspirin in the next lab period, set aside a small portion of the recrystallized product in a labeled test tube or beaker to dry until the next laboratory period.

2. **Tests on Aspirin**
    a. **Test with Bicarbonate**
        Acetylsalicylic acid molecules still contain the organic acid group (carboxyl) and will react with sodium bicarbonate to release carbon dioxide gas:

$$H^+(aq) + HCO_3^-(aq) \longrightarrow H_2O(l) + CO_2(g)$$

Add a very small portion of your aspirin (crude or purified) to a test tube. Add also a small portion of sodium bicarbonate. Add a small amount of water and note the evolution of carbon dioxide. This test indicates only that aspirin is an acid; it is not a specific test for aspirin.

b. Test with Iron(III)

If the synthesis of aspirin has not been effective, or if aspirin has decomposed with time, then free salicylic acid will be present. This would be harmful if ingested. The standard United States Pharmacopoeia test for the presence of salicylic acid is to treat the sample in question with a solution of iron(III). If salicylic acid is present, the phenolic functional group (–OH) of salicylic acid will produce a purple color with iron(III) ions. The *intensity* of the purple color is directly proportional to the amount of salicylic acid present.

Set up four test tubes in a rack. To the first test tube, add a small quantity of pure salicylic acid as a control. To the second and third test tubes, add very small portions of your crude and purified aspirin, respectively. To the fourth test tube, add a commercial aspirin tablet (if available).

Add 5 mL of distilled water to each test tube, and stir to mix. Add 8–10 drops of iron(III) chloride (ferric chloride) solution. The appearance of a pink or purple color in your aspirin or the commercial tablet sample indicates the presence of salicylic acid.

c. Melting Point Determination

After the crude and purified aspirin samples have dried for a week, determine their melting points by the capillary method described in Experiment 5. Compare the observed melting points with the literature value.

3. Preparation of Methyl Salicylate

Place about 1 g (roughly measured) of salicylic acid in a small Erlenmeyer flask and add 5 mL of methyl alcohol. Stir thoroughly to dissolve the salicylic acid.

Add 4–5 drops of 50% sulfuric acid (*Caution!*) and heat the flask in the 70°C water bath on the hotplate in the exhaust hood for approximately 5 minutes.

Pour the contents of the test tube into approximately 50 mL of warm water in a beaker, and cautiously waft some of the vapors toward your nose. There should be a pronounced odor of wintergreen.

EXPERIMENT 46

Ester Derivatives of Salicylic Acid

Pre-Laboratory Questions

1. From the formulas of salicylic acid and acetylsalicylic acid given in the Introduction to this experiment, determine their molar masses.

2. Suppose 3.50 g of salicylic acid is heated with an excess amount of acetic acid. What is the expected yield of acetylsalicylic acid? Show your calculation

3. Suppose for the experiment outlined in Question 2, only 1.82 g of aspirin is isolated from the reaction mixture. Calculate the percent yield for the process. Show your work.

4. Write the chemical equation for the preparation of methyl salicylate from salicylic acid and methyl alcohol

5. One unusual property of methyl salicylate is its ability (under the right conditions) to produce **triboluminescence.** Use a scientific encyclopedia or online reference to define triboluminescence and to describe the conditions under which methyl salicylate exhibits this phenomenon.

EXPERIMENT 46

Ester Derivatives of Salicylic Acid

Results/Observations

1. **Preparation of Aspirin**

    Observation on crude aspirin (appearance)

    _____

    _____

    _____

    _____

    Observation on recrystallized aspirin (appearance)

    _____

    _____

    _____

    _____

2. **Tests on Aspirin**

    Effect of bicarbonate _____

    Effect of iron(III) on crude aspirin _____

    Effect of iron(III) on purified aspirin _____

    Effect of iron(III) on salicylic acid _____

    Effect of iron(III) on aspirin tablet _____

    Melting points

        Crude aspirin _____ Purified aspirin _____

        Literature value _____ Reference _____

3. **Preparation of Methyl Salicylate**

    Observations on methyl salicylate _____

    _____

    Odor? _____

Questions

1. How does sulfuric acid catalyze the preparation of esters? Could some other acid be used?

2. In the aspirin synthesis, excess acetic anhydride was destroyed by the addition of water, which converts the anhydride into acetic acid. Write the balanced chemical equation for the reaction.

3. In the tests with Fe(III), which samples gave a purple color? What does the purple color signify? How did your crude aspirin and your purified aspirin compare when tested with Fe(III)? How did the commercial aspirin tablet compare? What can you conclude about the relative purities of the various samples tested?

4. Although aspirin's original primary use was as an analgesic and fever-reducer, it has largely been supplanted for those uses by other drugs. However, aspirin has found great usage in low-dosage therapies intended to prevent the formation of blood clots in the body. Use a scientific encyclopedia or online reference to discuss how aspirin is able to prevent blood clots from forming inside the body.

EXPERIMENT 47

Preparation of Fragrant Esters

Objective

Esters are an important class of organic chemical compounds. The bonding in esters is analogous to that found in some important classes of biological compounds. In this experiment you will synthesize some esters that have characteristic odors.

Introduction

When an organic acid, R–COOH, is heated with an alcohol, R'–OH, in the presence of a strong mineral acid, the chief organic product is a member of the family of organic compounds known as **esters**. The general reaction for the esterification of an organic acid with an alcohol is

$$R\text{–}COOH + HO\text{–}R' \rightleftarrows R\text{–}CO\text{–}OR' + H_2O$$

In this general reaction, R and R' represent hydrocarbon chains, which may be the same or different. As a specific example, suppose acetic acid, CH_3COOH, is heated with ethyl alcohol, CH_3CH_2OH, in the presence of a mineral acid catalyst. The esterification reaction will be

$$CH_3\text{–}COOH + HO\text{–}CH_2CH_3 \rightleftarrows CH_3\text{–}COO\text{–}CH_2CH_3 + H_2O$$

The ester product of this reaction ($CH_3\text{–}COO\text{–}CH_2CH_3$) is named *ethyl acetate*, indicating the alcohol and acid from which it is prepared. Esterification is an equilibrium reaction, which means that the reaction does *not* go to completion on its own. Frequently, however, the esters produced are extremely volatile and can be removed from the system by distillation. If the ester is not very easily distilled, it may be possible instead to add a drying-agent to the equilibrium system, thereby removing water from the system and forcing the equilibrium to the right.

Many of the odors and flavorings of fruits and flowers are due to the presence of esters in the essential oils of these materials. The table that follows lists some esters with pleasant fragrances, as well as indicating from what alcohol and which acid the ester may be prepared.

Table of Common Esters

Ester	Aroma	Constituents
n-propyl acetate	pears	*n*-propyl alcohol/acetic acid
methyl butyrate	apples	methyl alcohol/butyric acid
isobutyl propionate	rum	isobutyl alcohol/propionic acid
octyl acetate	oranges	*n*-octyl alcohol/acetic acid
methyl anthranilate	grapes	methyl alcohol/2-aminobenzoic acid
isoamyl acetate	bananas	isoamyl alcohol/acetic acid
ethyl butyrate	pineapples	ethyl alcohol/butyric acid
benzyl acetate	peaches	benzyl alcohol/acetic acid

A fruit or flower generally contains only a few drops of ester, giving a very subtle odor. Usually, the ester is part of some complex mixture of substances, which, taken as a whole, have the aroma attributed to the material. When prepared in the laboratory in relatively large amounts, the ester may seem to have a pronounced chemical odor, and it may be difficult to recognize the fruit or flower that has this aroma.

Safety Precautions	<ul><li>**Protective eyewear approved by your institution must be worn at all time while you are in the laboratory.**</li><li>**Most of the organic compounds used or produced in this experiment are highly flammable. All heating will be done using a hotplate, and no flames will be permitted in the laboratory.**</li><li>**Sulfuric acid is used as a catalyst for the esterification reaction. Sulfuric acid is dangers and can burn skin, eyes, and clothing. If it is spilled, wash *immediately* before the acid has a chance to cause a burn, and inform the instructor.**</li><li>**The vapors of the esters produced in this experiment may be harmful. When determining the odors of the esters, do not deeply inhale the vapors. Merely waft a small amount of vapor from the ester toward your nose.**</li><li>**NaOH solution is highly corrosive to skin and eyes. Wash immediately if spilled.**</li><li>**Dispose of all materials as directed by the instructor.**</li></ul>

Apparatus/Reagents Required

Hotplate; 50% sulfuric acid; assorted alcohols and organic acids, as provided by the instructor, for the preparation of fruit and flower aromas; methyl salicylate; 20% NaOH; disposable 4 mL plastic pipets with stem cut to 1inch

Procedure

1. **Preparation of Fragrant Esters**

   Set up a water bath in a 250-mL beaker on a hotplate in the fume exhaust hood. Most of the reactants and products in this choice are highly *flammable*, and no flames are permitted in the lab during this experiment. Adjust the heating control to maintain a temperature of around 70°C in the water bath.

   Some common esters, and the acids/alcohols from which they are synthesized, were indicated in the table in the Introduction to this experiment. Synthesize at least two of the esters, and note their aromas. Different students might synthesize different esters, as directed by the instructor, and compare the odors of the products.

   To synthesize the esters, mix 3-4 drops (or approximately 0.1 g if the acid is a solid) of the appropriate acid with 3-4 drops of the indicated alcohol on a clean, dry watch glass or in a small plastic beaker.

Add 1 drop of 50% sulfuric acid to the mixture on the watch glass (*Caution!*). Use the tip of a plastic pipet to stir the mixture on the watch glass, and then suck as much as possible of the mixture into the pipet.

Place the pipet, tip upward, into the warm-water bath, and allow it to heat for approximately 5 minutes.

Squirt the resulting ester from the pipet into a beaker of warm water, and cautiously waft the vapors toward your nose.

Remember that the odor of an ester is very concentrated. Several sniffs may be necessary for you to identify the odor of the ester. Record which esters you prepared and their aromas.

2.   Hydrolysis of an Ester

Esters may be destroyed by *reversing* the esterification reaction: water and an ester will react with one another (hydrolysis) to give an alcohol and an organic acid. This reaction is often carried out using sodium hydroxide as catalyst (in which case the sodium salt of the organic acid results, rather than the acid itself).

For a general ester, the reaction is:

$$R\text{--}CO\text{--}OR' + H_2O \longrightarrow R\text{--}COOH \ (Na^+R\text{--}COO^-) + HO\text{--}R'$$

Set up a water bath in a 250-mL beaker on the hotplate in the hood. Adjust the heating control to maintain a temperature of approximately 70°C in the water bath.

Obtain 10 drops of methyl salicylate in a clean test tube. Note the odor of the ester by cautiously wafting vapors from the test tube toward your nose.

Add 2–3 mL of water to the test tube, followed by 10–15 drops of 20% NaOH solution (*Caution!*).

Heat the methyl salicylate/NaOH mixture in the water bath for 15–20 minutes. After this heating period, again note the odor of the mixture. If the odor of mint is still noticeable, heat the test tube for an additional 10-minute period.

Write an equation for the hydrolysis of methyl salicylate.

EXPERIMENT 47

Preparation of Fragrant Esters

Pre-Laboratory Questions

1. Using R–COOH to represent an organic acid and R′–OH to represent an alcohol, write a general equation for the formation of an ester. What is the *other* product of the esterification reaction?

2. Why are no flames permitted in the laboratory during this experiment, and a hotplate used to heat the reaction mixture instead?

3. Write an equation describing the *hydrolysis* of methyl salicylate.

4. Choose four of the esters listed in the **Table of Common Esters** in the Introduction to this experiment and write equations below showing their synthesis from the appropriate alcohols and organic acids.

Name: _____ Section: _____

Lab Instructor: _____ Date: _____

EXPERIMENT 47

Preparation of Fragrant Esters

Results/Observations

1. **Preparation of Fragrant Esters**

   Which esters did you prepare?

   Ester #1 _____ Ester #2 _____

   Odor _____ Odor _____

   Formula _____ Formula _____

2. **Hydrolysis of Methyl Salicylate**

   Observations:

   _____

   _____

   _____

   _____

   _____

   Equation for the hydrolysis reaction:

   _____

Questions

1. Ordinarily, esterification reactions come to *equilibrium* before the full theoretical yield of ester is realized. Aside from the distillation discussed earlier, how might one experimentally shift the equilibrium of the esterification reaction so that a larger amount of ester might be isolated?

2. Using a handbook or encyclopedia of chemistry, give the structures and names of two additional fragrant esters that were not discussed in this experiment.

3. Use your textbook to write a definition for *hydrolysis*. Write an example of a hydrolysis reaction (other than the reaction performed in this experiment).

4. Esters are frequently used as additives in commercial products found in the home. Examine the labels of things you may have at home, such as shampoos, soaps, hand creams, and prepared foods, and find the names of two esters among the ingredients. List here the products and the esters you found.

EXPERIMENT 48

Proteins

Introduction

Proteins are long-chain polymers of the α-amino acids and make up about 15% of our bodies. Proteins have many functions in the body. Some proteins form the major structural feature of muscles, hair, fingernails, and cartilage. Other proteins help to transport molecules through the body, fight infections, or act as catalysts (enzymes) for biochemical reactions in the cells of the body.

$$H_2N-\underset{R}{\overset{H}{C}}-\underset{OH}{\overset{O}{C}}$$

An α-amino acid

Proteins have several levels of structure, each level of which is very important to a protein's function in the body. The **primary structure** of a protein is the particular sequence of amino acids in the polymer chain; each particular protein has a unique primary structure. The **secondary structure** of a protein consists of the basic arrangement in space of the overall chain of amino acids: some protein chains coil into a helical form, whereas other proteins chains may bind together to form a sheet of protein. The **tertiary structure** of a protein reflects how the protein chains (whether sheet or helix) fold in space as a consequence of the interactions of the side groups of the amino acids: for example, the helical secondary structures of some proteins fold into a globular (spherical) shape to enable them to travel more easily through the bloodstream.

The various levels of structure of a protein are absolutely crucial to the protein's function in the body. If any error occurs in the primary structure of the protein, a genetic disease may result (sickle-cell disease and Tay-Sachs disease are both due to errors in protein primary structures). If any change in the environment of a protein occurs, the tertiary and secondary structures of the protein may be changed enough to make the protein lose its ability to act in the body. For example, egg white is basically an aqueous solution of the protein albumin: when an egg is cooked, the increased temperature causes irreversible changes in the tertiary and secondary structures of the albumin, making it coagulate as solids. When the physical or chemical environment of a protein is modified, and the protein responds by changing its tertiary or secondary structure, the protein; we say the protein has been **denatured**. Denaturation of a protein may be reversible (if the change to the protein's environment is not too drastic) or irreversible (as with the egg white discussed above). Changes in environment that may affect a protein's structure include changes in temperature, changes in the pH (acidity) of the protein's environment, changes in the solvent in which the protein is suspended, and the presence of any other ions or molecules that may chemically interact with the protein (such as ions of the heavy metals).

Safety Precautions	• **Protective eyewear approved by your institution must be worn at all time while you are in the laboratory.**
	• **Sodium hydroxide, nitric acid, and hydrochloric acid solutions are corrosive to skin, eyes, and clothing. Wash immediately if these reagents are spilled and clean up all spills on the benchtop. Inform the instructor.**
	• **Lead acetate and copper(II) sulfate solutions are toxic if ingested. Wash after handling. Dispose of as directed by the instructor.**

Apparatus/Reagents Required

1% albumin solution; 1% alanine solution; nonfat milk; gelatin cubes; 10% NaOH solution; 3% $CuSO_4$ solution; 6 M nitric acid solution; 6 M NaOH solution; 5% lead acetate solution; 3 M HCl solution; 3 M NaOH solution; saturated NaCl solution

Procedure

1. Tests for Proteins

Samples of three proteinaceous materials will be available for testing: egg white (containing the protein *albumin*); nonfat milk (containing the protein *casein*), and gelatin. Samples of the free amino acid alanine will also be tested for comparison and contrast.

a.   The biuret test

The biuret test detects the presence of proteins and polypeptides but does *not* detect free amino acids (the test is specific for the *peptide linkage*).

Place about 10 drops of 1% albumin, nonfat milk, and 1% alanine solutions and a small cube of gelatin in separate clean, small test tubes.

Add 10 drops of 10% sodium hydroxide solution *(Caution!)* to each test tube, followed by 3–4 drops of 3% copper(II) sulfate solution.

Carefully mix the contents of each test tube with a clean stirring rod, being sure to clean the stirring rod when switching between solutions. Be sure to break up the gelatin sample as much as possible when mixing.

The blue color of the copper(II) sulfate solution will change to purple in those samples that contain proteinaceous material. Record which samples gave the purple color.

Why does alanine not give a positive biuret test?

b.   The xanthoproteic test

Two of the common amino acids found in proteins, tryptophan and tyrosine, contain substituted benzene rings in their side chains. When treated with nitric acid, a nitro group (–NO_2) can also be added to the benzene rings, which results in the formation of a yellow color in the protein (*xanthoproteic* means "yellow protein"). For example, if nitric acid is spilled on the skin, the proteins in the skin will turn yellow as the skin is destroyed by the acid.

Set up a 250-mL beaker containing approximately 100 mL of water on a wire gauze/ring stand, and heat to boiling.

Place about 10 drops of 1% albumin, nonfat milk, and 1% alanine solutions and a small cube of gelatin in separate clean, small test tubes.

Add 8–10 drops of 6 *M* nitric acid *(Caution!)* to each test tube.

Carefully mix the contents of each test tube with a clean stirring rod, being sure to clean the stirring rod when switching between solutions. Be sure to break up the gelatin sample as much as possible when mixing.

Transfer the test tubes to the boiling-water bath and heat the samples for 3-4 minutes.

Record which samples produce the yellow color characteristic of a positive xanthoproteic test. Which samples do not give a positive test?

c.   The lead acetate test for sulfur

In strongly basic solutions, the amino acid cysteine from proteins will react with lead acetate solution to produce a black precipitate of lead sulfide, PbS. Cysteine is found in many (but not all) proteins. The strongly basic conditions are necessary to break up polypeptide chains into individual amino acids (hydrolysis).

Set up a 250-mL beaker containing approximately 100 mL of water on a wire gauze/ring stand, and heat to boiling.

Place about 10 drops of 1% albumin, nonfat milk, and 1% alanine solutions and a small cube of gelatin in separate clean, small test tubes.

If you are willing to make the sacrifice, you may also test small samples of your hair (roll up 2 or 3 strands into a ball that will fit in a test tube) or some fingernail clippings.

Add 15–20 drops of 6 *M* sodium hydroxide *(Caution!)* to each sample.

Carefully mix the contents of each test tube with a clean stirring rod, being sure to clean the stirring rod when switching between solutions. Be sure to break up the gelatin sample as much as possible when mixing.

Transfer the test tubes to the boiling-water bath and heat the samples for 10-15 minutes. If the volume of liquid inside the test tubes begins to decrease during the heating period, add 5–10 drops of water to the test tubes to replace the volume.

After the heating period, allow the samples to cool. Then add 5 drops of 5% lead acetate solution to each test tube.

Record which samples produce a black precipitate of lead sulfide.

2. Denaturation of Proteins

Place 10 drops of 1% albumin solution in each of seven clean small test tubes. Set aside one sample as a control for comparisons.

Heat one albumin sample in a boiling-water bath for 5 minutes, and describe any change in the appearance of the solution after heating.

To the remaining albumin samples, add 3–4 drops of the following reagents (each to a separate albumin sample): 3 M hydrochloric acid; 3 M sodium hydroxide; saturated (5.4 M) sodium chloride solution; ethyl alcohol; 5% lead acetate solution.

Record any changes in the appearance of each albumin sample after adding the other reagent.

Repeat the tests above, using 10-drop samples of nonfat milk in place of the albumin solution.

EXPERIMENT 48

Proteins

Pre-Laboratory Questions

1. Use your textbook, an encyclopedia, or an online reference to explain why the amino acids found in animal proteins are referred to as *alpha* (α) amino acids.

2. Use your textbook to draw structural formulas for the following common α-amino acids.

Leucine   Glycine

Valine   Methionine

Phenylalanine   Serine

3.   Explain what is meant by the primary, secondary, and tertiary structures of a protein.

4.   What does it mean to say that a protein has been *denatured*? Give an example from everyday life of the denaturing of a protein.

EXPERIMENT 48

Proteins

Results/Observations

1. **Tests for Proteins**

   Biuret test: Albumin   _____

                        Milk   _____

                        Alanine   _____

                        Gelatin   _____

   Xanthoproteic test: Albumin   _____

                        Milk   _____

                        Alanine   _____

                        Gelatin   _____

   Lead acetate test: Albumin   _____

                        Milk   _____

                        Alanine   _____

                        Gelatin   _____

                        Other   _____

2. **Denaturation of Protein**

   Albumin: Heat   _____

              HCl   _____

              NaOH   _____

              NaCl   _____

              Alcohol   _____

              Lead   _____

Milk: Heat_____

               HCl   _____

               NaOH   _____

               NaCl   _____

               Alcohol _____

               Lead   _____

Questions

1. Why did the 1% alanine solution give a negative result in the biuret test?

2. Use a scientific encyclopedia or online reference to describe why amino acids are strongly affected by the pH of their surroundings.

3. Draw the structure of the simple dipeptides *gly-phe* and *phe-gly*. Circle the peptide bond in each structure.

EXPERIMENT 49

Enzymes

Objective

Enzymes are the essential catalysts in biological systems that make life possible. In this experiment you will examine the catalytic properties of several common enzymes.

Introduction

Enzymes are proteinaceous (protein-like) materials that function in the body as catalysts to control the speed of biochemical processes. Enzymes are said to be *specific* because a given enzyme typically acts on only a single type of molecule (or functional group): the molecule upon which an enzyme acts is referred to, in general, as the enzyme's *substrate*. Like other proteins, enzymes can be denatured by changes in their environment that cause the tertiary or secondary structure of the protein to deform.

The digestion of carbohydrates begins in the mouth, where the enzyme ptyalin is present in saliva. Ptyalin is a type of enzyme called an *amylase*, because the substrate it acts upon is the starch *amylose* (many enzymes are named after the substrate they act upon, with the ending of the substrate's name changed to -*ase* to indicate the enzyme). Ptyalin cleaves the long polysaccharide chain of amylose into smaller units called *oligosaccharides*. These oligosaccharides are then further digested into individual monosaccharide units in the small intestine.

Enzymes called *proteases* break down proteins into oligopeptides or free amino acids. For example, the protease called *bromelain* is present in fresh pineapple: if fresh pineapple is added to gelatin (a protein), the gelatin will not "set" because the protein is broken down. Meat tenderizers and several contact lens cleaners contain the enzyme *papain,* which is extracted from the papaya plant. Papain breaks down the proteins contained in muscle fibers of meat, making it more "tender," and dissolves the protein deposits that may cloud contact lenses.

Hydrogen peroxide is often used to clean wounds because oxygen is released when H_2O_2 breaks down (this is observed as bubbling):

$$2H_2O_2 \longrightarrow 2H_2O + O_2$$

Having an ample supply of oxygen available to a wound is thought to destroy, or slow the growth of, several dangerous anaerobic microorganisms (e.g., *Clostridium*). Hydrogen peroxide used to clean a wound decomposes as a result of the action of enzymes called catalases (and similar enzymes called peroxidases) found in the blood. Such enzymes are also found in abundance in potatoes and yeasts.

Safety Precautions	• Protective eyewear approved by your institution must be worn at all time while you are in the laboratory. • Lead acetate solution is toxic if ingested. Wash after handling. Dispose of as directed by the instructor. • Hydrogen peroxide solution may be corrosive to the eyes and skin. Wash immediately if spilled. Clean up spills on the benchtop.

Apparatus/Reagents Required

5% lead(II) acetate solution, 0.1 M iodine/potassium iodide, 6% hydrogen peroxide, meat tenderizer, contact lens cleaner, fresh and canned pineapple samples, potato, yeast suspension, crackers.

Procedure

1. **Action of ptyalin (salivary amylase)**

   Heat approximately 100 mL of water in a 250-mL beaker to a gentle boil.

   Using a small beaker, collect approximately 3 mL of saliva by letting the saliva flow freely from your mouth (do not spit).

   Obtain a small soda or "oyster" cracker, and crush it finely on a sheet of paper using the bottom of a small beaker.

   Label four clean test tubes as 1, 2, 3, and 4. Transfer a small amount (about the size of a match head) of the crushed cracker into each test tube.

   To the first test tube containing cracker, add 2 mL of distilled water as a control.

   To the second test tube containing cracker, add 1 mL of your saliva and 1 mL of distilled water.

   To the third test tube containing cracker, add 1 mL of your saliva, followed by 1 mL of 5% lead acetate solution. As a heavy metal, the lead(II) ion should denature the protein of the salivary enzyme.

   To the final test tube containing cracker, add 1 mL of your saliva and 1 mL of distilled water, and then place the test tube in the boiling-water bath for 5–10 minutes. Heating the saliva sample should denature the protein of the salivary enzyme. Carefully remove the test tube from the boiling-water bath and allow it to cool to room temperature.

   Set the four test tubes aside on the benchtop for at least 1 hour to allow the salivary enzyme to digest the starches in the cracker. Go on to the other portions of the experiment during the waiting period.

   After the test tubes have stood for at least1e hour, add 1 drop of 0.1 M I$_2$/KI solution to each of the test tubes. Iodine is a standard laboratory test for the presence of starch: if starch is present even in trace quantities in a sample, addition of I$_2$/KI solution will cause a deep blue-black color to appear. Record and explain your observations.

2. **Action of proteases.**

Your instructor will provide you with samples of prepared gelatin (a protein), either cast in the wells of a spot plate, or as small chunks that you can transfer to small test tubes. Although the gelatin is similar to the type used as a dessert, the gelatin has been made up double-strength to make it drier and easier to handle.

To one gelatin sample, add a few crystals of meat tenderizer.

To a second gelatin sample, add a few drops of the available commercial contact lens cleaner (record the brand and the name of the enzyme it contains, if available).

To a third gelatin sample, add a small piece of fresh pineapple (or juice extracted from fresh pineapple).

To a fourth gelatin sample, add a piece of cooked or canned pineapple.

Allow the gelatin samples to stand undisturbed for 30–60 minutes. Go on to the last part of the experiment while waiting.

After this time period, examine the gelatin samples for any evidence of breakdown of the gelatin by the proteases. Determine in particular whether the gelatin has seemed to melt or partially melt. Record your observations.

3. **Action of catalase**

Obtain two small slivers of freshly cut potato. Place one of the pieces of potato in a small test tube and transfer to a boiling-water bath for 5–10 minutes.

Place 10 drops of 6% hydrogen peroxide (*Caution!*) into each of three small test tubes.

To one test tube, add 8-10 drops of yeast suspension. Record your observations.

To the second test tube, add the small piece of uncooked, freshly cut potato. Record your observations.

To the third test tube, add the small piece of cooked potato. Why does the cooked potato not cause the hydrogen peroxide to decompose?

Name: _____ **Section:** _____

Lab Instructor: _____ **Date:** _____

EXPERIMENT 49

Enzymes

Pre-Laboratory Questions

1. Use your textbook, an encyclopedia, or an online reference to write a specific definition of the term *enzyme*.

2. Briefly explain the "lock-and-key" model for the action of enzymes.

3. Use your textbook, an encyclopedia, or an online reference to list the names and functions of five enzymes not mentioned in this experiment.

4. If the environment of an enzyme is changed from its normal state, what may happen to the enzyme? Why is this a problem in living creatures?

Name: ______________________________ **Section:** __________

Lab Instructor: ______________________________ **Date:** __________

EXPERIMENT 49

Enzymes

Results/Observations

1. **Action of ptyalin:** Results of I_2/KI test

   Control ______________________________

   Saliva ______________________________

   Saliva/lead ______________________________

   Saliva/heat ______________________________

2. **Action of proteases**: Observations on gelatin samples

   Meat tenderizer ______________________________

   Lens cleaner ______________________________

   Fresh pineapple ______________________________

   Cooked pineapple ______________________________

3. **Action of catalase:** Observations on hydrogen peroxide samples

   Yeast ______________________________

   Fresh potato ______________________________

   Cooked potato ______________________________

Questions

1. Why do the cooked pineapple and cooked potato samples appear to have no active enzymes?

2. List five ways in which an enzyme can be denatured.

3. Some antibiotic drugs work by *inhibiting* the enzymes of microorganisms. Use your textbook or an encyclopedia to explain what is meant by the *inhibition* of enzyme action.

4. Enzymes, like all materials containing proteins, are affected by the pH of their surroundings. For each enzyme, there is an optimum pH at which the enzyme is most effective. Many enzymes have their optimum pH in the range of pH 7.0 – 7.2, which is the normal pH of most cellular fluids. However the enzyme pepsin, found in the stomach, is most active around pH 2. Use a scientific encyclopedia or online reference to determine what process(es) pepsin catalyzes and why pepsin is most active at very low pH.

EXPERIMENT 50

Polymeric Substances 1:   Amorphous Sulfur

Objective

Polymeric substances contain molecules that consist of extremely long chains of repeating simpler units (monomers). In this experiment, you will investigate a polymeric allotrope of sulfur.

Introduction

The ordinary form/allotrope of sulfur at room conditions, *orthorhombic* sulfur, consists of discrete S_8 molecules. The eight sulfur atoms form a non-planar ring. Orthorhombic sulfur is brittle, hard, and pale yellow in color. It is insoluble in water but is relatively soluble in carbon disulfide, CS_2. When orthorhombic sulfur is heated to near its boiling point and then quickly quenched in cold water, a *polymeric* allotrope of sulfur, which has very different properties, is formed.

When sulfur is heated, the normal S_8 ring molecules split open, giving free chains of eight sulfur atoms. The terminal sulfur atoms of these chains tend to react with the terminal sulfur atoms of other chains, thereby building up still longer chains. The buildup of polymeric chains is observed during the heating of sulfur as the liquefied sulfur becomes thicker and very viscous ("sticky"). In contrast with this observation for heating sulfur, most other substances generally become thinner and more free-flowing when heated.

If the thick, sticky, liquefied sulfur is allowed to cool slowly, the original orthorhombic S_8 ring molecules re-form as the temperature drops. If the thick, sticky liquefied sulfur is *rapidly quenched* by cooling however, the polymeric allotrope can be "frozen" and examined. The polymeric sulfur is dark brown or black, and its properties are very different from those of orthorhombic sulfur. On standing for several days, the polymeric sulfur slowly converts into orthorhombic sulfur.

S_8 ring S_8 chain

Safety Precautions	• **Protective eyewear approved by your institution must be worn at all time while you are in the laboratory.** • **Carbon disulfide is very toxic, and its vapor is very volatile and forms explosive mixtures with air. Avoid contact with CS$_2$.** *Confine all use of CS$_2$ to the fume exhaust hood.* **No flames will be permitted in the room while anyone is using carbon disulfide.** *Check with the instructor before lighting any flames.* • **Hot oil is dangerous and can spatter and burn if moisture is present. Make sure the oil bath used is not cloudy before starting to heat.**

Apparatus/Reagents Required

Chunk sulfur, powdered sulfur, carbon disulfide, oil bath/hotplate

Procedure

Record all data and observations directly in your notebook in ink.

1. **Properties/Recrystallization of Orthorhombic Sulfur**

   Obtain a large chunk of sulfur and examine its color, hardness, and brittleness. To demonstrate the hardness and brittleness of the sulfur, crush the chunk with the metal base of a ringstand. Place a very small chunk of the crushed sulfur, or a small amount of powdered sulfur, in a small test tube and bring it to the exhaust hood.

   In the exhaust hood, add 10 drops of carbon disulfide (*Caution!*) to the sulfur, and stir carefully with a glass rod to dissolve the sulfur. Sulfur is slow to dissolve. If the sulfur has not dissolved after 4–5 minutes of stirring, add 5 drops more of carbon disulfide, and stir to dissolve the remaining sulfur.

   While it is still in the exhaust hood, pour the sulfur/carbon disulfide solution into a watchglass, allowing the CS$_2$ to evaporate slowly. Examine the crystals of orthorhombic sulfur with a magnifying glass and sketch their shape.

2. **Preparation of Polymeric Sulfur**

   Half fill a test tube with powdered sulfur. Using a clamp, set up the test tube in an oil bath (*Caution!*) on a hotplate in the exhaust hood. Begin heating the sulfur, noting changes in *color* and *viscosity* as the sulfur is heated. Use a glass rod to stir the sulfur during the heating.

   Prepare a beaker of ice-cold water for use in quenching the polymeric sulfur. When the sulfur has almost begun to boil, rapidly pour the thick molten sulfur into the beaker of cold water. Using forceps, remove the polymeric sulfur from the cold water. Examine the polymer's hardness and brittleness, and compare these properties with those of orthorhombic sulfur.

   In the exhaust hood, test the solubility of a tiny piece of the polymeric sulfur in 10 drops of carbon disulfide.

   Allow the polymeric sulfur to sit in your laboratory locker until the next lab period. At that time, re-examine the properties of the sulfur, testing for color, hardness, brittleness, and solubility. Have the properties of the polymeric sulfur changed?

Name: _____ **Section:** _____

Lab Instructor: _____ **Date:** _____

EXPERIMENT 50

Polymeric Substances 1: Amorphous Sulfur

Pre-Laboratory Questions

1. What, in general terms, is meant by the term *polymer*?

2. Describe how the polymeric form of sulfur forms from the normal S_8 molecules in which sulfur usually exists.

3. Sulfur has several allotropic forms in addition to the polymeric allotrope to be prepared in this experiment. Use your textbook or a chemical encyclopedia to find a representation of the *phase diagram* of sulfur in which the various forms of sulfur are indicated. Sketch the diagram here.

4. Carbon disulfide is used to dissolve sulfur in this experiment. Carbon disulfide is extremely flammable and its vapor is toxic. Discuss some safety precautions that must be taken when working with carbon disulfide.

EXPERIMENT 50

Polymeric Substances 1: Amorphous Sulfur

Results/Observations

Observations on color, hardness, brittleness of chunk sulfur

Observation on recrystallized orthorhombic sulfur (from CS_2)

Shape of crystals

Observation on color, hardness, and brittleness of plastic sulfur

Solubility in CS_2

Observation on plastic sulfur after standing 1 week

Questions

1. Most substances become more free-flowing when heated above their melting point. Why? Why did sulfur become *thicker* (that is, *less* free-flowing) when heated beyond its melting point?

2. Why was the polymeric sulfur insoluble in CS_2?

3. What changes occurred in the polymeric sulfur after standing for a week? Which form of sulfur appeared to be present after the waiting period?

EXPERIMENT 51

Polymeric Substances 2: Preparation of Nylon

Objective

Polymeric substances contain molecules that consist of extremely long chains of repeating simpler units (monomers). In this experiment, you will prepare a short length of nylon filament.

Introduction

In general, a polymeric substance contains long-chain molecules, in which a particular unit (monomer) is found *repeating* over and over again along the entire length of the polymer's chain. **Condensation polymers** are formed when two substances that ordinarily would not polymerize on their own are made to react, producing a monomer that is now capable of polymerizing. A small molecule, usually water, is also split out during the condensation.

Nylon is a condensation polymer of 1,6-hexanedioic acid (adipic acid) or one of its derivatives, with 1,6-hexanediamine (hexamethylenediamine). The monomer produced from the condensation of these two substances has a total of 12 carbon atoms, with 6 from each of the reacting species. The nylon resulting from the polymerization of this monomer is sometimes referred to as nylon-66 (other polymers result if other acids or amines are used). The condensation reaction results in the formation of an **amide link** between the acid and the amine, and nylon is therefore a polyamide.

$$n \left(\underset{\substack{|\\ H}}{H-N}-(CH_2)_6-\underset{\substack{|\\ H}}{N}-H \quad + \quad HO-\underset{\substack{\|\\ O}}{C}-(CH_2)_4-\underset{\substack{\|\\ O}}{C}-OH \right) \longrightarrow$$

hexamethylene diamine   adipic acid

$$* \left(\underset{\substack{|\\ H}}{N}-(CH_2)_6-\underset{\substack{|\\ H}}{N}-\underset{\substack{\|\\ O}}{C}-(CH_2)_4-\underset{\substack{\|\\ O}}{C} \right)_n * \quad + \quad n\,H_2O$$

nylon-66 repeating unit

As the condensation reaction occurs, it is possible to pull the nylon being produced into thin filaments (threads). Because of hydrogen bonding between polymeric molecules, the resulting nylon is very strong and stable.

Safety Precautions	• Protective eyewear approved by your institution must be worn at all time while you are in the laboratory.
	• The reagents used in this synthesis are toxic until mixed to form the nylon. Adipoyl chloride can burn skin, eyes, and clothing. Its vapor is harmful and irritating. Adipoyl chloride releases hydrogen chloride gas during the reaction. Work with the reagents in the fume exhaust hood. Wash immediately if spilled and clean up all spills.
	• The hexane used as solvent is volatile and highly flammable. No flames will be permitted in the laboratory. Work in the fume exhaust hood
	• Dispose of all material as directed by the instructor

Apparatus/Reagents Required

10% adipoyl chloride in hexane, 20% 1,6-hexanediamine in water, 1 M NaOH solution, tongs or forceps, watchglass

Procedure

Record all data and observations directly in your notebook in ink.

All portions of this experiment up to the washing of the nylon must be performed in the **exhaust hood!**

Have ready a 250-mL beaker full of cold water for washing the nylon after it is produced.

Place about 5 mL of 1 M NaOH on a watchglass. To the NaOH, add approximately 1 mL of 20% 1,6-hexanediamine solution. Stir with a stirring rod to mix the reagents.

Gently add 10 – 12 drops of the 10% adipoyl chloride/hexane solution to the surface of the aqueous reagents on the watchglass. Try to *float* the adipoyl chloride solution on top of the aqueous layer. Avoid bulk mixing of the two solutions; otherwise, the nylon will congeal into a blob rather than remaining threadlike.

Almost immediately, nylon will begin to form at the *interface* between the aqueous and nonaqueous solutions. With forceps or tongs, reach into the liquid on the watchglass and begin to pull on the film of nylon that has formed at the liquid junction. Pull the filament of nylon slowly and evenly from the mixture until you have a thread of nylon about 3 – 6 inches long.

Transfer the nylon filament to the beaker of rinsing water, and allow it to stand for several minutes. Dispose of the other reagents as directed by your instructor.

Take the beaker with the nylon filament to your bench, rinse the nylon with several additional portions of water, and examine its color, hardness, and strength when pulled.

Name: ______________________ **Section:** __________

Lab Instructor: ______________________ **Date:** __________

EXPERIMENT 51

Polymeric Substances 2: Preparation of Nylon

Pre-Laboratory Questions

1. What, in general terms, is a **condensation polymer**?

2. Use a handbook of chemical data or online source to find the following physical properties of nylon-66:
   a. Density ______________________
   b. Color ______________________
   c. Melting point ______________________
   Reference ______________________

3. Nylon-66 is a *polyamide*. What functional group(s) is(are) present in a molecule that make(s) the molecule an *amide*? Draw the structure of a representative example of such a molecule.

4. Write a chemical equation/structure showing the formation of a unit of nylon-66 from molecules of 1,6-hexanediamine and 1,6-hexanedioic acid.

5. Use your textbook to draw the structures of the repeating unit of four other common condensation polymers. Give the name of each polymer and indicate the reagents from which the polymer could be synthesized.

EXPERIMENT 51

Polymeric Substances 2: Preparation of Nylon

Results/Observations

Observation on adding adipoyl chloride solution to 1,6-hexanediamine solution

Observation on pulling nylon filament from the reaction mixture

Observation on color, hardness, and strength of nylon filament

Questions

1. Why did the adipoyl chloride solution *float* on the surface of the 1,6-hexanediamine solution?

2. It was mentioned that nylon-66 is not the only nylon that can be synthesized. Use your textbook or a chemical encyclopedia to find the formulas of some other nylons. Enter those formulas here.

3. Use a scientific encyclopedia or online reference to indicate the repeating monomer unit for each of the following important polymers.

    a.   polyethylene

    b.   polyvinyl chloride

    c.   polytetrafluoroethylene (Teflon®)

EXPERIMENT 52

Qualitative Analysis of Selected Cations and Anions

Objective

Behaviors of selected cations and anions will be investigated, with known and unknown samples being tested for the purpose of identification. Elimination and confirmation tests will be used to divide the samples tested into groups on the basis of specific chemical properties.

Introduction

Over the years, chemists have devised means to identify the component ions present in inorganic compounds. Unfortunately, a lot of these tests involve reagents that are highly toxic, many of which are known carcinogens and/or mutagens. The good news is that modern electronic instrumentation has given us the means to identify the elements and ions present without resorting to classical analytical schemes that face the investigator with dangerous health risks. In the previous four experiments, methods were outlined for the separation and identification of various cations.

In those classical analytical schemes, the cations were divided into particular groups on the basis of chemical properties shown by the members of each group. For example, cations were placed in Group I if their chloride salts were insoluble in acid. Cations whose sulfides were insoluble (which essentially includes all but the cations of Group 1A) were divided into subgroups according to whether their sulfide salts were soluble in acidic systems. By repeated separation techniques, the cations can be identified and confirmed one by one until all the cations in a mixture have been determined. A similar progressive method can be used to identify the components of a mixture of anions. The anions are separated into smaller groups from the general mixture and then confirmed one by one.

For our purposes, a series of simple solubility tests will be run on four members of the alkaline earth family, Group 2A on the periodic table, then a streamlined scheme will be used to identify the anions present in solid samples. Individual solid samples containing a single anion will be tested, and the normal confirmatory test for the particular anion will be demonstrated. Then unknown samples will be tested, with each unknown again containing only a single cation or anion. Preliminary screening tests will be performed on the known and unknown samples. On the basis of these preliminary tests, the traditional anion groups are constituted.

Safety Precautions	• Protective eyewear approved by your institution must be worn at all time while you are in the laboratory.
	• All the solid samples in this experiment should be assumed to be toxic. Wash after handling them.
	• Barium, nitrite, molybdenum, and permanganate compounds are toxic. Wash after handling.
	• Sulfuric and nitric acids are damaging to skin, eyes, and clothing. Wash immediately if the acids are spilled and inform the instructor.
	• SO_2 and NO_2 gases are toxic and irritating to the respiratory system. Only the smallest possible quantities of these gases should be inhaled in the confirming tests for the sulfite and nitrate ions.
	• Methylene chloride is toxic and its vapors are harmful.
	• Ammonia is a strong cardiac stimulant and its vapors are extremely irritating to the respiratory tract. Use in the fume exhaust hood.
	• Silver nitrate will stain the skin and clothing if spilled. Wash immediately. The stain on skin will take several days to wear off
	• Dispose of all materials as directed by the instructor.

Apparatus/Reagents Required

Group 2A Cations: 0.10 M solutions of $Mg(NO_3)_2$, $Ca(NO_3)_2$, $Sr(NO_3)_2$, and $Ba(NO_3)_2$; 0.5 M $(NH_4)_2MoO_4$, 0.1 M $(NH_4)_2C_2O_4$, 0.1 M Na_2SO_4, and 0.1 M $NaOH$; unknown, containing one of the Group 2A nitrates.

Selected Anions: Known solid samples of the anions (sulfate, sulfite, carbonate, nitrate, phosphate, chloride, bromide, iodide), two unknown samples (each containing one of the anions), 0.1 M barium chloride, 6 M nitric acid, 6 M sulfuric acid, 0.1 M silver nitrate, 0.5 M ammonium molybdate reagent, 1 M iron(II) sulfate, 1 M potassium nitrite, methylene chloride, 0.1 M potassium permanganate, 6 M aqueous ammonia

Procedure

Record all data and observations directly in your notebook in ink.

A. The Cations of Group 2A

1. The Analytical Scheme

Place 1-2 mL each of 0.1 M solutions of $Mg(NO_3)_2$, $Ca(NO_3)_2$, $Sr(NO_3)_2$, and $Ba(NO_3)_2$ respectively in four separate 100-mm ("4-inch") test tubes. The tubes must be scrupulously clean; contamination of reagents can make it difficult or even impossible to interpret results. There is no need to dry the inside of the tubes; just shake out any excess water.

494

Add an equal volume of 0.5 M ammonium molybdate, $(NH_4)_2MoO_4$, to each one and look for formation of a precipitate in each tube. Precipitates may be slow to form, so be patient. If no precipitation appears within a couple of minutes, you can assume that none will occur and proceed with your experiment. If a precipitate does form, note its color, whether it is fast or slow to appear, and whether it is heavy (thick) or light (just a light cloudiness). Record your observations in the Data Table for Part A.

Dispose of the contents of your tubes in the waste container provided for this part of this experiment. Clean your tubes as described above if they are needed for the following steps.

Repeat the previous step, starting with fresh 1-2 mL quantities of each of the four alkaline earth metal solutions. This time, in place of the ammonium molybdate, add 1-2 mL of 0.1 M ammonium oxalate, $(NH_4)_2C_2O_4$, to each of the tubes. As before, agitate the tubes to facilitate mixing, then allow the tubes to stand for 2-3 minutes. Record in the Data Table whether or not a precipitate forms, how quickly it forms, its color, and appearance (heavy or light). Dispose of the contents of your tubes in the appropriate container.

Using clean tubes and 1-2 mL of each of the four cation solutions, test each with 0.1 M sodium sulfate, Na_2SO_4. Record your observations in the Data Table, then dispose of the contents of your tubes and proceed to the final test.

Using 1-2 mL samples of the four Group 2A cations, in separate, clean test tubes, test each one with 1-2 mL of 0.1 M sodium hydroxide, NaOH.

Dispose of the contents of your tubes and thoroughly clean all of your test tubes following the procedure described at the start of this procedure section.

2. **Analysis of the Unknown**

Starting with four clean tubes, place 1-2 mL of your assigned unknown in each tube. To identify the group 2A cation, add 1-2 mL of a different test reagent to the four tubes: $(NH_4)_2MoO_4$ in the first, $(NH_4)_2C_2O_4$ in the second, and so on. By comparing these results with your data table, you are to determine which of the four Group 2A cations was present in your unknown.

B. Analysis of Selected Anions

The anions may be quickly divided into three subgroups by some simple preliminary tests. Apply the tests to approximately 0.1-g samples of each of the known anion samples and to each of the unknown samples.

The known anion samples are sodium or potassium salts, ensuring that the anion samples will be soluble in water. You will test the following anions: sulfate, SO_4^{2-}; sulfite, SO_3^{2-}; carbonate, CO_3^{2-}; nitrate, NO_3^-; phosphate, PO_4^{3-}; and the halides (Cl^-, Br^-, and I^-). In addition to the tests on the known anion samples, two unknown samples will be determined.

1. **Anions Whose Barium Salts Are Insoluble**

Set up a rack of test tubes, and label each of the test tubes for one of the anions to be tested. Also provide labeled test tubes for each of your unknown anion samples.

Place a few crystals of the appropriate anion or unknown sample into its respective test tube. Add about 1 mL of distilled water, and swirl each test tube to dissolve the salt.

Add 10 drops of 0.1 M barium chloride to each test tube, and record which anions produce insoluble barium salts.

Add 10–15 drops of 6 M nitric acid to each test tube, and swirl the test tubes. Record which anions give precipitates with barium ion even in acidic solution. Discard the samples in the appropriate waste container – barium salts are toxic and require special handling and disposal.

2. **Anions That Form Volatile Substances When Acidified**

   Set up a rack of test tubes, and label each of them for one of the anions to be tested. Also provide labeled test tubes for each of your unknown anion samples. Set up a boiling-water bath for use in heating the samples.

   Place a few crystals of the appropriate anion salt or unknown sample into its respective test tube. Add about 1 mL of distilled water, and swirl each test tube to dissolve the salt.

   Treat each sample of anion/unknown separately to avoid confusing the samples. Add 10 – 12 drops of 6 M sulfuric acid to the sample, and note whether any gas is evolved from the sample. Cautiously waft the vapors from the test tube toward your nose to detect any characteristic odor.

   Heat each sample one at a time in the boiling-water bath to see whether any gas/odor is evolved under these conditions. Record which samples produced a volatile product/odor. Remove the samples to the fume hood and discard the samples in the appropriate waste container.

3. **Anions That Form Insoluble Silver Compounds**

   Set up a rack of test tubes, and label each of the test tubes for one of the anions to be tested. Also provide labeled test tubes for each of your unknown anion samples.

   Place a few crystals of the appropriate anion or unknown sample into its respective test tube. Add about 1 mL of distilled water, and swirl each test tube to dissolve the salt.

   Add 10 drops of 0.1 M silver nitrate solution to each test tube. Record which samples precipitate with silver ion.

   Add 10–15 drops of 6 M nitric acid to each sample. Swirl the test tubes, and record which samples produce, with silver ion, a precipitate that is then insoluble in acid.

4. **Confirmation Tests for the Anions**

   The preliminary tests given in Parts 1, 2, and 3 should have indicated to you some general properties that can be used to distinguish among the anions. The results of these tests should give you some indication as to what anions your two unknown samples contain. If the results of the preliminary tests are not clear to you, consult with your instructor before proceeding.

   a. **Sulfate Ion**

      Sulfate ion is precipitated by barium ion, and the precipitate is insoluble in acid. Other substances may precipitate *initially* with barium ion, but such precipitates are subsequently soluble in acid.

      Place a small portion of the sulfate known and a small portion of any unknown believed (from the preliminary tests) to be sulfate into separate clean test tubes. Add 10 drops of water and swirl to dissolve the samples.

      Add 10 drops of 0.1 M barium chloride solution, followed by 10 drops of 6 M nitric acid. If the unknown sample does contain sulfate ion, the precipitate formed with barium ion will not dissolve when acid is added.

   b. **Phosphate Ion**

      Phosphate ion is one of the ions that is precipitated by barium ion, but the precipitate is subsequently soluble in acid. Phosphate ion similarly is precipitated by silver ion, but as before, the precipitate subsequently dissolves in acid. Phosphate ion does not generate a volatile product when acidified and boiled. A specific reagent, ammonium molybdate, is used to confirm the presence of phosphate. Phosphate forms a characteristic yellow precipitate with this reagent.

Place a small portion of the phosphate known and of any unknown believed (from the preliminary tests) to be phosphate into separate clean test tubes. Add 10 drops of water and 5 drops of 6 *M* nitric acid to each test tube. Swirl the test tubes to dissolve the samples.

Add 10 drops of 0.5 *M* ammonium molybdate reagent to each sample, and transfer the test tubes to a boiling-water bath. The appearance of a yellow precipitate confirms the presence of phosphate.

The precipitate is sometimes slow in forming if the sample is not saturated, but generally the sample will turn yellow as the ammonium molybdate is added.

Continue heating the samples in the boiling-water bath for 5 minutes; then remove and transfer to an ice bath. Check each sample carefully for the yellow precipitate. (There may be only a small amount of precipitate present if the solution was not truly saturated.)

c. Nitrate Ion

Nitrate ion does not precipitate with barium ion or with silver ion and does not generate a volatile product when acidified and heated. A specific test reagent is used for nitrate ion, consisting of acidified iron(II) sulfate. Nitrate ion will oxidize iron(II) to iron(III), and brown nitrogen(IV) oxide gas will be evolved if the mixture is heated.

Place a small portion of the nitrate known and of any unknown that is believed (from the preliminary tests) to be nitrate into separate clean test tubes. Dissolve the samples in 15–20 drops of 6 *M* sulfuric acid.

Add 10–15 drops of 1 *M* iron(II) sulfate to each sample and transfer to the boiling-water bath. The evolution of brown NO_2 gas confirms nitrate in the sample.

d. Sulfite Ion

Sulfite ion precipitates with barium ion, but the precipitate is subsequently soluble in acid. Sulfite ion may also precipitate with silver ion, but the precipitate will dissolve after a few minutes as complexation occurs. Sulfite ion, however, generates sulfur dioxide gas when acidified and heated, and the gas may be identified by its characteristic odor.

Place a small portion of the sulfite known and of any unknown believed (from the preliminary tests) to be sulfite ion into separate clean test tubes. Add 10–15 drops of distilled water, and swirl to dissolve.

Add 10 drops of 6 *M* sulfuric acid and note whether a gas or odor is evolved from the sample. Heat the sample in the boiling-water bath for 30 seconds and check again for the characteristic odor of sulfur dioxide.

e. Carbonate Ion

Carbonate precipitates with barium ion and silver ion, but the precipitates are soluble in acid. Carbonate ion evolves colorless, odorless carbon dioxide gas when acidified. If a more specific test for carbonate is desired, the gas evolved from the carbonate when acidified can be bubbled into limewater, from which insoluble calcium carbonate will precipitate.

Place a small portion of the carbonate known and of any unknown that is believed to be carbonate ion into separate clean test tubes. Add 5–6 drops of 6 *M* sulfuric acid. The vigorous evolution of a colorless, odorless gas can be taken as a positive test for carbonate ion.

f. Halide Ions

Halide ions (chloride, bromide, iodide) do not form precipitates with barium ion and generally do not evolve a volatile product when acidified with a non-oxidizing acid. The halides do precipitate with silver ion, however.

Place a small portion of the chloride, bromide, and iodide known samples into separate clean test tubes (label the test tubes). Also place a small portion of any unknown believed to be one of the halides into a clean test tube. Dissolve each of the samples in 10–15 drops of distilled water.

To the iodide known sample and the unknown sample, add 1 mL of methylene chloride and 20–30 drops of 1 M potassium nitrite solution. Stopper the samples and shake. Nitrite ion will oxidize iodide ion to elemental iodine, which is preferentially soluble in the methylene chloride layer.

The iodide known sample will develop a purple color in the organic layer, as will the unknown if it contains iodide. Save the unknown sample if iodide is not present.

Place a small portion of the bromide known sample into a clean test tube. If the unknown tested earlier did not contain iodide ion, it may contain bromide or chloride ion. With a medicine dropper, remove the upper (aqueous) layer from the unknown sample that was tested for iodide ion, and transfer the upper layer to a clean test tube.

Add 1 mL of methylene chloride to each test tube, followed by 2 drops of 0.1 M potassium permanganate. Stopper the test tubes and shake. Permanganate ion will oxidize bromide ion to elemental bromine, which is preferentially soluble in the methylene chloride added.

The bromide known sample will develop a red color in the methylene chloride layer, as will the unknown if the unknown contains bromide ion. Save the unknown sample if bromide ion is not present.

Place a small sample of the chloride known sample into a clean test tube. If the unknown tested earlier did not contain bromide ion, then remove the upper (aqueous) layer from the sample with a medicine dropper, and transfer to a clean test tube.

Add 10–15 drops of 0.1 M silver nitrate to each sample. The appearance of a white precipitate is taken as an initial indication that chloride ion is present. Centrifuge the mixture and decant the supernatant liquid. Wash the precipitate with 20–30 drops of distilled water, centrifuge, and discard the rinsings.

To confirm chloride ion, dissolve the precipitate in 15–20 drops of 6 M aqueous ammonia. Then add 6 M nitric acid dropwise to the solution, which will cause the precipitate of silver chloride to reappear.

Name: ______________________________ **Section:** __________

Lab Instructor: ______________________________ **Date:** __________

EXPERIMENT 52

Qualitative Analysis of Selected Ions

Pre-Laboratory Questions

1. Use a chemical reference or a table of solubility rules to predict which Group 2A cations are likely to form precipitates with the anions being used in Part A of this experiment.

2. Describe how each of the following anions is confirmed.

   a. nitrate ion, NO_3^-

   ____

   b. sulfite ion, SO_3^{2-}

   ____

   d. phosphate ion, PO_4^{3-}

   ____

   e. carbonate ion, CO_3^{2-}

   ____

3.   This experiment provides methods for the detection of a few of the more common anions, but many equally common anions are omitted from the analysis for brevity. Consult a textbook of qualitative analysis or a handbook, and indicate briefly how the presence of the following anions could be detected.

   a.   acetate ion, $C_2H_3O_2^-$

   b.   sulfide ion, S^{2-}

   c.   chromate ion, CrO_4^{2-}

   d.   nitrite ion, NO_2^-

Name: ______________________________ Section: __________

Lab Instructor: ______________________________ Date: __________

EXPERIMENT 52

Qualitative Analysis of Selected Anions

Results/Observations

Part A. The Cations of Group 2A

Reagents	Mg^{2+}	Ca^{2+}	Sr^{2+}	Ba^{2+}	Unknown #
0.02 *M* $(NH_4)_2MoO_4$					
0.1 *M* $(NH_4)_2C_2O_4$					
0.1 *M* Na_2SO_4					
0.1 *M* NaOH					

Identity of Unknown: __________________

Part B. Analysis of Selected Anions

Identification numbers of unknowns __________________ __________________

1. Which known/unknown samples form precipitates with Ba^{2+}?

2. Which known/unknown samples evolved a gas when acidified?

3. Which known/unknown samples form precipitates with Ag^{+}?

4. Confirmatory Tests: Observations

   a. Sulfate

b. Phosphate

c. Nitrate

d. Sulfite

e. Carbonate

f. Chloride

g. Bromide

h. Iodide

5. Which anions do your unknowns contain? Give your reasoning.

Questions

1. Why is nitrite ion used in the oxidation of iodide ion, whereas permanganate is used for bromide ion?

2. Write an equation showing the oxidation of iodide ion by nitrite ion.

Plotting Graphs of Experimental Data

Very often the data collected in the chemistry laboratory are best presented pictorially, by means of a simple graph. This section reviews how graphs are most commonly constructed, but you should consult with your instructor about any special considerations he or she may wish you to include in the graphs you will be plotting in the laboratory.

The use of a graphical presentation allows you to display clearly both experimental data and any relationships that may exist among the data. Graphs also permit interpolation and extrapolation of data, to cover those situations that were not directly investigated in an experiment and to allow predictions to be made for such situations. In order to be meaningful, your graphs must be *well planned*. Study the data before attempting to plot them. Recording all data in tabular form in your notebook will be a great help in organizing the data.

The particular sort of graph paper you use for your graphs will differ from experiment to experiment, depending on the precision possible in a particular situation. For example, it would be foolish to use graph paper with half-inch squares for plotting data that had been recorded to four-significant-figure precision. It is impossible to plot such precise data on paper with such a large grid system. Similarly, it would not be correct to plot very roughly determined data on very fine graph paper. The fine scale would imply too much precision in the determinations.

A major problem in plotting graphs is deciding what each scale division on the axes of the graph should represent. You must learn to scale your data, so that the graphs you construct will fill virtually the entire graph paper page. Scale divisions on graphs should always be spaced at very regular intervals, usually in units of 10, 100, or some other appropriate power of 10. Each grid line on the graph paper should represent some readily evident number (for example, not 1/3 or 1/4 unit). It is not necessary or desirable to have the intersection of the horizontal and vertical axes on your graph always represent the origin (0, 0). For example, if you were plotting temperatures from 100°C to 200°C, it would be silly to start the graph at zero degrees. The axes should be labeled in ink as to what they indicate, and the major scale divisions should be clearly marked. By studying your data, you should be able to come up with realistic minimum and maximum limits for the scale of the graph and for what each grid line on the graph paper will represent. Consult the many graphs in your textbook for examples of properly constructed graphs.

By convention, the **horizontal** (x) **axis,** formally known as the abscissa, of a graph is used for plotting the quantity that the experimenter has varied during an experiment (independent variable). The **vertical** (y) **axis**, formally known as the ordinate, is used for plotting the quantity that is being measured in the experiment (as a function of the other variable). For example, if you were to perform an experiment in which you measured the pressure of a gas sample as its temperature was varied, you would plot temperature on the horizontal axis, since this is the variable that is being controlled by you, the experimenter. The pressure, which is measured and which results from the various temperatures used, would be plotted on the vertical axis.

When plotting the actual data points on your graph, use a sharp, hard pencil. Place a single, small, round dot to represent each datum. If more than one set of data is being plotted on the same graph, small squares or triangles may be used for the additional sets of data. If an estimate can be made of the probable magnitude of error in the experimental measurements, this can be indicated by **error bars** above and below each data point (the size of the error bars on the vertical scale of the graph should indicate the magnitude of the error on the scale). To show the relationship between the data points, draw the best possible straight line or smooth curve through the data points. Straight lines should obviously be inked in with a ruler, whereas curves should be sketched using a drafter's French curve.

If the data plot for an experiment gives rise to a straight line, you may be asked to calculate the **slope** and **intercept** of the line. Although the best way to determine the slope of a straight line is by the techniques of numerical regression, if the data seem reasonably linear, the slope may be approximated by

$$\text{slope} = (y_2 - y_1)/(x_2 - x_1)$$

where (x_1, y_1) and (x_2, y_2) are two points on the line. Once the slope has been determined, either intercept of the line may be determined by setting y_1 or x_1 equal to zero (as appropriate) in the equation of a straight line:

$$(y_2 - y_1) = (\text{slope})(x_2 - x_1)$$

If you or your university or college has computers available for your use, simple computer programs for determining the slope and intercept of linear data are commonly available. Spreadsheet programs also have the ability to create "x-y" graphs. Use of computer plots may not be permitted; check with your instructor. The idea is for *you* to learn how to plot data.

Remember always *why* graphs are plotted. Although it may seem quite a chore when first learning how to construct them, graphs are intended to simplify your understanding of what you have determined in an experiment. Rather than just a list of numbers that may seem meaningless, a well-constructed graph can show you instantaneously whether your experimental data exhibit consistency and whether a relationship exists among them.

APPENDIX B

Errors and Error Analysis in General Chemistry Experiments

The likelihood of error in experimental scientific work is a fact of life. No measurement technique is perfect. Errors in experiments do happen—no matter how much effort is expended to prevent them. Naturally, however, the scientist must take every reasonable precaution to exclude errors due to sloppiness, poor preparation, and misinterpretation of results. Other errors, though, are unavoidable.

For such unavoidable errors, an estimate of the magnitude of the error, and how that error will affect the results and conclusions of the experiment, must be included as part of the data treatment for the experiment. Errors in experimental work can generally be classified as either of two types, determinate or indeterminate. **Systematic** errors may include personal errors by the experimenter, errors in the apparatus or instrumentation used for an experiment, and errors in the method used for the experiment. **Random** errors are introduced because no measurement can be made with truly absolute certainty, and because, for a series of duplicate measurements, there will always be minor fluctuations in the results obtained.

Personal errors are very common among beginning students who have not yet mastered the various procedures and techniques used routinely in the chemistry laboratory. Personal errors can be eliminated by close attention to published procedures and to the advice and suggestions of your instructor. Most experiments are repeated several times in an attempt to provide a set of data, so that personal errors can be recognized if the results are not reproducible—that is, if the same measurement does not consistently give the "same" answer.

Errors due to the apparatus or instruments used for an experiment are less easy to detect or correct. However, some major sources of such error can be eliminated. Be sure all glassware you use is clean. This applies especially to volumetric glassware used for transferring or containing specific volumes of liquids or solutions. Oftentimes, volumetric glassware is calibrated before use to ensure that the glassware has been manufactured correctly and has not suffered any loss in precision during previous use. When making mass determinations, always use the same laboratory balance throughout an experiment. When mass determinations made in the laboratory are done by a difference technique, errors due to poorly calibrated balances will cancel out; this also applies to the "tare" feature common to electronic balances, so long as all measurements are done using the same instrument. In many experiments in this manual, you are asked to consult your instructor before starting a procedure so that the instructor can locate any error in the construction of the apparatus.

Errors in experimental procedures occur because no experimental procedure is perfect, and no author of a procedure can anticipate every situation or eventuality among users of the procedure. Even for the most common experimental techniques, procedures are constantly being rewritten and improved. As you perform the experiments in this manual, be aware of suggestions for improving the procedures or for clearing up any ambiguity in them. If you believe you have a better way of doing something, consult with your instructor before trying the new procedure. Be sure you note any changes in procedure in your notebook.

Although errors can be anticipated, and sometimes corrected or allowed for before they occur, they will still occur in virtually all experiments. For numerical data, standard methods have been developed for indicating the magnitude of the error associated with a series of measurements. Such methods of error analysis are based on the mathematical science of statistics; these mathematical tools can provide insights about the precision of an experiment or experimental method.

Two words that must be carefully distinguished from one another are used in discussions of error: *accuracy* and *precision.* The **accuracy** of a measurement represents how closely the measurement agrees with the true or accepted value for the property being measured. For example, the true boiling point of water under 1 atmosphere barometric pressure is 100.0°C. If the thermometer in your laboratory locker gives a reading of 100°C in boiling water, you can conclude that your thermometer is fairly accurate. The **precision** of a series of measurements reflects how closely the members of the series agree with each other. If you weighed a series of similar coins on a laboratory balance, and each coin's mass was indicated as 2.65 grams on the balance, you would conclude that the coins had been manufactured to have this precise mass. However, precision in a series of measurements does not necessarily mean that the measurements are accurate. In the preceding example of the coins, even though the balance used indicates the mass of each coin to be 2.65 grams, if the balance is malfunctioning, the mass determined may be completely inaccurate.

Generally, if you have made a valiant attempt to exclude personal, apparatus, and procedural errors from an experiment, you may assume that precision in a series of measurements does at least hint that the measurements are also accurate. Expressing the degree of precision in a series of measurements may be done by several standard methods. To demonstrate these methods, consider the following set of data.

Suppose an experiment has been performed to determine the percent by mass of sulfate ion in a sample. To indicate the precision of the method, the experimental determination was repeated four times, with the following results:

Sample	% Sulfate
A	44.02
B	44.11
C	43.98
D	44.09

The **mean** or average value of these determinations is obtained by dividing the sum of the four determinations by the number of determinations.

Mean = (44.02 + 44.11 + 43.98 + 44.09)/4 = 44.05

The **deviation** of each individual result can then be calculated, and, from these, the **average deviation** from the mean (without regard to the mathematical sign of the individual deviations).

Sample	% Sulfate	Deviation from Mean
A	44.02	−0.03
B	44.11	+0.06
C	43.98	−0.07
D	44.09	+0.04
Mean	44.05	

The average deviation from the mean is then given by

(|−0.03| + |+0.06| + |−0.07| + |+0.04|)/4 = 0.05   (Note the use of absolute values, denoted by ||.)

More commonly, average deviations from the mean value of a series of duplicate measurements are expressed on a relative basis, with reference to the mean value. For a mean value of 44.05 and an average deviation from the mean of 0.05, as in the preceding example, the **relative precision** may be expressed as a percentage:

$(0.05/44.05) \times 100 = 0.1\%$

The lower the relative deviation from the mean, the more precise the set of measurements. The precision of an experiment is most commonly expressed in terms of the average or relative average deviation from the mean in the situation where only a few measurements have been performed.

Another common method of expressing the precision of a large set of n measurements is the **standard deviation**, s.

$$s = \sqrt{\frac{d_1^2 + d_2^2 + d_3^2 + \mathrm{K} + d_n^2}{n-1}}$$

where each d represents the deviation of an individual measurement from the mean, and n is the total number of measurements.

For the data given earlier, we have

Sample	% Sulfate	Deviation from Mean	Deviation Squared
A	44.02	0.03	0.0009
B	44.11	0.06	0.0036
C	43.98	0.07	0.0049
D	44.09	0.04	0.0016
	Mean = 44.05	Mean deviation = 0.05	

the standard deviation, s, would be given by

$$s = \sqrt{\frac{0.0009 + 0.0036 + 0.0049 + 0.0016}{(4-1)}} = 0.06$$

and the results of the experiment would be reported as 44.05 ± 0.06 for the percentage of sulfate ion in the sample. As before, the smaller the standard deviation, the more precise the determination.

Often, one measurement in a series will appear to differ greatly from the other measurements. For example, suppose a fifth determination of the percentage of sulfate were made in the experiment discussed earlier (Sample E below).

Sample	% Sulfate
A	44.02
B	44.11
C	43.98
D	44.09
E	42.15

Obviously, samples A, B, C, and D seem to "agree" with one another, but sample E seems entirely different (maybe some sort of personal error occurred during the experiment). Your first tendency is probably to discard result E; but you need a mathematical method for deciding when a result of a measurement should be rejected as being completely in error. There are various methods for doing this; a very common method is called the *Q-test*, which is applied to the *most deviant result* in a series of measurements. The function Q is defined as follows:

$$Q = \frac{\text{difference between deviant result and nearest neighbor}}{\text{range of all measurements in the series}}$$

For the data given, sample E (42.15) is the most deviant, and sample C (43.98) is its nearest neighbor (closest value to the deviant value). The range of the measurements is the difference between the highest value (sample B, 44.11) and the lowest value (sample E, 42.15). Q would then be given by

$$Q = \frac{43.98 - 42.15}{44.11 - 42.15} = \frac{1.83}{1.96} = 0.93$$

The calculated value of Q (0.93) is then compared to a table listing standard Q values. If the calculated experimental value of Q is *larger* than the table value of Q, the deviant result should be rejected and not included in subsequent calculations (such as that of the mean or standard deviation). The following short table of standard "Q-90" values indicates when a deviant result may be rejected with 90% confidence that the result is in error.

Number of Measurements	Q-90
2	—
3	0.94
4	0.76
5	0.64
6	0.56
7	0.51
8	0.47
9	0.44
10	0.41

In the example, there were five measurements made, and the calculated value of Q (0.93) was considerably larger than the table value for five measurements (0.64). This indicates that the results of sample E should be omitted from any subsequent calculations.

Vapor Pressure of Water at Various Temperatures

Temperature, °C	Pressure, mm Hg	Temperature, °C	Pressure, mm Hg
0	4.580	31	33.696
5	6.543	32	35.663
10	9.209	34	39.899
15	12.788	36	44.563
16	13.634	38	49.692
17	14.530	40	55.324
18	15.477	45	71.882
19	16.478	50	92.511
20	17.535	55	118.03
21	18.650	60	149.37
22	19.827	65	187.55
23	21.068	70	233.71
24	22.377	75	289.10
25	23.756	80	355.11
26	25.209	85	433.62
27	26.739	90	525.77
28	28.349	95	633.91
29	30.044	100	760.00
30	31.823		

Density of Water at Various Temperatures

Temperature, °C	Density, g/mL	Temperature, °C	Density, g/mL
0	0.99987	26	0.99681
2	0.99997	28	0.99626
4	1.0000	30	0.99567
6	0.99997	32	0.99505
8	0.99988	34	0.99440
10	0.99973	36	0.99371
12	0.99952	38	0.99299
14	0.99927	40	0.99224
16	0.99897	45	0.99025
18	0.99862	50	0.98807
20	0.99823	55	0.98573
22	0.99780	60	0.98324
24	0.99732		

APPENDIX E

Solubility Infomation

Solubility Rules for Ionic Compounds in Water

- Nearly all compounds containing Na^+, K^+, and NH_4^+ are readily soluble in water.

- Nearly all compounds containing NO_3^- are readily soluble in water.

- Most compounds containing Cl^- are soluble in water, with the common exceptions of $AgCl$, $PbCl_2$, and Hg_2Cl_2.

- Most compounds containing SO_4^{2-} are soluble in water, with the common exceptions of $BaSO_4$, $PbSO_4$, and $CaSO_4$.

- Most compounds containing OH^- ion are *not* readily soluble in water, with the common exceptions of the hydroxides of the alkali metals (Group 1) and $Ba(OH)_2$.

- Most compounds containing other ions not mentioned above are *not* appreciably soluble in water.

Solubility Product Constants

Ag_2CO_3	8.1×10^{-12}	CdS	1.0×10^{-28}	$MgCO_3$	1.0×10^{-15}
Ag_2CrO_4	9.0×10^{-12}	CoS	$5. \times 10^{-22}$	$MgNH_4PO_4$	2.5×10^{-13}
Ag_2S	1.6×10^{-49}	$Cr(OH)_3$	6.7×10^{-31}	MnS	2.3×10^{-13}
Ag_2SO_4	1.2×10^{-5}	$Cu(OH)_2$	1.6×10^{-19}	$Ni(OH)_2$	1.6×10^{-16}
Ag_3PO_4	1.8×10^{-18}	CuS	8.5×10^{-45}	NiS	$3. \times 10^{-21}$
AgBr	5.0×10^{-13}	$Fe(OH)_2$	1.8×10^{-15}	$Pb(OH)_2$	1.2×10^{-15}
AgCl	1.6×10^{-10}	$CaSO_4$	6.1×10^{-5}	$Pb_3(PO_4)_2$	$1. \times 10^{-54}$
AgI	1.5×10^{-16}	CdS	1.0×10^{-28}	$PbBr_2$	4.6×10^{-6}
AgOH	2.0×10^{-8}	CoS	$5. \times 10^{-22}$	$PbCl_2$	1.6×10^{-5}
$Al(OH)_3$	2.0×10^{-32}	$Cr(OH)_3$	6.7×10^{-31}	$PbCO_3$	1.5×10^{-15}
$Ba(OH)_2$	5.0×10^{-3}	$Cu(OH)_2$	1.6×10^{-19}	$PbCrO_4$	2.0×10^{-16}
$Ba_3(PO_4)_2$	$6. \times 10^{-39}$	CuS	8.5×10^{-45}	PbI_2	4.6×10^{-6}
$BaCO_3$	1.6×10^{-9}	$Fe(OH)_2$	1.8×10^{-15}	PbS	$7. \times 10^{-29}$
$BaCrO_4$	8.5×10^{-11}	$Fe(OH)_3$	4.0×10^{-38}	$PbSO_4$	1.3×10^{-8}
$BaSO_4$	1.5×10^{-9}	FeS	3.7×10^{-19}	$Sn(OH)_2$	5×10^{-26}
$Ca(OH)_2$	1.3×10^{-6}	Hg_2Cl_2	1.1×10^{-18}	$Sn(OH)_4$	$1. \times 10^{-56}$
$Ca_3(PO_4)_2$	1.3×10^{-32}	HgS	1.6×10^{-54}	SnS	$1. \times 10^{-26}$
CaC_2O_4	2.3×10^{-9}	$KClO_4$	$1. \times 10^{-2}$	$Zn(OH)_2$	4.5×10^{-17}
$CaCO_3$	8.7×10^{-9}	$KHC_4H_4O_6$	$3. \times 10^{-4}$	$ZnCO_3$	2.1×10^{-10}
$CaSO_4$	6.1×10^{-5}	$Mg(OH)_2$	8.9×10^{-12}	ZnS	2.0×10^{-25}

Properties of Substances

Introduction

As a convenience to the user, some properties of the various chemical substances employed in this manual are listed in the table that follows. This table by no means lists all notable properties of the substances mentioned, and should it serve only as a quick reference to the substances. *Additional information about the substances to be used in an experiment should be obtained before use of the substances.* Sources of such additional information include the lab instructor or professor, the Material Safety Data Sheet (MSDS) for the substance, the label on the substance's bottle, and a handbook of dangerous or toxic substances.

In the following table, corrosiveness and toxicity risks are indicated as L (low), M (moderate), or H (high).

Substance	Flammability	Corrosiveness	Toxicity	Notes
acetaldehyde	Yes	M	M	1, 4
acetic acid	Yes	H	H	1, 4
acetic anhydride	Yes	H	H	
acetone	Yes	M	H	
acetylacetone	Yes	M	H	1, 4
adipoyl chloride	Yes	H	H	1, 4, 7
alanine	No	L	L	
aluminon reagent	No	M	M	
aluminum	Yes*	L	L	
ammonia	No	M	H	1, 4
ammonium carbonate	No	L	M	2
ammonium chloride	No	L	L	
ammonium molybdate	No	M	H	1
ammonium nitrate	No	M	H	3
ammonium sulfate	No	L	M	
ammonium sulfide	No	M	H	4
ammonium thiocyanate	No	L	H	
asparagine	Yes*	L	L	
barium chloride	No	M	H	
barium nitrate	No	M	H	3
Benedict's reagent	No	M	M	
benzoic acid, 2-amino-	Yes	M	H	
benzyl alcohol	Yes	L	M	
boric acid	No	L	M	
boron	Yes*	M	H	
bromine	No	H	H	1
butyl alcohol, *n-*	Yes	M	M	
butylamine, *n-*	Yes	M	M	1, 4
butyric acid	Yes	M	M	4
calcium	Yes*	M	L	7
calcium acetate	Yes	H	H	1, 4
calcium carbonate	No	L	L	2
calcium chloride	No	M	M	

Substance	Flammability	Corrosiveness	Toxicity	Notes
calcium hydroxide	No	M	M	
carbon	Yes*	L	M	
carbon disulfide	Yes	H	H	1, 4
chlorine	Yes*	H	H	1, 4
chromium(III) chloride	No	H	H	1
cobalt(II) chloride	No	M	H	
copper	Yes*	L	M	
copper(II) oxide	No	M	M	
copper(II) sulfate	No	M	M	
cyclohexane	Yes	L	H	
cyclohexene	Yes	M	H	4
dichlorobenzene, 1, 4-	Yes	M	H	
diethyl ether	Yes	M	H	
dimethylglyoxime	Yes	M	M	
dry ice	No	H	L	5
Eriochrome Black-T	No	M	H	
ethyl alcohol (ethanol)	Yes	L	H	6
ethyl ether	Yes	M	H	
ethylene diamine	Yes	M	H	1, 4
ethylene diamine tetraacetic acid (EDTA)	Yes*	M	H	
formaldehyde	Yes	M	H	1, 4
fructose	Yes*	L	L	
gelatin	Yes*	L	L	
glucose	Yes*	L	L	
glycine	Yes*	L	L	
graphite	Yes*	M	M	
hexane	Yes	M	H	
hexanediamine	Yes	H	H	4
hexene, 1-	Yes	M	M	
hydrochloric acid	No	H	H	1, 4
hydrogen	Yes	L	M	
hydrogen peroxide, 3%	No	H	M	3
iodine	No	H	H	
iron(II) sulfate	No	L	L	
iron(III) chloride	No	M	M	
iron(III) nitrate	No	M	M	3
isoamyl alcohol	Yes	L	M	
isobutyl alcohol	Yes	M	M	
isoleucine	Yes*	L	L	
isopropyl alcohol	Yes	L	M	
lead	Yes*	L	M	
lead acetate	Yes*	M	H	
lithium	Yes	H	H	7
lithium chloride	No	M	H	
magnesium	Yes	L	M	
magnesium chloride	No	L	M	
manganese(II) chloride	No	M	M	3
manganese(IV) oxide	No	M	M	3
mercury	No	H	H	
mercury(II) chloride	No	H	H	
methyl alcohol (methanol)	Yes	H	H	

Substance	Flammability	Corrosiveness	Toxicity	Notes
methyl salicylate	Yes	M	H	
methylene chloride	Yes	H	H	1, 4
n-butyl alcohol	Yes	M	M	
n-butylamine	Yes	M	M	4
n-octyl alcohol	Yes	M	M	
n-pentyl alcohol	Yes	M	M	
n-propyl alcohol	Yes	L	M	
naphthalene	Yes	M	H	
nickel(II) chloride	No	M	H	
nickel(II) sulfate	No	M	H	
ninhydrin	Yes	H	H	
nitric acid	No	H	H	1, 3, 4
nitrogen	No	L	L	
oleic acid	Yes	M	L	
oxygen	Supports	L	L	
pentane	Yes	M	H	
pentanedione, 2,4-	Yes	M	H	1, 4
pentyl alcohol, n-	Yes	M	M	
phenanthroline, 1,10-	Yes	M	M	
phosphoric acid	No	H	H	
phosphorus, red	Yes	H	H	
phosphorus, white	Yes	H	H	
potassium	Yes	H	H	7
potassium bromide	No	L	L	
potassium carbonate	No	L	L	
potassium chloride	No	L	L	
potassium chromate	No	H	H	3
potassium dichromate	No	H	H	3
potassium ferrocyanide	No	M	H	
potassium hydrogen phthalate	Yes*	M	M	4
potassium hydroxide	No	H	H	
potassium iodate	No	H	H	3
potassium iodide	No	M	M	
potassium nitrate	No	M	H	3
potassium oxalate	Yes*	M	H	
potassium permanganate	No	H	H	3
potassium sulfate	No	L	L	
proline	Yes*	L	L	
propionic acid	Yes	H	H	
propyl alcohol, n-	Yes	L	M	
salicylic acid	Yes*	M	M	
serine	Yes*	L	L	
silicon	No	M	M	
silver acetate	Yes*	M	M	
silver chromate	No	H	H	3
silver nitrate	No	M	M	
sodium	Yes	H	H	7
sodium acetate	Yes*	M	M	
sodium bicarbonate	No	L	L	
sodium bromide	No	L	L	
sodium chloride	No	L	L	
sodium chromate	No	H	H	3

514

Substance	Flammability	Corrosiveness	Toxicity	Notes
sodium hydrogen sulfite	No	M	M	4
sodium hydroxide	No	H	H	
sodium hypochlorite	No	H	H	1, 4
sodium iodide	No	M	M	
sodium nitrate	No	H	H	3
sodium nitrite	No	H	H	3
sodium nitroprusside	No	M	M	
sodium silicate	No	L	L	
sodium sulfate	No	L	L	
sodium sulfite	No	M	M	3, 4
sodium thiosulfate	No	M	M	
starch	Yes*	L	L	
strontium chloride	No	M	M	
sulfur	Yes	L	L	
sulfuric acid	No	H	H	3, 7
tartaric acid	Yes*	M	L	
thioacetamide	Yes*	H	H	4
tin	No	L	L	
tin(II) chloride	No	M	M	3
toluene	Yes	H	H	
zinc	Yes*	L	L	
zinc sulfate	No	L	M	

*Will burn if deliberately ignited, but has very low vapor pressure and is not a fire hazard under normal laboratory conditions. Such substances are more precisely termed "combustible" rather than flammable.

Notes:

1. This substance is a respiratory irritant. Confine use of the substance to the exhaust hood.

2. This substance vigorously evolves carbon dioxide if acidified.

3. This substance is a strong oxidant or reductant. Although the substance itself may not be flammable, contact with an easily oxidized or reduced material may cause a fire or explosion. Avoid especially contact with any organic substance.

4. This substance possesses or may give rise to a stench. Confine use of this substance to the exhaust hood.

5. Handle dry ice with tongs to avoid frostbite.

6. Ethyl alcohol for use in the laboratory has been denatured. It is *not* fit to drink and *cannot be made* fit to drink.

7. This substance reacts vigorously with water and may explode.

Concentrated Acid and Base Reagent Data

Acids

Reagent	HCl	HNO₃	H₂SO₄	CH₃COOH
Formula mass of solute	36.46	63.01	98.08	60.05
Sp. gr of concentrated reagent	1.19	1.42	1.84	1.05
% Concentration (w/w)	37.2	70.4	96.0	99.8
Molarity	12.1	15.9	18.0	17.4
Volume (mL) needed for 1 L of 1 M solution	82.5	63.0	55.5	57.5

Bases

Reagent	NaOH	NH₃(aq)
Formula mass of solute	40.00	35.06[1]
Sp. gr. of concentrated reagent	1.54	0.90
% Concentration (w/w)	50.5	56.6
Molarity	19.44	14.5
Volume (mL) needed for 1 L of 1 M solution	51.5	69.0

Data on concentration, density, and amount required to prepare solution will differ slightly from batch to batch of concentrated reagent. Values given here are typical.

---

[1] As NH₄OH